JONATHAN KRALL

Interplanetary Magnetohydrodynamics

International Series on Astronomy and Astrophysics

1. E.N. Parker, *Spontaneous Current Sheets in Magnetic Fields with Applications to Stellar X-rays*
2. C.F. Kennel, *Convection and Substorms: Paradigms of Magnetospheric Phenomenology*
3. L.F. Burlaga, *Interplanetary Magnetohydrodynamics*

Interplanetary Magnetohydrodynamics

L.F. Burlaga

New York   Oxford
OXFORD UNIVERSITY PRESS
1995

Oxford University Press

Oxford New York
Athens Auckland Bangkok Bombay
Calcutta Cape Town Dar es Salaam Delhi
Florence Hong Kong Istanbul Karachi
Kuala Lumpur Madras Madrid Melbourne
Mexico City Nairobi Paris Singapore
Taipei Tokyo Toronto

and associated companies in
Berlin Ibadan

Copyright © 1995 by Oxford University Press, Inc.

Published by Oxford University Press, Inc.,
200 Madison Avenue, New York, New York 10016

Oxford is a registered trademark of Oxford University Press

All rights reserved. No part of this publication may be reproduced,
stored in a retrieval system, or transmitted, in any form or by any means,
electronic, mechanical, photocopying, recording, or otherwise,
without the prior permission of Oxford University Press.

Library of Congress Cataloging-in-Publication Data
Burlaga, L. F.
Interplanetary magnetohydrodynamics / L.F. Burlaga.
p.   cm. — (International series on astronomy and astrophysics; 3)
Includes bibliographical references and indexes.
ISBN 0-19-508472-1
1. Interstellar matter. 2. Magnetohydrodynamics.
I. Title. II. Series.
QB790.B79   1995
523.01'86—dc20   94-24056

1 3 5 7 9 8 6 4 2
Printed in the United States of America
on acid-free paper

*Dedicated to my Mother
and the Memory of my Father*

PREFACE

In situ measurements of the interplanetary medium have been made in the region between the orbit of Mercury and the orbit of Pluto. The Mariner II spacecraft made the first comprehensive measurements within 1 AU. Helios 1 and 2 moved from the Earth at 1 AU to 0.3 AU. The IMPs, ISEE-3, the Vela spacecraft, and the Soviet spacecraft such as Prognoz made detailed studies at 1 AU. Pioneer 10, Pioneer 11, Voyager 1, and Voyager 2 are moving far beyond the orbit of Neptune. These remarkable voyages are historic achievements, representing the accomplishments of thousands of people and costing hundreds of millions of dollars. The analysis of data from these spacecraft provides a detailed picture of our Sun's exotic and beautiful environment. The time is ripe for a book that summarizes some of the essential results of these measurements and their interpretations.

The interplanetary medium is a remarkable physical system, which has served as a laboratory for the study of turbulent, supersonic, ideal magnetohydrodynamic flows. In situ observations of the particles and fields provided detailed measurements of the magnetic field and plasma on scales from less than the thermal proton gyroradius to greater than 40 AU. These observations confirmed the existence of many MHD phenomena that were predicted, including collisionless fast and slow shocks, pressure balanced structures, tangential and rotational discontinuities, force-free field configurations, and MHD waves. Numerous detailed studies provided a wealth of information about these MHD phenomena, leading to a deeper understanding of them. The observations of the outer heliosphere, beyond 1 AU, provided evidence of new nonlinear MHD processes, such as the merging of shocks and the merging of interaction regions resulting in period doubling, the formation of large-scale structures, and memory loss. The interplanetary data have also revealed the existence of intermittent MHD turbulence, multifractal fluctuations in the large-scale magnetic field and plasma, and a vortex street of astronomical dimensions.

This book provides an observational and theoretical account of many of the fundamental MHD structures and processes in the interplanetary medium. The book emphasizes the fundamental forms and motions in the solar wind—"things with reason infused," as Leonardo da Vinci would call them. It illustrates the universal forms and motions, gives simple formulas describing them, and explains how they are created and destroyed by the basic forces and physical processes.

The book focuses on the fundamental structures and dynamical processes in the interplanetary medium that have been subjects of the author's research—the things the author feels most competent to write about. Special efforts were made to include numerous references to the work of the many people who contributed substantially to these topics. These references should quickly guide the specialist to the papers not discussed in the book. However, no attempt is made to review all the observations and theories of interplanetary phenomena that have been produced during the last 25 years; such a bibliography would itself fill a book of this size. The discussion is confined almost completely to papers published prior to mid-1993. Some important topics have been omitted, such as MHD turbulence and the interplanetary ejecta other than magnetic clouds. Most of the figures in this book are from the author's collection, but many other figures of importance are referenced and discussed. Lengthy mathematical derivations and detailed descriptions of models are omitted, but references are made to many of the papers on these subjects as well.

The author is grateful to many individuals who have influenced this work either directly or indirectly. Prof. E.N. Parker read the entire manuscript and provided many valuable comments. Critical comments on individual chapters of the book were provided by C. Farrugia, R. Lepping, F. McDonald, K. Ogilvie, J. Piragglia, and A. Roberts. I am particularly indebted to Drs. N.F. Ness and K.W. Ogilvie with whom I worked closely throughout my career. Collaborations with Dr. F.B. McDonald on the relation between interplanetary magnetohydrodynamics and cosmic rays have been very stimulating and productive. Most of my papers were written as collaborative efforts with scientists throughout the world. Many experimenters provided their data unselfishly for specific studies without participating directly as authors. Conversations and correspondence with numerous scientists have helped to shape my thoughts. Debates on controversial issues have also been important in clarifying my views and searching for unambiguous solutions to fundamental problems.

The book is written such that it can be understood by advanced graduate students and scientists who are not specialists in interplanetary physics. It should be of interest to scientists in many disciplines: astrophysics; solar and coronal structure and dynamics; planetary, magnetospheric, and ionospheric phenomena; cosmic ray physics; magnetohydrodynamics; and of course interplanetary and heliospheric physics. The book will also be of interest to those who are investigating nonlinear phenomena such as intermittent turbulence, multifractals, fractals, the formation of large-scale structures, period doubling, and catastrophe theory.

Davidsonville, MD L.F.B.

CONTENTS

1 Introduction, 3

    1.1 The Solar Wind, 3
    1.2 Geometrical and Topological Properties of the Solar Wind, 5
    1.3 Approach, 6
    1.4 Coordinates, 7
    1.5 Basic Equations, 9
    1.6 Interplanetary Spacecraft and Trajectories, 13

2 Large-Scale Magnetic Field, 14

    2.1 The Solar Activity Cycle, 14
    2.2 Sectors and the Heliospheric Current Sheet, 14
    2.3 The Spiral Magnetic Field, 27

3 Large-Scale Plasma, 34

    3.1 Typical Plasma Characteristics at 1 AU, 34
    3.2 Solar Cycle Variations at 1 AU, 37
    3.3 Relations Among Fields at 1 AU, 38
    3.4 Latitude Variations, 39
    3.5 Radial Variations, 42

4 Pressure Balanced Structures, 45

    4.1 Basic Concept, 45
    4.2 Tangential and Rotational Discontinuities, 50
    4.3 Directional Discontinuities, 54
    4.4 Stream Interfaces, 58
    4.5 Kelvin–Helmholtz Instability and Shear Layers, 60
    4.6 Magnetic Holes, 62

5 Shocks, 70

    5.1 Fast MHD Shocks, 70
    5.2 Slow MHD Shocks, 85

6 Magnetic Clouds and Force-Free Magnetic Fields, 89

    6.1 Interplanetary Ejecta, 89
    6.2 Magnetic Clouds, 91
    6.3 Magnetic Flux Tube Model of Magnetic Clouds, 94
    6.4 Force-Free Field Models of Locally Cylindrical Magnetic Clouds, 96

6.5 Motions of Magnetic Clouds, 98
6.6 Force-Free Tori and Spheroids, 106
6.7 Internal Structure, 109
6.8 Relation to Shocks, 113

7 Corotating Streams and Interaction Regions, 115

7.1 Corotating Streams and Interaction Regions <1 AU, 115
7.2 Compound Streams Near 1 AU, 129
7.3 Structure of Interaction Regions and Streams >1 AU, 133

8 Merged Interaction Regions, 138

8.1 Definition and Classification of Merged Interaction Regions, 138
8.2 Formation of a Corotating Merged Interaction Region, 139
8.3 Quasi-Periodic Corotating MIRs, 151
8.4 Local MIRs, 162
8.5 Global MIRs, 165

9 Large-Scale Fluctuations, 169

9.1 Introduction, 169
9.2 Spectral Signatures of Large-Scale Fluctuations, 171
9.3 Multifractal Fluctuations, 181
9.4 Intermittent Turbulence and Multifractal Velocity Fluctuations, 182
9.5 Multifractal Magnetic Field Strength Fluctuations, 192

10 Heliospheric Vortex Street, 201

10.1 Observations, 201
10.2 Conventional Flow Models, 203
10.3 Vortex Street Models, 204

References, 215

Author Index, 243

Subject Index, 247

Interplanetary Magnetohydrodynamics

1

Introduction

1.1 The Solar Wind

This book is concerned with the magnetohydrodynamic processes in the interplanetary medium, the region near the solar equatorial plane extending from approximately 0.5 AU to 40 AU. The interplanetary medium lies between the sun and the interstellar medium. The observations that are discussed in this book were made by instruments on spacecraft moving through the interplanetary medium. Thus, our view of the interplanetary medium based on these observations is that of one who is immersed in the medium that he seeks to understand. There is another view of the interplanetary medium, the view that would be seen by an observer standing far beyond the interplanetary medium, say two or three hundred astronomical units from the sun. This section provides a brief overview of the interplanetary medium from the latter point of view. The sun and the interstellar medium compete for control of the interplanetary medium. The sun that we see during the day has a sharp boundary, the photosphere, having a temperature of 4000 K. However, anyone who has seen a solar eclipse knows that the sun reaches out to at least 10 solar radii. The solar "corona" is hot, of the order of a million degrees (van de Hulst, 1953; Billings, 1959). One of the great unanswered questions is how the sun produces such a hot corona (Parker, 1961; Scudder, 1992). The interstellar medium is relatively cool, the temperature being only of 10,000 K. The matter in the corona is fully ionized, consisting primarily of protons and electrons. The local interstellar medium is only weakly ionized, consisting primarily of neutral hydrogen. The density of the solar corona is high, of the order of 10^8 cm^{-3}. The density of the interstellar medium is low, of the order of 0.1 cm^{-3}. The magnetic field in the corona is approximately 10^5 nT, while that in the interstellar medium is perhaps 0.5 nT.

One might expect the interstellar material and magnetic field to move into the solar system until they reach a point where a pressure equilibrium with the extended corona is attained. Parker (1958) showed that if the corona were static, its pressure very far from the sun would be orders of magnitude larger than the pressure of the interstellar gas. No equilibrium between the sun and the interstellar medium is possible for a static corona!

Parker (1958) proposed the existence of a solar wind, an extension of the corona into the interplanetary medium, moving supersonically through the realm of the planets. Evidence for such a wind can be seen by observing that the tail of a comet is directed away from the sun regardless of the comet's direction of motion. The existence of a solar wind was also inferred by scientists who noted changes in the geomagnetic field following great solar flares by a day or so. Parker showed that the solar wind is a consequence of the pressure gradient in the solar corona and the sun's gravitational field (see Parker, 1963).

Spacecraft observations demonstrate that the solar wind extends beyond the interplanetary medium, at least to 50 AU. Parker (1961) proposed that the supersonic solar wind flow would end in a huge standing shock wave, allowing the solar material to decelerate to subsonic speeds. The location of this termination shock is a subject of great interest. Current estimates place it at approximately 100 AU (Suess, 1990), although there is considerable uncertainty in this number.

Beyond the termination shock, the solar plasma continues to flow away from the sun for some distance. This plasma, which was hot, dense, and supersonic near the sun is now cool and rarefied as a result of expansion, and it moves slowly after passing through the termination shock. At some sufficiently large distance from the sun, the pressure of this solar plasma is balanced by the pressure of the interstellar material at a boundary called the heliopause (Suess, 1990). Beyond this point, the interstellar medium rules supreme.

If the sun were immobile, the heliopause would be approximately spherical. However, the sun is moving through the interstellar medium at a speed of approximately 23 km/s. The distance from the sun to the heliopause in the direction of this motion is of the order of 150 AU. There, the solar wind plasma cannot advance further in that direction, so it is deflected by the interstellar medium as it approaches the heliopause. Ultimately, the solar wind escapes in the opposite direction in a long tail whose length and properties are poorly known.

Thus, the sun wins control of the interplanetary medium by means of a supersonic solar wind formed by the corona and the sun's gravitation. However, the sun's control of the interplanetary medium is not complete. Neutral interstellar material enters the solar system; it has been detected at the orbit of earth. The sun tries to exclude this material by ionizing it, thermalizing it, and carrying the "pickup ions" away by means of the magnetic fields in the solar wind. The ionization occurs by photoionization and by charge exchange between the solar wind protons and interstellar neutral atoms. Near the orbit of Pluto, the pressure of these pickup protons can exceed that of the solar wind protons and electrons. Thus, the interstellar medium can exert some influence on solar plasma beyond the interplanetary medium, but the sun's control remains dominant out to the termination shock.

Cosmic rays from the interstellar medium also penetrate the interplanetary medium. Their motions are strongly influenced by the solar wind and its magnetic field. The intensity of cosmic rays varies with the changes in solar activity, but cosmic rays are never excluded totally. This is significant, because cosmic rays played a major role in the evolution of life on earth by causing genetic mutations. Since cosmic rays have little influence on the solar wind, they are not considered in this book. However, observations from interplanetary spacecraft have contributed much to our understanding of the physics of cosmic rays.

1.2 Geometrical and Topological Properties of the Solar Wind

The solar wind is a supersonic, fully ionized plasma moving approximately radially away from the sun. The sun is the most important organizing factor in the heliosphere, the region extending from the sun to the termination shock. To lowest order the sun is a point source of plasma, a singularity in the velocity field. This singularity alters the topology of the space so that it is no longer simply a Euclidean space $E3$, but rather $E3$ minus a point. To this order the heliosphere has spherical symmetry. Since the streamlines are straight lines through the origin (the sun) in this approximation, the space is a projective space, with no distinction between points on opposite sides of the sun. The basic gas dynamic models of the solar wind are based on this symmetry (Parker, 1958, 1963). According to this model the solar wind velocity is constant and radial, independent of both time and position in the interplanetary medium. As a consequence of the radial expansion, a basic transformation in the global solar wind is a (multiplicative) central dilation with the sun as the invariant point.

The sun is rotating in the same sense as the planets about an axis that is fixed as far as interplanetary physics is concerned. The sun's rotation assigns an orientation to the space around it. The sun's equatorial plane divides the space into two parts, a northern hemisphere and a southern hemisphere. In this approximation the symmetry is axial and invariant with respect to rotations about the solar rotation axis. The solar rotation axis is inclined 7.25° with respect to the normal to the ecliptic plane in which the planets move, allowing a spacecraft in the ecliptic to sample a small range of latitudes as it moves around the sun. As a consequence of the solar rotation about an axis, another basic transformation in the solar wind is a two-dimensional rotation with the solar rotation axis as an invariant line.

The sun has a magnetic field, and the solar wind tends to carry the field outward, because the solar wind is a highly conducting plasma in which the field is "frozen in." In its simplest state the magnetic field of the solar wind is an extended dipole, with the dipole axis close to the sun's rotation axis and a plane of symmetry passing through the sun. To zeroth order the magnetic field lines are radial near the sun (at 1.5–2.5 solar radii), extending away from (toward) the sun above the plane of symmetry and toward (away from) the sun below the plane of symmetry, depending on the phase of the approximately 22-year magnetic solar cycle. The transition from toward-fields to away-fields in the interplanetary medium is usually a relatively thin region, not necessarily planar, called the heliospheric current sheet. Thus, the helisopheric current sheet divides the heliosphere into two basic parts, one hemisphere with "positive magnetic polarity" (away-fields) and another with "negative magnetic polarity" (toward-fields).

The two basic transformations of the solar wind, a central dilation owing to the radial expansion of the solar wind and a rotation owing to the rotation of the sun, give the fundamental transformation group of the solar wind, the spiral similarity transformation. The spiral similarity leaves the origin (the sun) invariant, and the fundamental geometrical form is an Archimedean spiral (see Fig. 10.4 in Parker, 1963). The spiral appears as the geometry of a streamline relative to a frame corotating with the sun and the geometry of a magnetic field line in either a stationary or a corotating stream.

The basic global structure of the interplanetary medium, discussed in Chapter 2, is a consequence of the symmetries and the singularity discussed above. Most of the interesting dynamical processes in the interplanetary medium are smaller scale structures that are the result of departures from these simple symmetries. The departures include temporal and radial variations in the speed that break the dilational symmetry and azimuthal variations in the speed that break the rotational (axial) symmetry.

1.3 Approach

We shall treat the interplanetary medium as an MHD flow. The basic fields of interest are two vector fields (the velocity field and the magnetic field) the three scalar fields (the density, the proton temperature, and the electron temperature). Each of these fields is a function of position and time as well as various parameters depending on the problem under consideration.

From a mathematical point of view, the aim of interplanetary science is to obtain a description of each of the basic fields throughout the interplanetary medium and as a function of time for at least one 22-year magnetic cycle. A focus on a particular field (e.g., the magnetic field or the velocity field) leads one to identify certain structures such as sectors, corotating streams, and the heliospheric current sheet. The most important characteristics of the basic fields are topological properties such as singularities in the vector fields (e.g., the sun and the heliospheric current sheet) and critical points in the scalar fields (e.g., maxima in the density field, minima in the temperature field, and inflection points in the measured speed profile). Also important are discontinuities (abrupt transitions) in the fields such as shocks and boundaries. Finally, various symmetries in the fields are important in identifying structures of specific types and describing them analytically. The mathematical emphasis in this book is on singularities, critical points, discontinuities, symmetries, and simple analytic relations.

From the point of view of physics, the most important objective of interplanetary science is to determine and understand the relations among the basic fields and their changes. This leads us to identify and analyze physical objects such as shock waves, MDH waves and discontinuities, force-free fields (magnetic clouds), and vortices.

Interplanetary science also has a phenomenological aspect, which deals with specific properties of the interplanetary medium that are unique to our star, in contrast to the universal mathematical and physical properties discussed above. Properties such as the streams and interaction regions that corotate with the period of the sun's rotation, the solar cycle variation of the basic fields, the shape of the heliospheric current sheet, and the temporal variations of mass flux and magnetic flux are basic. They provide initial conditions, boundary conditions and constraints, and a general context for studies of specific physical phenomena and processes. Chapters 2 and 3 emphasize the basic phenomenological properties of the interplanetary magnetic field and plasma, respectively.

1.4 Coordinates

To specify the position of a point in the interplanetary medium, it is necessary to introduce an inertial coordinate system. There is no unique inertial coordinate system, but a natural choice is the inertial heliographic coordinate system illustrated in Fig. 1.1. The position of a point is given by a vector whose components are defined by an orthonormal triad of basis vectors centered at the sun. The axis of rotation of the sun singles out one line, and the sun's sense of rotation defines a natural orientation. Thus, the Z_{IHG} axis is chosen to point northward along the solar rotation axis. The solar rotation also defines a plane, the solar equatorial plane. The intersection of this plane with the ecliptic plane defines a line, the longitude of the ascending node, which gives the $\hat{X}_{IHG}$ basis vector. The longitude of the ascending node drifts slowly with time (about $1°/72$ years). The $\hat{X}_{IHG}$ basis vector is defined to be along the direction of the ascending node in 1900. The $\hat{Y}_{IHG}$ basis vector is chosen to complete a right-handed triad. Thus the position of a point in the interplanetary medium is given by a vector whose components are (X_{IHG}, Y_{IHG}, Z_{IHG}) in the inertial heliographic coordinate system.

In the space surrounding the sun, as in any mathematical manifold, there

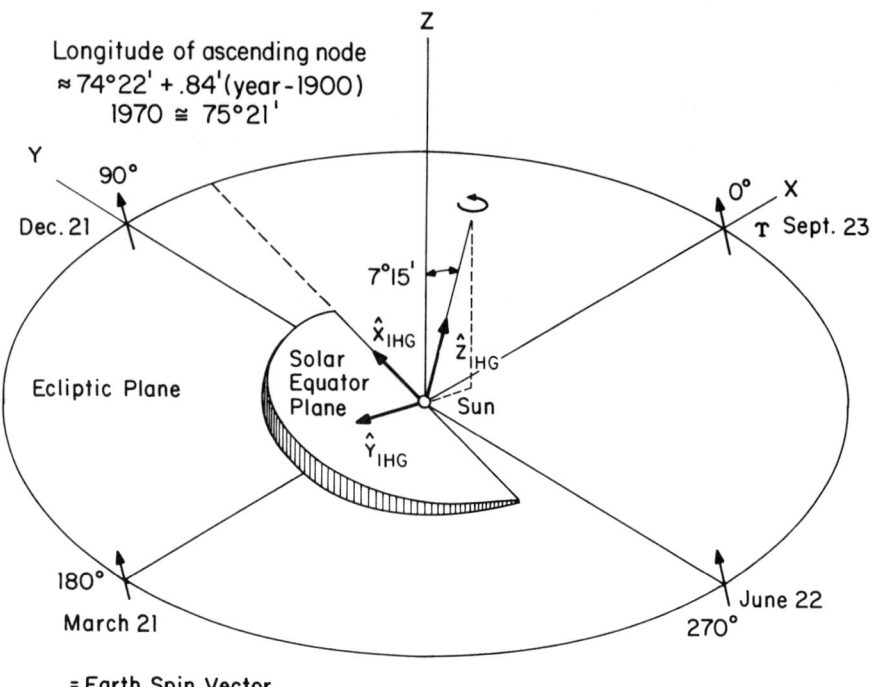

Fig. 1.1. Inertial heliographic coordinates.

is a vector space at each point (the "tangent space") containing all possible vectors at that point. To identify a specific vector at a point, such as the measured magnetic field vector or the velocity vector at a given time, it is necessary to introduce a coordinate system at that point. In fact, one must define a coordinate system at every point in the manifold in order to define a vector field. There is no unique way to define such a coordinate system. A coordinate system that is frequently used, the "heliographic coordinate system," is defined in Fig. 1.2. The coordinate system is determined by a frame consisting of three orthogonal unit vectors whose origin is at the point in question. One basis vector, $\hat{X}_{HG}$, is directed radially away from the sun. The second basis vector, $\hat{Y}_{HG}$, is perpendicular to $\hat{X}_{HG}$ and to $\hat{Z}_{IHG}$ and is directed in the sense of the motion of the planets. The third basis vector, $\hat{Z}_{HG}$, forms a right-handed triad. The Cartesian components of a magnetic field vector $\mathbf{B}(x)$ at a point x are (X_{HG}, Y_{HG}, Z_{HG}) in heliographic coordinates.

It is frequently meaningful to define a vector in terms of its magnitude and direction. The magnitude of a vector is independent of the coordinate system. We define the elevation angle of a vector with respect to the $X_{HG}-Y_{HG}$ plane as δ, δ increasing as the vector rotates increasingly northward. We define the direction of the component of $\mathbf{B}$ in the $X_{HG}-Y_{HG}$ plane by the angle λ, which is zero for a field directed away from the sun

Introduction

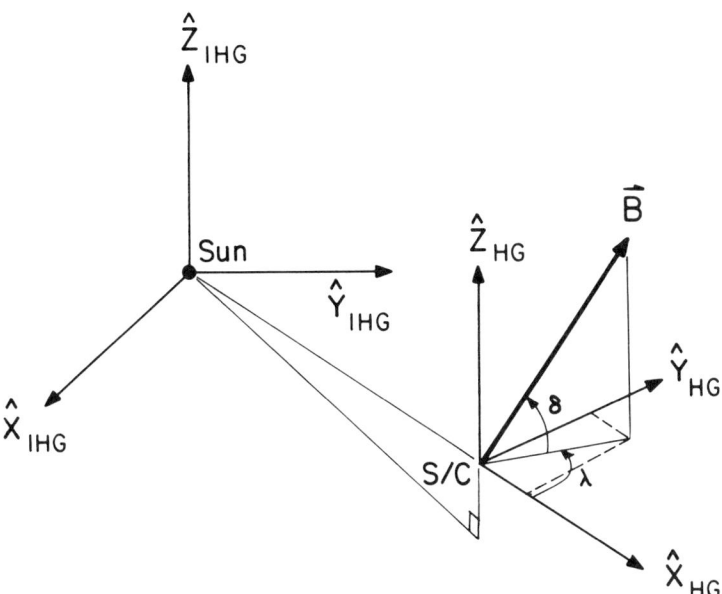

Fig. 1.2. Heliographic coordinates.

and increases as the vector moves toward Y_{HG} (see Fig. 1.2). Thus, $\delta = \sin^{-1}(B_Z/B)$ and $\lambda = \tan^{-1}(B_Y/B_X)$. A position vector relative to the inertial heliographic coordinate system can also be defined in terms of a magnitude and direction ($\delta_{IHG}, \lambda_{IHG}$) in the same way.

1.5 Basic Equations

A vast variety of interplanetary structures and processes, on scales ranging from 50 km to more than 50 AU ($=7.5 \times 10^9$ km), can be derived from the equations of MHD. The basic measured quantities are two vector fields (the magnetic field **B** and the velocity field **V**) and three scalar fields (the number density N of the protons, the proton temperture T, and the electron temperature T_e). Since there are nine unknown functions, we need nine equations in general. Fewer equations suffice when there are symmetries in a particular problem.

The equation of motion in vector form provides three differential equations relating spatial and temporal changes in **V** to spatial changes in **B** and the pressure p', which is the sum of the electron pressure p_e and the proton pressure p, neglecting the contributions of the minor ions that are present. The equation of motion is

$$\rho \partial_t \mathbf{V} + \rho(\mathbf{V} \cdot \nabla)\mathbf{V} = -\nabla p' + (\mathbf{J} \times \mathbf{B}) + \rho \nu \nabla^2 \mathbf{V} \qquad (1.1)$$

The left-hand side represents inertial effects. The Lagrangian acceleration of a volume element, $d\mathbf{V}/dt$, is written in Eulerian form as the sum of the term $\partial_t \mathbf{V}$ (the partial derivative of $\mathbf{V}$ with respect to t), representing the nonstationarity of the flow, and a convective term $(\mathbf{V} \cdot \nabla)\mathbf{V}$. The mass density $\rho = NM + N_e m$, where N (N_e) is the number density of protons (electrons) and M (m) is the mass of a proton (electron). To the extent that minor ions can be neglected, charge neutrality gives $N = N_e$, hence $\rho = N(M + m) \simeq NM$.

For purely inertial motion

$$\partial_t \mathbf{V} + (\mathbf{V} \cdot \nabla)\mathbf{V} = 0 \tag{1.2}$$

This equation can be written in Lagrangian form as $d\mathbf{V}/dt = 0$, which says that the velocity of a volume element is constant. The solution of equation (1.2) describes the basic kinematic effects, which are important for understanding the flow even when forces are not negligible. The mathematical subject of interplanetary kinematics has not been developed systematically, although it is significant in its own right. We shall discuss several topics in which a kinematic description provides important insights and useful results. These topics include the large-scale spiral magnetic field, the geometry of the heliospheric current sheet, the expansion of magnetic clouds, and turbulence. Unfortunately, kinematic descriptions have been applied by some people to dynamical problems where they are not appropriate, which has led to criticisms of kinematical analyses in general. There is considerable scope for further developments in interplanetary kinematics, particularly regarding the effects of complex motions such as mixing and turbulence on interplanetary magnetic fields.

The first two terms on the right-hand side of equation (1.1) are the basic forces involved in interplanetary dynamics: the pressure gradient force $\nabla p'$, and the magnetic force $\mathbf{J} \times \mathbf{B}$. We shall show that a remarkable variety of physical processes is associated with these two forces. In writing $4\pi \mathbf{J} = \nabla \times \mathbf{B}$, we assume that the displacement current is zero, which is a very good approximation in the interplanetary plasma. The plasma has unit magnetic permeability, so that $\mathbf{B} = \mathbf{H}$. The thermal pressure p' (the thermal energy density), is given in terms of the measured quantities N, T_e, and T by the ideal gas law

$$p' = p_e + p = Nk(T_e + T) \tag{1.3}$$

to the extent that minor ions can be neglected. In some cases it is necessary to consider the pressure of alpha particles, but this pressure is generally small because the abundance of alpha particles is typically only 5% of that of the protons (Neugebauer, 1981). Interstellar pickup protons can be the dominant pressure at approximately 35 AU (Burlaga et al., 1994), but this will not be considered.

The magnetic force $\mathbf{J} \times \mathbf{B}$ can be expressed as the sum of two forces, a magnetic tension $(\mathbf{B}\cdot\nabla)\mathbf{B}/4\pi$ and a gradient in the magnetic pressure $-\nabla(\mathbf{B}\cdot\mathbf{B})/8\pi$

$$\mathbf{J}\times\mathbf{B} = \frac{(\mathbf{B}\cdot\nabla)\mathbf{B}}{4\pi} - \frac{\nabla(\mathbf{B}\cdot\mathbf{B})}{8\pi} \tag{1.4}$$

A particularly important class of equilibria is that for which the magnetic force is zero and all other terms of equation (1.1) are either zero or negligible. These are called "force-free" magnetic field configurations. A force-free configuration actually represents a balance between two forces: the gradient of the magnetic pressure $-\nabla(\mathbf{B}\cdot\mathbf{B})/8\pi$ and the magnetic tension $(\mathbf{B}\cdot\nabla)\mathbf{B}/4\pi$. The magnetic tension is related to the curvature of the magnetic field lines, which behave like elastic bands in this respect.

The viscous force, $\rho v \nabla^2 \mathbf{V}$, is negligible everywhere in the solar wind, except within very thin layers with large velocity shear, such as in shock waves, and possibly other very small regions. Thus, the equation of motion in the interplanetary medium is nondissipative and the motion is reversible everywhere, except at shocks and at certain singularities on very small scales, such as those associated with turbulence and the Kelvin–Helmholtz instability. However, these exceptions are of fundamental significance for the structure and dynamics of the interplanetary medium, especially beyond the orbit of Mars. For example, the temperature increase produced locally by a nearly discontinuous change at a shock can affect the temperature of a very large region behind the shock, and turbulent heating becomes increasingly important at larger distances from the sun.

Certain aspects of the solar wind, particularly within 2 AU, can be described in terms of Hamiltonian dynamics. Some elegant and powerful formulations of ideal MHD in terms of Hamiltonian systems have been published (e.g., Holm and Kupershmidt, 1980) and will probably be of significance in future analytical studies of the solar wind. The association of the basic equations of physics with Lie groups (e.g., Morrison and Greene, 1980; Marsden, 1982) is a development that is intrinsically beautiful and promises to be useful in the discovery of analytical solutions. These approaches are beyond the scope of this book, but they offer a promising area of research in heliospheric physics.

A constraint expressing the conservation of mass (the continuity equation) is

$$\partial_t \rho + \nabla \cdot (\rho \mathbf{V}) = 0 \tag{1.5}$$

in Eulerian form. The continuity equation simply states that the change in density at a point is equal to the divergence of the flux of mass, $\nabla \cdot (\rho \mathbf{V})$ at that point. In Lagrangian form, the continuity equation is

$$\frac{1}{\rho}\frac{d\rho}{dt} = -\nabla \cdot \mathbf{V} \tag{1.6}$$

which says that the relative change in density following a volume element $(1/\rho)\,d\rho/dt$, is the negative of the divergence of the velocity. Thus, the density field is determined when the velocity field is known. In some cases the continuity equation is automatically satisfied by introducing a velocity potential.

Another fundamental conservation law, the conservation of magnetic flux as a consequence of the absence of magnetic charge, is given by the equation (Gauss's law)

$$\nabla \cdot \mathbf{B} = 0 \qquad (1.7)$$

which is one of the Maxwell's field equations. This equation implies that magnetic field lines, which are always defined as the integral curves of the vector field, either extend indefinitely, are closed, or terminate at singular points. Two magnetic field lines can cross only at a singular point where $\mathbf{B} = 0$. Two regions with magnetic field lines pointing in opposite directions can be separated by a singular surface where $\mathbf{B} = 0$. Thus regions where $\mathbf{B}$ is zero or small (such as magnetic holes and the heliospheric current sheet) are of special importance in interplanetary physics.

Three additional partial differential equations are given in vector form by

$$\partial_t \mathbf{B} = \nabla \times (\mathbf{V} \times \mathbf{B}) \qquad (1.8)$$

which is derived from the Maxwell's equation expressing Faraday's law $\nabla \times \mathbf{E} = -\partial_t \mathbf{B}$ and the assumption of infinite electrical conductivity. The latter gives $\mathbf{E} = 0$ in a frame moving with the plasma. Thus, $\mathbf{E} = -(\mathbf{V} \times \mathbf{B})$ in an inertial frame, by the Lorentz transformation for electric fields.

An energy equation is needed to complete the set of MHD equations. The transport of "thermal" energy in a low density nonuniform magnetoplasma is poorly understood. Moreover, the temperatures are not well defined, in general. The uncertainty in the temperature measurements is probably no greater than 50% in most cases and rarely greater than a factor of 2.

Given the uncertainties in the definition of temperature in the interplanetary medium, the difficulties in measuring the temperature, and the incomplete understanding of thermal energy transport in a low density collisionless plasma, it is reasonable to make a simple assumption relating the thermal pressure and temperature. One can assume that the proton and electron pressures each obey the ideal gas law: $p = NkT$ and $p_e = NkT_e$. Everywhere, except at certain singularities such as shocks, one can assume polytropic laws for the protons and electrons:

$$p = A(S)\rho^\gamma \qquad (1.9)$$

$$p_e = B(S)\rho^{\gamma e} \qquad (1.10)$$

where S is the entropy. The polytropic exponents γe and γ must be determined experimentally.

1.6 Interplanetary Spacecraft and Trajectories

It is convenient to distinguish between "near-Earth" spacecraft and deep-space probes. The near-Earth spacecraft provide (1) observations of the temporal variations of the solar wind over more than two solar cycles, (2) baseline data that are essential for determining radial gradients from the data acquired by deep-space probes, and (3) "intensive observations" involving high time resolution data from advanced instrumentation. The deep-space spacecraft provide exploratory observations of the regions far from Earth.

The deep-space probes can be classified as "inner-heliosphere probes," "outer-heliosphere probes," and "deep-space probes near 1 AU." The inner-heliosphere probes are the spacecraft that explored the region between 1 AU (the orbit of Earth) and 0.29 AU (the closest distance to the sun reached by a spacecraft to date). The outer-heliosphere probes are the spacecraft that explored the region from 1 AU to beyond 50 AU. The deep-space probes near 1 AU are those whose positions remained between 0.8 AU and 1 AU, but moved far from the earth.

The spacecraft Luna 2, Luna 3, and Venus 3, launched by the Soviet Union between 1959 and 1961, and the American spacecraft Explorer 10 launched in 1961, were perhaps the first spacecraft to enter interplanetary space, but the time spent by these spacecraft in the interplanetary medium was very brief. Explorer 10 was the first spacecraft to measure the basic plasma parameters: the number density $N = 6/cm^3$, the relatively low speed $V \approx 280$ km/s, and the proton temperature $T = (3-8) \times 10^5$ K.

The first of the inner-heliosphere probes was Mariner 2, which was also the first spacecraft to make continuous measurements of the interplanetary medium over an extended interval, beginning with its launch in 1962. Helios 1 made continuous measurements of the interplanetary medium between 1 AU and 0.31 AU, and Helios 2 made continuous measurements of the interplanetary medium between 1 AU and 0.29 AU. Helios 1 was launched on December 10, 1974. Helios 2 was launched on January 15, 1976. A plot of the orbits of the Helios probes is shown in Fig. 1.1 of the Introduction by Marsch and Schwenn to the book edited by Schwenn and Marsch (1990). The Pioneer–Venus orbiter provided valuable interplanetary measurements from 1 to 0.6 Au.

The region between 1 AU and the orbit of Pluto (and beyond) was explored by Pioneer 10, Pioneer 11, Voyager 1, and Voyager 2. A plot of the trajectories of these remarkable spacecraft may be found in Gazis et al. (1992) and Barnes (1990). Pioneer 10 was launched in 1972 and continues to make measurements. Pioneer 11 was launched in 1973, but it will soon cease to return useful data. Voyagers 1 and 2 were launched on September 5, 1977, and August 20, 1977, respectively. They can provide exploratory measurements out to at least 100 AU, if they stay healthy and continue to be tracked.

2

Large-Scale Magnetic Field

2.1 The Solar Activity Cycle

The structures of the solar wind and the interplanetary magnetic field vary systematically over a period of 11 years in relation to the solar activity cycle. The solar activity cycle is traditionally measured by the sunspot number. The sunspot number from 1962 to 1993 is shown in Fig. 2.1.

Four phases of the solar cycle can be identified: solar minimum, the ascending phase, solar maximum, and the descending phase. The three solar minima observed during the space age to date are shown in Fig. 2.1: 1964–65, 1976, and 1986. All the minima have similar values for the sunspot number. Three solar maxima observed during the space age to date are also shown in Fig. 2.1: a broad maximum centered around 1969, a larger maximum in 1979, and a comparable maximum in 1989. An ascending phase extends from a solar minimum to the following solar maximum, and the descending phases extends from a solar maximum to the following solar minimum.

The cosmic ray intensity is related to solar activity, with a tendency to be low when solar activity is high and high when solar activity is low (Forbush, 1954, 1958). However, the relation between cosmic ray intensity and solar activity is more complicated than a simple inverse relation. Explaining the solar cycle variation of the cosmic ray intensity is one of the outstanding problems of heliospheric research. The cosmic ray intensity is closely related to various MHD structures in the solar wind.

2.2 Sectors and the Heliospheric Current Sheet

2.2.1 Sectors and the Solar Magnetic Field

The existence of sectors at 1 AU was demonstrated by Ness and Wilcox (1964) and Wilcox and Ness (1965) in the IMP-1 magnetometer data of Ness et al. (1966). The sectors were observed at the end of 1963 and the

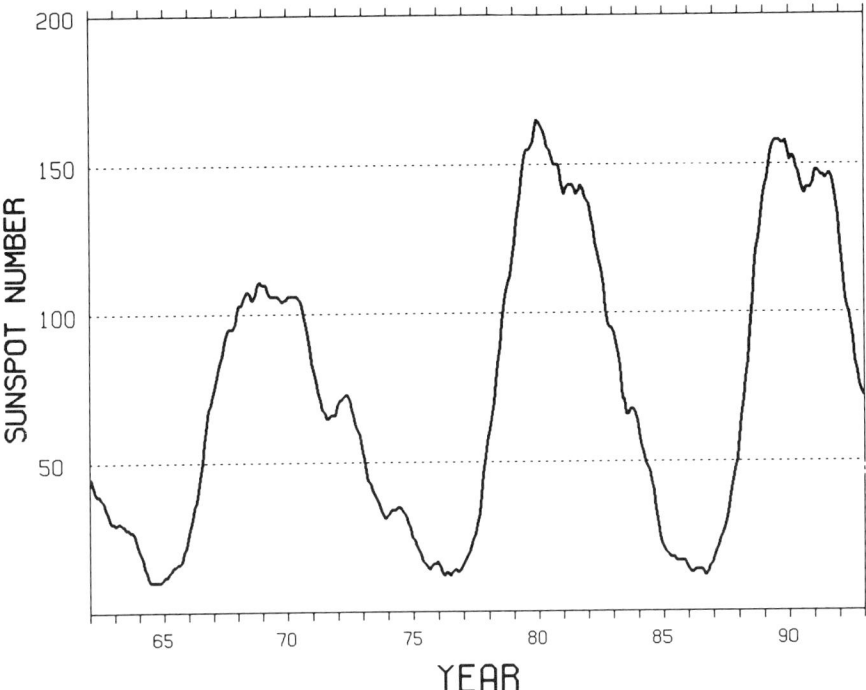

Fig. 2.1. Sunspot number versus time. (Y.C. Whang, private communication.)

beginning of 1964, near the solar minimum at the beginning of solar cycle 20. Wilcox and Ness define a "sector" as a region in which the polarity of the magnetic field is constant for at least 4 days while the region rotates past a spacecraft. They observed a "four-sector pattern," which means that four sectors of alternating positive and negative polarity were observed during a solar rotation. The pattern was stable, and it was observed repeatedly with a recurrence period of 27.0 ± 0.1 days.

The sector pattern is a mapping of the lowest order multipole components of the photospheric magnetic field. Models of this mapping are usually carried out in two parts: (1) a mapping from the photosphere to a spherical source surface centered at the sun with a radius of 1.5–2.5 solar radii, and (2) a mapping from the source surface to the interplanetary medium.

The mapping from the photosphere to the source surface was originally developed on the basis of a potential field model by Schatten et al. (1969) and Altschuler and Newkirk (1969). These authors solved Laplace's equation between the photosphere and a "source surface" located between $1.5 R_s$ and $2.5 R_s$. The photospheric magnetic field observations were taken at the inner boundary, and the magnetic field was assumed to be radial at the source surface. This solution provides the polarity of the magnetic field at the source surface, which serves as the boundary condition for a mapping into the interplanetary medium.

The most important feature of the magnetic field on the source surface is the "neutral line" (e.g., see Hoeksema, 1986). Above the neutral line (i.e., at higher latitudes) the magnetic field is directed away from (or toward) the sun, and below the neutral line the magnetic field is directed toward (or away from) the sun. As one moves along the equator, the polarity changes discontinuously upon crossing the neutral line. Since the magnetic field has different directions on opposite sides of the neutral line, the magnitude of the field is theoretically zero at the neutral line, hence the name. The neutral line is a curve on which the magnetic field is singular.

A relation between the neutral line on the source surface and the maximum brightness contour of the K-coronameter observations was found by Hansen et al. (1974) and Howard and Koomen (1974). Pneuman et al. (1978) noted a discrepancy between the position of the neutral line and the position of the maximum brightness line. They suggested that the potential field model was in error owing to the neglect of the polar field. Burlaga et al. (1981a) showed that the maximum brightness contour of the K-coronameter observations was more consistent with the Helios observations of the sectors than the neutral line, which tended to extend to higher latitudes than the maximum brightness contour (see the discussion of Fig. 2.2 in Section 2.2.2). These results support the suggestion of Pneuman et al. that one must include a contribution of the polar field in the calculation of the neutral line by the potential field method. Such a contribution is now routinely included in the neutral line models (Hoeksema et al., 1983; Wilcox and Hundhausen, 1983; Bruno et al., 1984).

2.2.2 Heliospheric Current Sheet

Schulz (1973) proposed that the sector pattern is related to a near-equatorial, warped (nonplanar) surface, the heliospheric current sheet (HCS), which in turn is related to the magnetic equator separating the two solar hemispheres of opposite polarities. The HCS is a surface determined by mapping the neutral line from the sun to the heliosphere. If the magnetic field in the northern hemisphere of the sun is positive, a spacecraft observes a sector with positive polarity when it is above the HCS and a sector with negative polarity when it is below the HCS. Schulz interpreted a two-sector pattern observed near the ecliptic as the result of a tilted solar magnetic dipole (Pneuman and Kopp, 1970), in which case the HCS near the sun is a rotating plane. This model is in contrast to an earlier model in which sectors and sector boundaries were thought to extend to high latitudes, like the segments of an orange. Schulz interpreted the four-sector pattern as the result of a warp in a quasi-equatorial current sheet owing to a quadrupole contribution of the solar magnetic field.

The model of Schulz (1973) was confirmed by two complementary sets of observations made near the solar minimum at the beginning of solar cycle 21. Pioneer 11 observations made at 16°N and 4 AU during February 1976 showed a single polarity for more than 1 solar rotation, hence no sector structure (Smith et al., 1978). This observation demonstrates that the sector structure did not extend above 16°N, thereby providing definitive evidence against the "orange segment" hypothesis, near solar minimum. The calculated "tilt" of the HCS (the maximum latitudinal extent of the neutral line on the source surface) in June 1976 was 14° so that the absence of the sector pattern at Pioneer 11 is consistent with the model of Schulz. Similar results were obtained by Voyager 1 at the beginning of solar cycle 22 (Burlaga and Ness, 1993a), when the spacecraft was above the latitudinal extent of the HCS inferred from the position of the neutral line. The actual position of the HCS was not measured by either Pioneer 11 or Voyager 1 during the intervals when they observed the single polarity for several solar rotations. Thus, these observations are consistent with the existence of an HCS with small amplitude, but Pioneer 11 and Voyager 1 did not observe the position of the HCS directly.

Direct evidence for the existence of a near-equatorial HCS that is an extrapolation of the neutral line at the source surface was derived from Helios 1 and 2 observations made between 0.3 AU and 1 AU. Helios 1 and 2 were both in the ecliptic plane, but because their longitudes differed, at times their heliographic latitudes differed by several degrees. Figure 2.2 (top panel) shows that Helios 1 observed positive magnetic polarities at less than 6°N between Carrington longitude 0° and 100° while Helios 2 observed negative magnetic polarities at a few degrees south in the same longitude range on Carrington rotation (CR) 1639. This means that the heliospheric current sheet was between the latitudes of Helios 1 and 2 at that time. The position of the HCS determined from the white light maximum brightness curves at 1.5 solar radii is consistent with the position of the HCS being between Helios 1 and Helios 2.

The neutral line computed without correction for the polar fields, shown by the dotted lines in Fig. 2.2, is not consistent with the Helios data. This is definitive evidence that sun's polar fields must be considered in computing the neutral line, as suggested by Pneuman et al. (1978). The observation of mixed polarities by Voyager 1 near 2 AU, corresponding to a time when the maximum white light brightness contour was in the ecliptic over a wide range of longitudes (Behannon et al., 1983), is additional direct evidence for the existence of a near-equatorial HCS that is an extension of the neutral line at the source surface, if one assumes that the neutral line coincides with the maximum white light brightness contour.

The form of the HCS in the corona at this time (Fig. 2.3, bottom panel) is obtained by taking the position of the maximum brightness contour as the footprint of the HCS at $1.5R_s$ and extrapolating it radially to $5R_s$. The HCS is asymmetric, extending to larger latitudes in the south than in the north.

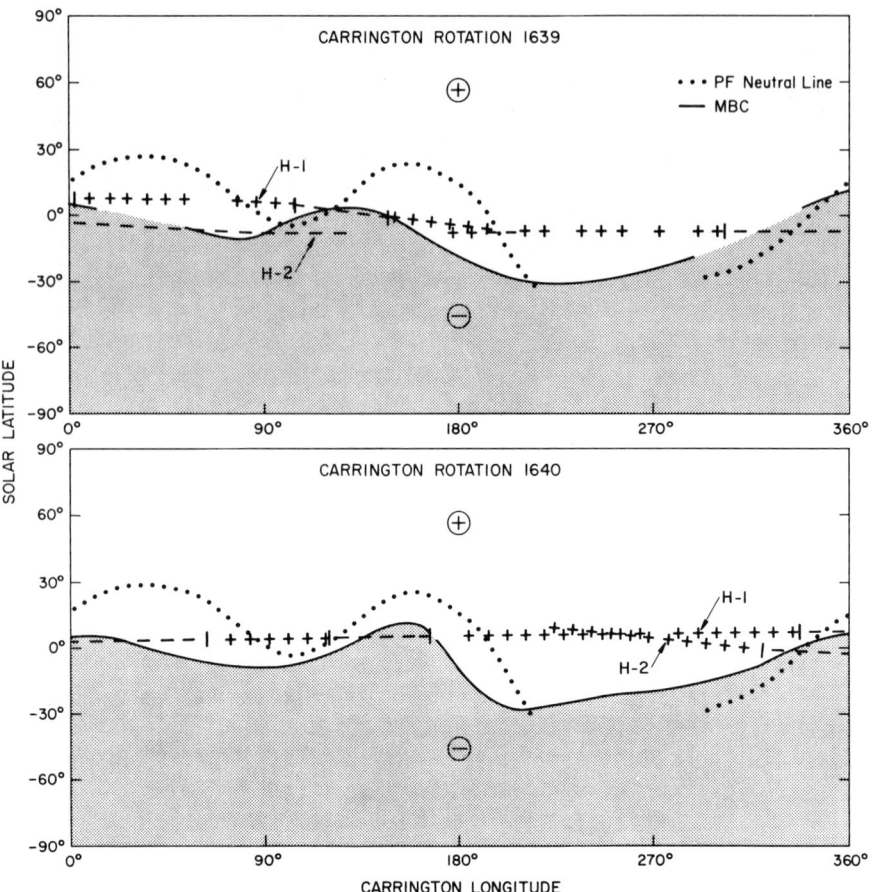

Fig. 2.2. Sector polarities observed by Helios 1 and 2 (plus and minus signs), the neutral line computed from the potential field model (dotted curve), and the maximum white light brightness contour (solid curve). (L.F. Burlaga, A.J. Hundhausen, and X.-P. Zhao, *J. Geophys. Res.*, **86**, 8893, 1981a, copyright by the American Geophysical Union.)

Such an asymmetry is generally ignored in discussions of the HCS near solar minimum, but it can be significant. In particular, it implies that the model of the HCS as a rotating plane in the corona near solar minimum is only an approximation. The rotating plane model of the HCS has another important shortcoming. It does not indicate that the inclination of the footpoints of the HCS (maximum brightness contour) with respect to the solar equator can be large locally even if the effective tilt of the HCS is small, as illustrated by the lower panel of Fig. 2.2.

Figures 2.2 and 2.3 imply that a spacecraft in the solar equator should observe a four-sector pattern at the time under consideration, consistent

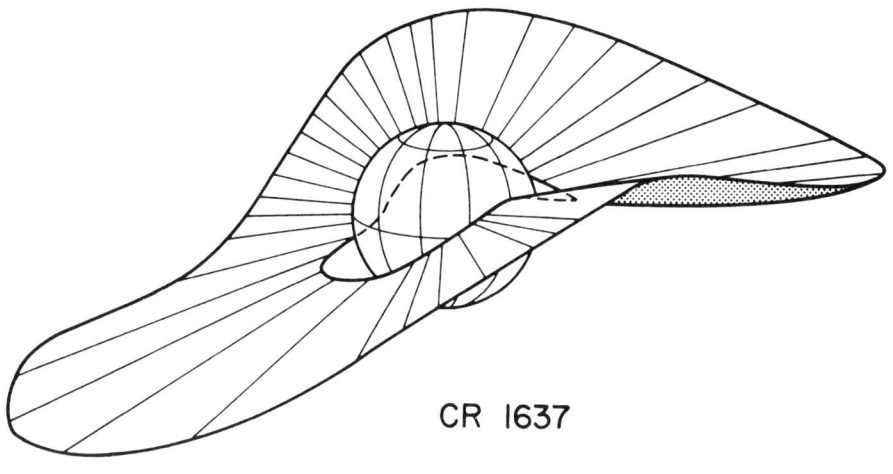

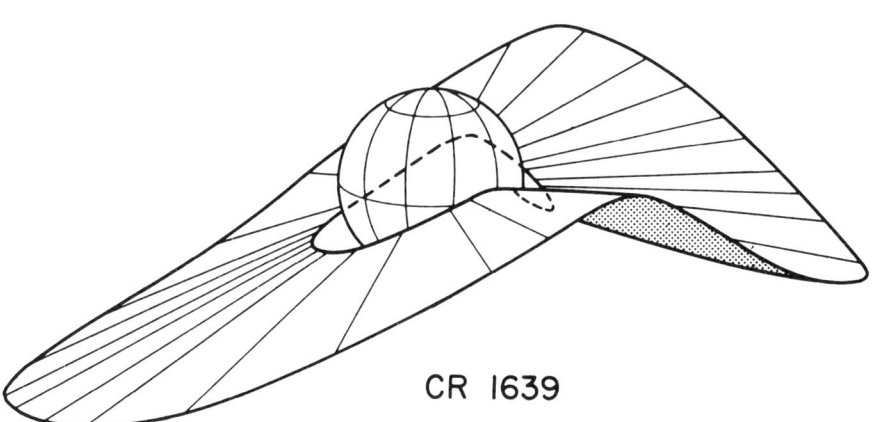

Fig. 2.3. The heliospheric current sheet near the sun. (L.F. Burlaga, A.J. Hundhausen, and X.-P. Zhao, *J. Geophys. Res.,* **86,** 8893, 1981a, copyright by the American Geophysical Union.)

with observations (Villante et al., 1979, 1982; Bruno et al., 1984). To first approximation (neglecting the asymmetry of the HCS) the four-sector pattern is the result of warps of the HCS produced by a quadrupole component of the solar magnetic field amounting to 17% of the dipole component from May 1976 to May 1977 (Bruno et al., 1982), consistent with the qualitative picture the Schulz (1973) expressed earlier.

The extension of the HCS into the interplanetary medium is a surface determined to first approximation by a kinematic mapping from the neutral line on the source surface to the interplanetary medium. This mapping is usually constructed by assuming that the solar wind velocity is constant and radial on the neutral line (maximum brightness contour) and independent of the distance from the sun. The footpoints of the HCS (the neutral line) are assumed to corotate rigidly with the sun. A highly idealized but frequently reproduced form of the heliospheric current sheet for the case in which the HCS is a rotating plane near the sun with a tilt angle of approximately 20° was drawn by Thomas and Smith (1981). Numerous other examples of the shape of the HCS for different tilt angles and different neutral lines have been published in many papers, including Svalgaard and Wilcox (1974, 1978), Alfvén (1977), Akasofu and Fry (1986), Fry and Akasofu (1986), and Hundhausen (1977).

The shape of the HCS can be very complex. There is generally a north–south asymmetry; the amplitude can be small while the inclination near the ecliptic is large, giving a square wave character; the amplitude and spacing between successive maxima may vary; there can be four sectors instead of two; there can be intermediate-scale warps superimposed on the HCS shown in Fig. 2.3; and there will be small-scale fluctuations because the HCS is embedded in a turbulent medium. Moreover, the shape of the HCS probably varies with distance from the sun owing to temporal variations and transient ejecta, as discussed below. Two essential features of the HCS are its latitudinal extent and the number of sectors that it defines in the equatorial plane.

The foregoing definition of the HCS assumes that the neutral line is a closed curve and that the solar and interplanetary conditions are stationary. Under such conditions the HCS can be viewed as a set that is the product $N \times S$ of two curves: the neutral "line" N and a curve S corresponding to an Archimedean spiral S originating from a point on N. Such a surface is analogous to a ruled surface, where the directrix is the neutral curve and a spiral takes the place of a line as the generator. The differential geometry of this surface has not been studied systematically, to the author's knowledge, although such a study would be straightforward and the results would be useful. From this construction of the HCS it is obvious that contrary to the statements of some authors, there is no meridional (N–S) component of the magnetic field at the HCS for a spiral magnetic field, even when the surface is highly warped.

2.2.3 Temporal and Latitudinal Variations of the Polarity

The observations discussed in the preceding section refer to the conditions near solar minimum. When the solar activity increases, the form of the neutral line changes, as discussed by many authors, including Hundhausen (1977), Villante et al. (1982), and Hoeksema et al. (1982). Figure 2.4, based on Fig. 2 in Hoeksema (1992), illustrates the temporal evolution of the neutral line at the source surface for representative solar rotations for each

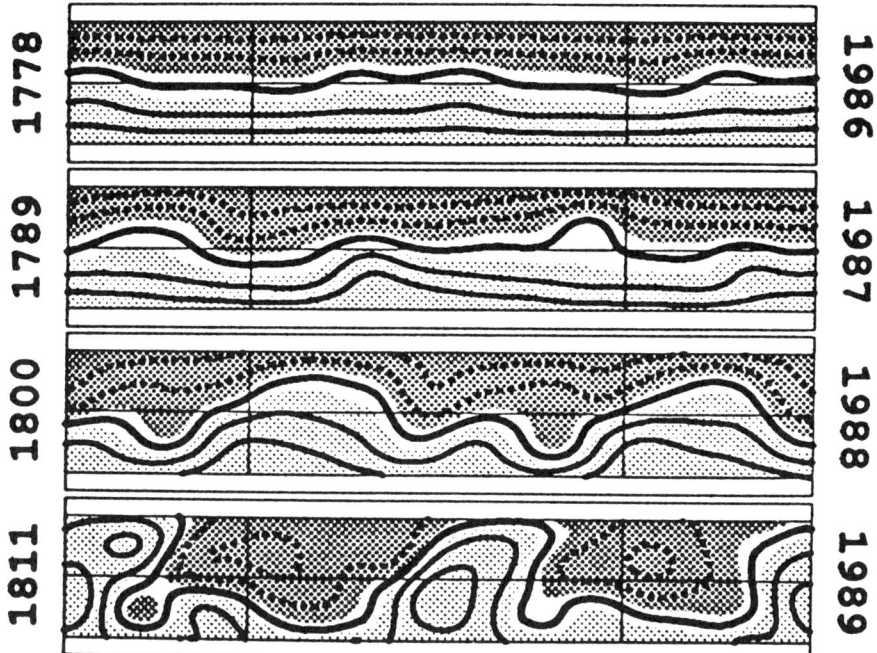

Fig. 2.4. Evolution of the neutral line from 1986 through 1989. [L.F. Burlaga and N.F. Ness, *J. Geophys. Res.*, **98,** 17451, 1993a, copyright by the American Geophysical Union. (After Hoeksema, 1992).]

of four years from solar minimum in 1986 through 1989 near the solar maximum of solar cycle 22. Note that a four-sector pattern is observed at the equator near solar minimum when the neutral line is near the equator, owing to quadrupole distortions as discussed in the preceding section. When solar activity increases, the latitudinal extent of the neutral line increases and a two-sector pattern is observed at the equator.

A similar evolution of the neutral line from a four-sector configuration to a two-sector configuration was observed on the preceding solar cycle, solar cycle 21 from 1976 to 1980. A four-sector pattern was observed at Voyager 1 from day 268, 1977, to day 200, 1978, and a two-sector pattern was observed thereafter until ≈ day 51, 1979, when the spacecraft was located near 5 AU (Burlaga et al., 1984a). Essentially the same pattern was observed at 1 AU (Sheeley and Harvey, 1981). Thus, the evolution from a four-sector pattern to a two-sector pattern was a consequence of the temporal evolution of the neutral line. Smith (1981) and Hakamada and Akasofu (1982) suggested that stream interactions might alter the latitudinal extent of the HCS.

The distributions of the azimuthal magnetic field directions observed by Voyager 1 and Voyager 2 from 1986 through 1989 are shown in Fig. 2.5.

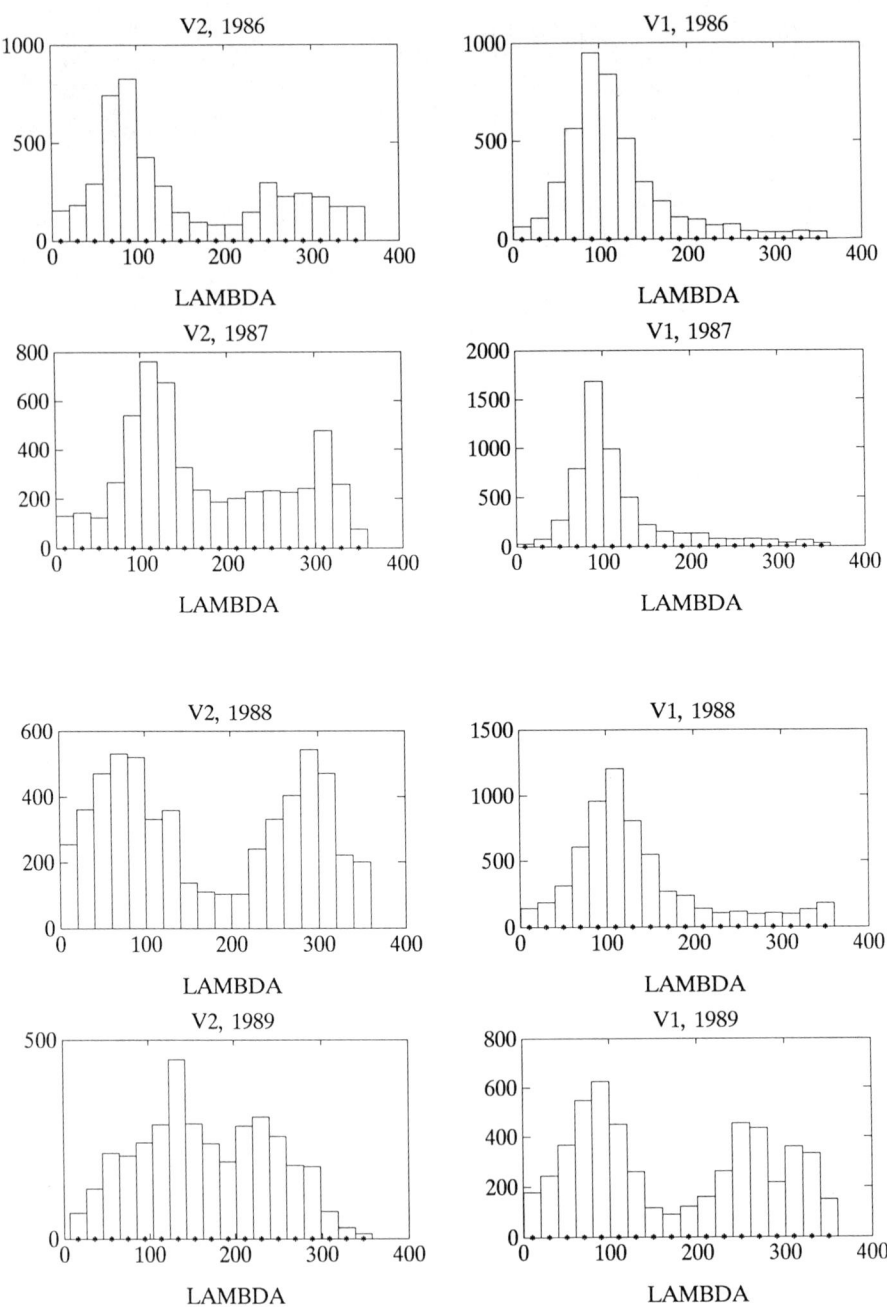

Fig. 2.5. Distribution of magnetic field directions at two latitudes. (L.F. Burlaga and N.F. Ness, *J. Geophys. Res.*, **98**, 17451, 1993a, copyright by the American Geophysical Union.)

Throughout the interval, Voyager 2 observed both positive and negative polarities near the equator, as expected. Voyager 1 at approximately 25°N to 30°N observed predominantly magnetic fields directed toward the sun, for the three years from 1986 through 1988, because it was above the HCS and because the dominant polarity of the sun's northern hemisphere during this phase of the solar cycle was negative. A similar distribution was observed by Pioneer 11 during 1986 at ≈ 20 AU and 15°N (Smith et al., 1988), which was also above the HCS at this time. During 1988, the amplitude of the HCS was approaching the latitude of Voyager 1, giving more hours with positive polarities than observed during the preceding two years, but Voyager 1 continued to observe predominantly negative polarity. During 1989, Voyager 1 (at 30°N and ≈ 40 AU) observed both positive and negative polarities, because the amplitude of the neutral line and that of the HCS were significantly above the latitude of Voyager 1 (see Fig. 2.4).

The agreement between the observed polarity distributions and those inferred from the latitudinal extent of the neutral line on the source surface from solar minimum to the year before solar maximum suggests that the average latitudinal extent of the HCS was relatively unperturbed between the sun and 40 AU, when the sun was not extremely active. Thus, the latitudinal variation of the yearly polarity distributions can be understood in terms of a kinematic mapping of the neutral line, except near solar maximum. Near solar maximum the interplanetary medium is very disturbed and the sector pattern is not well-defined.

The first indication of a possible latitudinal variation in the polarity pattern was published by Rosenberg and Coleman (1969), who observed that from 1964 to 1968 the dominant polarity between 0.7 and 1.5 AU varied periodically during the year. There were more days of negative polarity when the spacecraft were above the equator than below it. The dominant polarity was inward at heliographic latitudes above the equator, where the sun's dipole field had negative polarity. The dominant polarity was outward at heliographic latitudes below the heliographic equator, where the dipole field had positive polarity. This result was confirmed by Rosenberg et al. (1977) using simultaneous measurements at two different latitudes from Pioneer 10 at 1.0–4.3 AU and Helios 1 and 2 at 1.0 AU. These investigators found that the greater the difference in heliographic latitude between the spacecraft, the greater the difference in percentage positive polarity per solar rotation, for the period 1972–73. In this case the dominant polarity in the northern hemisphere was outward, because the sun's polarity reversed between the time of these observations and those reported by Rosenberg and Coleman (1969). The latitude dependence of the polarity was also studied by Rosenberg and Coleman (1980). Although the effect observed by Rosenberg and Coleman does not demonstrate the existence of an HCS, it finds a natural explanation in terms of an HCS with the properties discussed above.

2.2.4 Radial Variations of the Sector Structure and HCS

Evidence for a radial variation of the polarity pattern and the sector pattern because 1 AU and 10 AU was first presented by Behannon et al. (1989). The percent agreement between the polarities observed by the Voyager spacecraft and the corresponding polarities of the solar field decreased when the Voyager spacecraft moved from about 4 AU to 7 AU. This decrease is not the result of a temporal variation, because no such decrease was observed by the Pioneer–Venus orbiter at 0.6 AU during the same time interval.

When the neutral line and HCS are close to the equator, a four-sector pattern or a six-sector pattern is observed near the sun. The corresponding sector pattern observed in the interplanetary medium is very sensitive to small temporal variations in the shape of the neutral line and to small perturbations of the HCS in the interplanetary medium. Such was the case during 1986 and 1987, which might account in part for the complex sector pattern observed by Voyager 1 and Voyager 2 beyond 20 AU during this period. (The Voyager 2 pattern is shown in Fig. 2.6). It is very difficult to separate radial variations from temporal variations in such a situation.

When the amplitude of the neutral line is large and the neutral line makes a large angle with the equator where it intersects the equator, and when a two-sector pattern is observed near the sun, a constant-speed extrapolation of the neutral line implies that the two-sector pattern should be observed throughout the interplanetary medium, regardless of small perturbations that might be present. Such a neutral line and two-sector pattern were observed near the sun during 1988 and 1989 (Fig. 2.4). However, Voyager 2 at 20–30 AU did not observe a quasi-periodic sector pattern during 1988 and 1989 (Fig. 2.6). There was significant evolution of the sector pattern between the sun and 25 AU during 1988 and 1989.

The degradation of the sector pattern between the sun and 25 AU during 1988 and 1989 implies that the heliospheric current sheet was disturbed in an irregular way relative to the form expected from the simple extrapolation of the neutral line. The solar activity was high during 1988 and 1989, so that the interplanetary medium was disturbed by transient ejecta. The magnetic fields in these ejecta are typically not along the spiral direction, and they are likely to contribute regions of mixed polarity. Moreover, the transient ejecta will interact with sectors that might be present near the sun, causing displacements of the sector boundaries in an irregular way. For example, an ejection moving only 50 km/s faster than a sector boundary would displace the sector boundary about 2.5 AU between the sun and 20 AU, causing a change in the arrival time of the sector boundary at 20 AU equal to 10 days. Thus, transient ejecta, occurring randomly when the sun is active, can destroy a quasi-stationary sector pattern that might otherwise be present, without necessarily destroying the sectors themselves. In this way one can reconcile the presence of a two-sector pattern at the source surface with the absence of a recurrent two-sector pattern at Voyager 2 between 20 AU and 30 AU during 1988 and 1989 when the solar activity was high.

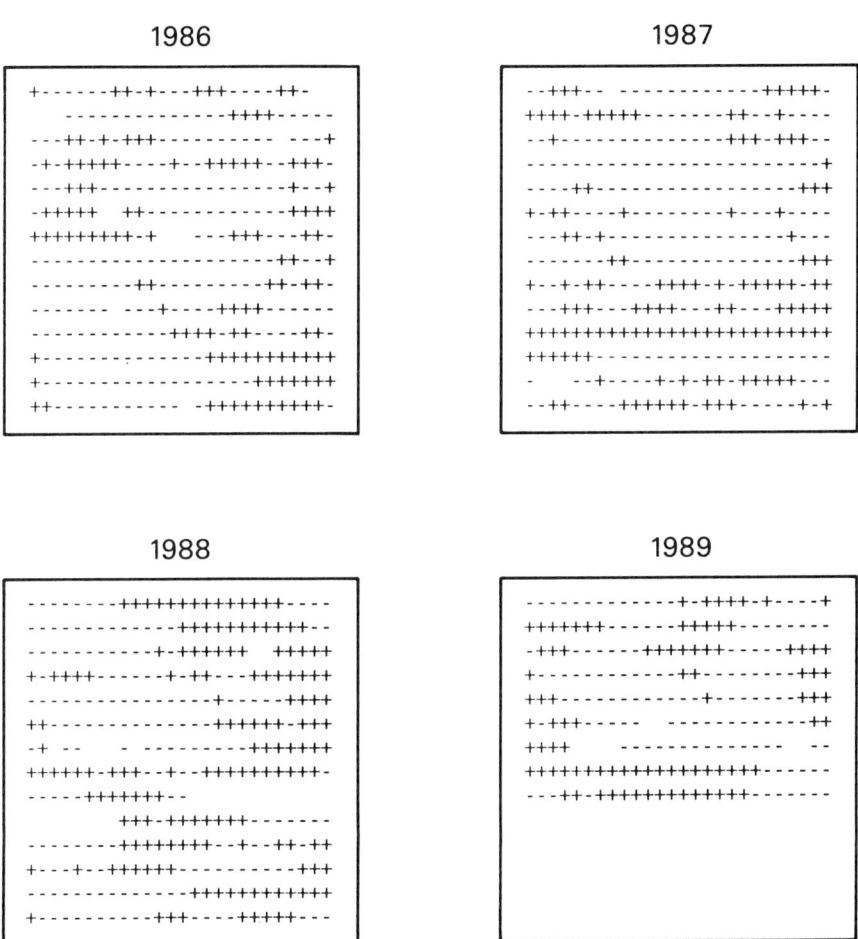

Fig. 2.6. Sectors in the outer heliosphere during ascending solar activity. (L.F. Burlaga and N.F. Ness, *J. Geophys. Res.*, **98**, 17451, 1993a, copyright by the American Geophysical Union.)

The elementary kinematic argument given above shows that transients can significantly modify the HCS and sector structure when the sun is active. Three more elaborate kinematic models of distortions of the HCS are worthy of note. They are not based on the idea that the HCS is perturbed by transients, and they do not account for the observations described above. These models are significant insofar as they demonstrate the sensitivity of the shape of the HCS to velocity perturbations.

The kinematic effects of a stationary latitudinal gradient of the solar wind speed on the shape of the HCS were considered by Suess and Hildner (1985), who found very large distortions of the HCS. Although this model is instructive and demonstrates the sensitivity of the HCS shape to latitudinal variation in the speed of the HCS, its relevance is questionable, because the assumption that the speed of the HCS is a strong function of latitude is probably not valid in general. Typically, the solar wind speed is minimum near a sector boundary (the HCS) at all latitudes during relatively quiet times, and the speed of the HCS can be constant at all latitudes even if the solar wind speed between the sector boundaries is latitude dependent. In another kinematic model, Suess et al. (1986) determined the fluctuations in the speed from observations made by Voyager when the HCS was nearly parallel to the ecliptic and very close to Voyager, and they used this ecliptic speed profile to calculate the distortions of an HCS that is initially perpendicular to the ecliptic. Both the assumption that the Voyager speeds measured near the HCS are the same as the speeds on the HCS and the assumption that speed variations in the ecliptic are representative of those normal to the ecliptic are questionable. Nevertheless, the authors demonstrate the sensitivity of the shape of the HCS to small variations in the velocity along the HCS.

Another kinematic model of the HCS representing an important class of distortions of the HCS (LeRoux and Potgieter, 1992) is based on the observation that the neutral line can evolve significantly during a year, which is the time for the HCS to move from 1 AU to 100 AU. This evolution causes distortions of the HCS on a scale of 100 AU relative to the profile of a stationary HCS, even if the solar wind speed is constant everywhere on the HCS, independent of both distance and latitude. The amplitude of the curve formed by the intersection of the HCS with a meridian plane during the ascending phase of solar cycle 22 may vary nonlinearly with increasing distance from the sun. In this case the HCS depends on the radial distance, but the radial dependence is a consequence of time variations of the neutral line at the source, rather than variations in the speed.

2.2.5 Solar Cycle Variations of Sectors and Neutral Line

The evolution of the neutral line over a whole solar cycle from October 1979 (CR 1687) through February 1989 (CR 1812) is discussed by Hoeksema and Suess (1990). The neutral line is most complex, evolves most rapidly, and extends to the poles when the polar fields reversed from CR 1687 to CR 1697, centered about CR 1692 at solar maximum. Multiple neutral lines were observed on CR 1697. The polar field reversal did not occur by the rotation of a dipole axis through the equator, as proposed by Saito et al. (1989). Rather, the polar fields fade and then grow with the opposite polarity (Hoeksema and Suess, 1990). This point was also demonstrated by Smith and Thomas (1986). Shortly after solar maximum during 1981, there was a simple four-sector pattern in the equatorial plane. During 1982 a two-sector pattern was observed, and during 1983 a four-sector pattern was observed again near the equator. The two-sector pattern with a relatively large amplitude of the neutral line appeared again in 1984, and a two-sector pattern with a smaller amplitude of the neutral line was observed during 1985. Finally, from solar minimum in late 1986 to mid-1987 the neutral line was very close to the equator with warps less than 10° at times, as discussed above. The pattern just described is seen again in reverse form on a shorter time scale from solar minimum to solar maximum. In particular, when the solar activity increased from mid-1987 through February 1989, the latitudinal extent of the neutral line increased, reaching near to the poles during 1989.

The model of the HCS as a tilted rotating plane near the sun is not valid near solar maximum, and it is only an approximation at other parts of the solar cycle. Nevertheless, this approximation is useful insofar as it allows one to describe the temporal variations of the HCS in terms of a single parameter, the tilt angle. The tilt angle exceeds 40° from the beginning of 1978 through most of 1984 and again from the beginning of 1988 through 1992 (Hoeksema, 1992).

2.3 The Spiral Magnetic Field

2.3.1 Parker's Model of Magnetic Field Directions

Parker (1958, 1963) predicted that the interplanetary magnetic field components in spherical coordinates should vary with distance R from the sun and heliographic latitude θ as follows:

$$B_R(R, \theta, \phi) = B_1\left(\theta, \phi - \frac{R\Omega}{V}\right)\left(\frac{R_1}{R}\right)^2 \tag{2.1}$$

$$B_\theta(R, \theta, \phi) = 0 \tag{2.2}$$

$$B_\phi(R, \theta, \phi) = B_1\left(\theta, \phi - \frac{R\Omega}{V}\right)\left[\left(\frac{R_1\Omega}{V}\right)\left(\frac{R_1}{R}\right)\cos(\theta)\right] \tag{2.3}$$

Here B_1 is the magnetic field strength at a reference radius R_1 which is usually chosen to be 1 AU; Ω is the rotation rate of the sun, and V is the solar wind speed in km/s. These equations are referred to as the "spiral field model" and "Parker's model." They are derived from equations (1.7) and

28 INTERPLANETARY MAGNETOHYDRODYNAMICS

(1.8) in spherical coordinates on the following assumptions: (1) Ω is a constant, and (2) **V** is radial and independent of R and ϕ but can be a function of latitude. These equations are a consequence of the fact that the sun is a singularity in the heliosphere, which to lowest order is a particular kind of a point defect (Burlaga and Ness, 1993b). The theory of defects (Mermin, 1979; Michel, 1980) might prove to be useful in the analysis of complicated vector fields in the solar wind.

The integral curves of the magnetic field **B** are Archimedean spirals, $R = R_1(V/\Omega R_1)(\phi - \phi_0)$. The coefficient $(V/\Omega R_1)$ is in fact approximately equal to 1, although one must be careful in using this approximation when high accuracy is required. The form of an Archimedean spiral on a scale of 40 AU is shown in Fig. 2.7. The magnetic field lines are nearly radial close to the sun, and they are nearly perpendicular to the radial direction beyond $\approx$5–10 AU.

Near the sun the dominant component of the magnetic field is B_R,

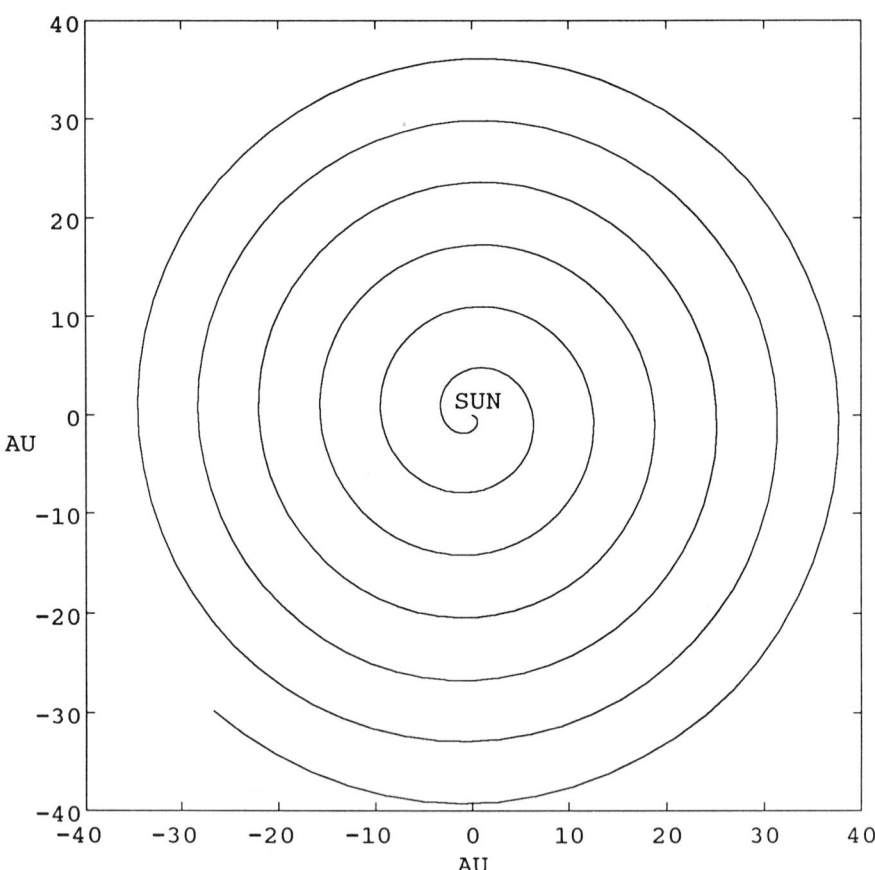

Fig. 2.7. Archimedean spiral from the sun to 40 AU.

which decreases rapidly with increasing distance from the sun as R^{-2} (see the review by Mariani and Neubauer, 1990). Beyond ≈ 3 AU the fluctuations in the radial component of the magnetic field owing to waves and turbulence become larger than the radial component of the magnetic field predicted by Parker's model (Burlaga et al., 1982b), so that it is no longer meaningful to compare the observations of $B_r(R)$ with equation (2.1).

Parker's model implies that $B_\theta = 0$. The angle ϕ_s between the magnetic field and the radial direction (the "spiral angle") is

$$\tan \phi_s = \frac{R\Omega}{V} \qquad (2.4)$$

where ϕ_s is measured clockwise from the outward radial direction when viewed from the northern hemisphere (i.e., $\phi_s = -\lambda$).

Parker's model of the global magnetic field direction is strongly supported by observations made within 10 AU (Behannon, 1978; Thomas and Smith, 1980; Burlaga et al., 1984a; Klein et al., 1987; Behannon et al., 1989). Even out to 30 AU, the magnetic field direction can be described to good approximation by Parker's model (Burlaga and Ness, 1993b). The most probable value of B_θ is 0° within ±5°. The other components of the magnetic field direction are nearly normal to the radial direction when the sun is not very active and there is a well-defined peak in the angle distribution. Deviations from the spiral direction are observed at large distances when the sun is active.

Bieber et al. (1993) analyzed 23 years of spacecraft observations spanning 27 AU and found that the spiral north of the HCS is more tightly wound than the spiral south of the HCS. The effect is small, $\approx 2°$, and one must be concerned about the experimental uncertainties, particularly in the early measurements at 1 AU and in the measurements of weak magnetic fields in the outer heliosphere, but the analysis cited was very careful. Smith and Bieber suggested that the effect could be the result of a nonzero azimuthal component at the source boundary. A kinematic analysis by Burlaga and Barouch (1976) showed that the azimuthal component of the interplanetary magnetic field is sensitive to a small nonzero azimuthal component of the magnetic field near the sun.

2.3.2 Radial Variations of Magnetic Field Strength

Parker's equations for the components of **B** given by equations (2.1)–(2.3) imply that the magnitude of the magnetic field strength in HG coordinates should vary as

$$B(R, t, \delta) = A(t, \delta) R^{-2} \left[1 + \left(\frac{419.5 R \cos(\delta)}{V(t, \delta)} \right)^2 \right]^{1/2} \qquad (2.5)$$

where R is the distance from the sun in AU, t is time, δ is the heliographic latitude, $V(t, \delta)$ is the bulk speed in km/s, and $A(t, \delta)$ is a measure of the source field strength which is independent of R. The rotation period of the source of the magnetic fields is assumed to be 26 days, leading to the factor 419.5.

When the speed is a constant independent of time and δ,

$$B(R) = A(t, \delta)R^{-2}\left[1 + \left(\frac{419.5}{V}\right)^2 (R\cos(\delta))^2\right]^{1/2} \quad (2.6)$$

The magnetic field strength, at a given distance and near the equatorial plane where $\delta = 0$, varies with time depending on the magnetic field strength near the sun. This temporal variation is measured by $A(t, \delta)$ which is related to the temporal variations of the magnetic field strength at 1 AU. Close to the sun, $B(R, t)$ varies approximately as $1/R^2$, essentially independent of the speed.

Far from the sun, equation (2.5) gives approximately

$$B(R, t, \delta) = \frac{419.5 A(t, \delta)}{RV(t, \delta)} \cos(\delta) \quad (2.7)$$

This shows that the magnetic field strength varies inversely with R, because the azimuthal component of the magnetic field is dominant, but it also varies inversely with the speed and directly as the temporal variations of the source magnetic field strength. Thus, any discussion of the radial variation of the magnetic field strength must correct for temporal variations in the source field strength (King, 1979; Burlaga et al., 1982b; Slavin et al., 1984), and it must consider the solar wind speed, which can vary with both latitude and time (Burlaga et al., 1984a; Klein et al., 1987).

Good agreement between the Voyager observations and the prediction of Parker's model for the magnetic field strength given by equation (2.5) was found by Burlaga et al. (1984a) and Klein et al. (1987). However, the analyses of the Pioneer 10 and 11 data obtained from 1972 to 1982 suggested that the magnetic field strength falls off faster with R than predicted by equation (2.5) (Smith and Barnes, 1983; Slavin et al., 1984; Thomas et al., 1986; Winterhalter et al., 1988, 1990; Smith, 1989, 1990, 1993; Winterhalter and Smith, 1989). This apparent departure from the model of Parker was described as a "flux deficit" by Thomas et al. (1986), who suggested that it was caused by meridional flux transport. The subsequent papers by Winterhalter and colleagues referenced above endorsed this explanation of the observations. Thomas et al. (1986) reported that the deficit is as large as 25% at 10 AU. Winterhalter et al. (1988) reported a deficit of 29% at 20 AU. Winterhalter et al. (1990), concluded that the deficit is "approximately 1%/AU" out to 20 AU.

Theoretically, small meridional flows can be produced by several mechanisms. The magnetic field pressure varies as $\cos^2(\delta)$, so that it is a maximum at the equatorial plane and decreases above and below the equatorial plane. This gives a pressure gradient that would tend to drive a flow away from the ecliptic, but the effect is much too small to explain the alleged flux deficit. An MHD model of Suess et al. (1985) predicts relatively large meridional flow velocities that could produce a flux deficit as large as 1%/AU, but Pizzo and Goldstein (1987) pointed out an error in the calculation. Correcting for this error, the meridional flow predicted by the model of Suess et al. is too small to account for the flux deficit. Pizzo and Goldstein (1987) offered an alternative explanation for the alleged flux deficit based on three-dimensional corotating streams. They argued that for suitable 3-D stream configurations, the pressure gradients in the corotating interaction regions at the front of the streams could drive a meridional flow of the required order of magnitude. However, the model requires very special stream configurations, and it does not explain why a flux deficit is observed during times when corotating streams are not the dominant feature of the interplanetary medium. There is no compelling theoretical model for meridional flows throughout the solar cycle of the order of magnitude required to explain a flux deficit of 1%/AU or more.

A quantitative analysis of the data from Pioneers 10 and 11 and Voyagers 1 and 2 obtained between 1 AU and 19 AU for the period 1972 through 1985 was carried out by Burlaga and Ness (1993b) with the aim of determining the magnitude of the alleged flux deficit and its uncertainty as a function of distance from the sun and time. They solved equation (2.5) for A, which gives

$$A(R, t, \delta) = B(R, t, \delta)R^2\left[1 + \left(\frac{419.5R\cos(\delta)}{V(t, \delta)}\right)^2\right]^{-1/2} \qquad (2.8)$$

They computed $A = A(R, t, \delta)$ from observations of $B(R, t, \delta)$ and $V(R, t, \delta)$, using the Pioneer and Voyager data, and they compared it with $A_1 = A(1\text{AU}, t, 0)$ computed from spacecraft at 1 AU. According to Parker's model, $A(R, t, \delta) = A_1(1\text{AU}, t, 0)$ for spacecraft in the equatorial plane. Thus, any flux deficit would be measured by

$$D = \frac{A - A_1}{A_1} \qquad (2.9)$$

Assuming that $D = 0$ at 1 AU, as it must be by definition, Burlaga et al. found the results for $D(R)$ shown in Fig. 2.8 for the Voyager and Pioneer data. A linear least squares fit to the Voyager data gives $D(R) = (0.0001 \pm 0.0021)R$, and a similar fit to the Pioneer data gives $D(R) = (-0.0015 \pm 0.0023)$. Neither set of data shows a significant flux deficit within the errors, and a flux deficit as large as 1%/AU is ruled out.

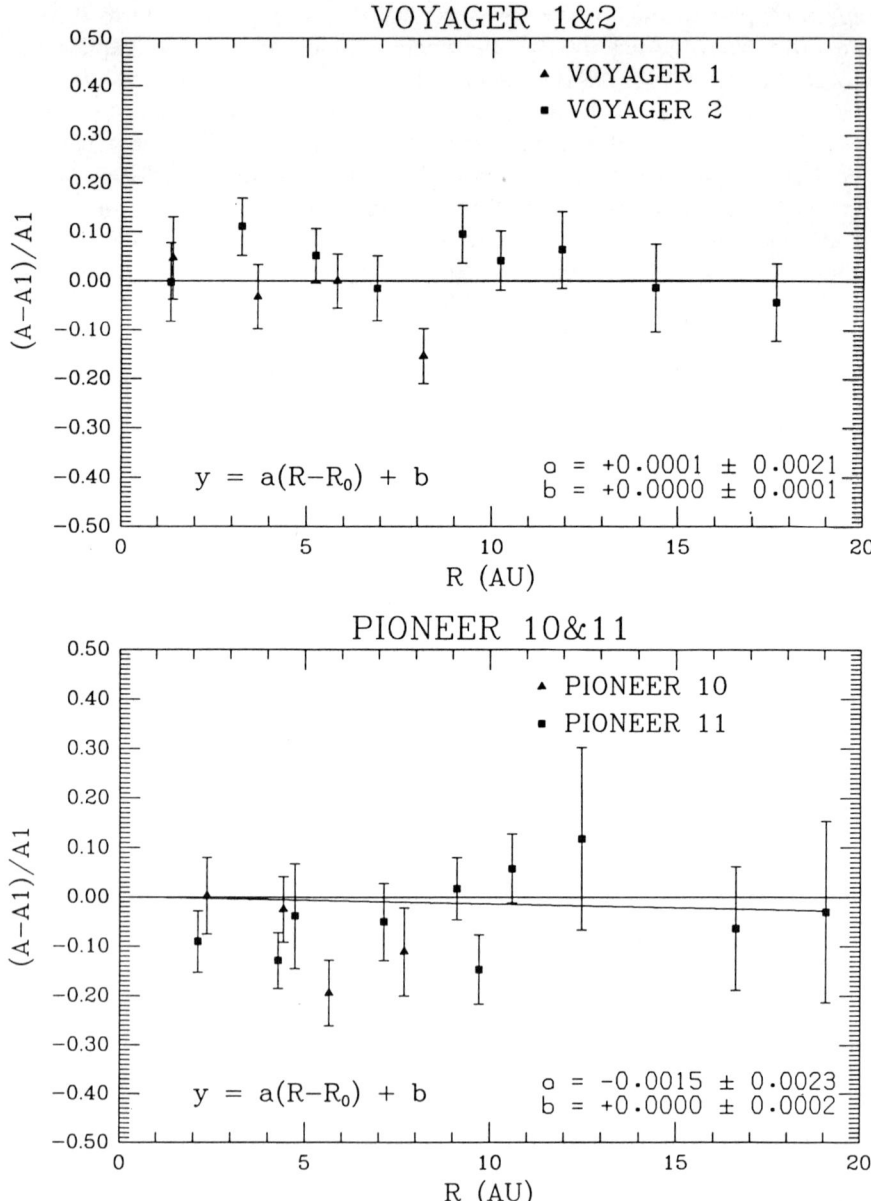

Fig. 2.8. Magnetic flux deficit? (L.F. Burlaga and N.F. Ness, *J. Geophys. Res.*, **98**, 3539, 1993b, copyright by the American Geophysical Union.)

Parker's model for the radial variation of the magnetic field strength provides a very satisfactory description of the observations between 1 AU, despite its simplicity and the complexity of the solar wind. The model works well because the main assumption of the model (radial solar wind velocity is independent of distance from the sun) is valid in the region and time period analyzed above; this will be demonstrated in Chapter 3 in the discussion of radial variations of the plasma (Section 3.5).

2.3.3 Latitudinal Variations of Magnetic Field Strength

Relatively little is known about the latitudinal variations of the magnetic field. The Voyager observations indicate that the magnetic field strength is weaker at higher latitudes than near the ecliptic (Burlaga et al., 1987b; Klein et al., 1987; Burlaga and Ness, 1993b). Luhman et al. (1988) suggested a latitude dependence in B_R proportional to $(1 - 0.8 \sin \delta)$ from 1979 to 1981 and during the first half of 1984, but proportional to $(1 + 0.8 \sin \delta)$ during 1982.

The magnetic field strength can vary with latitude at a given distance R from the sun for at least three reasons. Parker predicted a $\cos \delta$ dependence for the spiral magnetic field model assuming constant speed and source field strength, independent of latitude. The magnetic field strength at distance R will decrease with latitude if the magnetic field strength near the sun decreases with increasing heliographic latitude. Finally, the magnetic field strength far from the sun will decrease with increasing latitude if the solar wind velocity increases with latitude (see equation 2.5). Burlaga and Ness (1993b) inferred that the source magnetic field strength at 27°N was greater than the source field strength at 1°N during 1986 when the sun was relatively quiet, whereas the source field strength at 30°N was less than that at 3°N during 1988 and 1989 when the sun was active.

3

Large-Scale Plasma

3.1 Typical Plasma Characteristics at 1 AU

The state of the plasma at 1 AU is given by the bulk speed V, the proton number density N, the proton temperature T, and the electron temperature T_e. This section describes typical values of these basic quantities, the uncertainties in the measurements, and some fundamental parameters derived from them. Since the values vary with solar activity, this section aims only to present representative values, including both some early results of historical interest and some more modern results.

Vela 3 and IMP-3 observations of the state of the interplanetary medium at 1 AU just after solar minimum, from 1965 to 1967, were discussed by Ness et al. (1971) (referred to as NHB throughout this section. Heos-1 observations approaching solar maximum, from December 1968 to December 1969, were summarized by Formisano et al. (1974) (referred to as FMA throughout this section). The IMP-7 and IMP-8 observations made at 1 AU from 1972 to 1976 and the Helios observations made between 0.3 AU and 1 AU during the declining phase of solar activity from December 1974 to December 1976 are reviewed by Schwenn (1990).

The average speed V calculated by NHB and FMA is 412 and 409 km/s, respectively. The rms of the speed distribution is 78 and 72 km/s for the NHB data and the FMA data, respectively. The agreement between these two sets of measurements is good, as one expects for measurements of the solar wind speed at a given epoch of the solar cycle, because the speed is the most accurately determined plasma parameter. Schwenn (1990) gives a higher average speed, 468 km/s, but this was measured at a different phase of the solar cycle. It is convenient to think of the solar wind speed as approximately 0.25 AU/day, which corresponds to 434 km/s.

The average density N reported by NHB and FMA is 7.0 and 4.2 cm^{-3}, respectively. The difference in these two numbers might be the result of experimental errors. Absolute measurements of the density are very difficult to obtain, and variations of 30% from one instrument to another are not uncommon. The rms of the density distribution, which is a relative measurement, is 3.3 and 3.5 cm^{-3} for the NHB and FMA data, respectively. Schwenn (1990) gives 6.1 cm^{-3} for the average density later in the solar cycle.

The proton and electron temperatures are calculated from the respective velocity distributions. When computing the temperature it is often assumed that the temperature distribution is isotropic, but this is usually not the case. The temperature along the magnetic field direction is often different from the temperature perpendicular to the magnetic field. In some cases the temperature is calculated by fitting a Maxwellian distribution to the main part of the velocity distribution, thereby obtaining a "core temperature." In other cases the temperature is calculated from the second moment of the velocity distribution. The "moment temperature" is generally higher than the core temperature. Both approaches are sensitive to the presence of a non-Maxwellian tail that is usually present. The tail of the proton distribution function occasionally contains a bump (a "beam"). The tail in the proton distribution functions measured by electrostatic analyzers is contaminated by a contribution from helium ions when the temperature is high. The tail in the electron distribution can contain a highly directed component called the "strahl," discovery in the Helios plasma data (see Schwenn, 1990, for a discussion of the strahl and other references).

The average proton temperature reported by NHB and FMA is 8.0×10^4 and 7.4×10^4 K, respectively. The rms value of the proton temperature distribution is 4.0×10^4 and 6.0×10^4 K, respectively. The average proton temperature given by Schwenn (1990) for the declining phase of the solar cycle is significantly higher, 1.2×10^5 K.

The first direct measurements of T_e were reported during 1968, but the results varied widely. Perhaps the first definitive determinations of the electron temperature are those of Montgomery et al. (1968) and Burlaga (1968). Montgomery et al. measured the electron temperature directly with a special instrument. They found that the electron temperatures for 27 hours of data sampled during May and June 1967 by Vela 4 at 1 AU ranged between 7×10^4 and 2×10^6 K. The principal source of error is the distortion of the electron distribution caused by the electric fields associated with a spacecraft potential, resulting from the photoemission of electrons from the spacecraft. Burlaga (1968) determined the electron temperature indirectly from pressure balanced structures. He found $T_e = 1.5 \times 10^5$ K, consistent with the results of Montgomery et al. (1968). Both papers show that the electron temperature is higher than the proton temperature. Montgomery et al. found that the electrons are always hotter than the protons, T_e being 1.5–5 times T. They also observed that the electron temperature fluctuated much less than the proton temperature on the scale of a solar rotation. The mean electron temperature over long intervals was found to be 1.82×10^5 K (Serbu, 1972) and $(1.55 \pm 0.03) \times 10^5$ K (Scudder et al., 1973). Schwenn (1990) gives $T_e = 1.4 \times 10^5$ K for the electron temperature from the Helios data, which is comparable to the proton temperature during the declining phase of the solar cycle but higher than T during the rest of the solar cycle.

In summary, the electron temperature at 1 AU is approximately 1.5×10^5 K, it tends to be larger than the proton temperature, and it varies much less than the proton temperature.

The non-Maxwellian tails of the proton and electron distribution functions carry energy analogous to a heat flux, although the flux is not necessarily proportional to a temperature gradient as it is in ordinary fluids. The proton heat flux is 1.3×10^{-4} erg/cm^2/s and the electron heat flux is much larger, 4.3×10^{-3} erg/cm^2/s (Schwenn, 1990). Hundhausen et al. (1971) reported a proton thermal energy flux of about 10^{-5} erg/cm^2/s. Montgomery et al. (1968) estimated that the electron thermal energy flux varied between 5×10^{-3} and 2×10^{-2} erg/cm^2/s, while Ogilvie et al. (1971) determined that the electron "heat flux" is $<10^{-2}$ erg/cm^2/s. Representative values of the electron heat flux obtained by Feldman et al. (1975) are approximately $(8 \pm 5) \times 10^{-3}$ erg/cm^2/s. Obviously, the proton and electron heat fluxes are difficult to measure. Nevertheless the important point is clear: electron thermal energy is transported very efficiently from the sun through the interplanetary medium, while the proton thermal energy is not transported efficiently. This implies that protons tend to cool more rapidly than the electrons with increasing distance from the sun and expansion of the plasma. Thus, the electrons are hotter than the protons at 1 AU and beyond.

The average magnetic field strength reported by NHB and FMA is 5.2 and 6.0 nT, respectively, and the corresponding rms value of the distribution of magnetic field strengths is 2.4 and 2.5 nT, respectively. Since the earth's magnetic field strength is of the order of 100,000 nT, one can understand why measuring magnetic fields on a spacecraft in interplanetary space is a great challenge. Nevertheless, the magnetic field strength is typically known to ±0.1 nT at 1 AU. Measuring the magnetic fields in the outer heliosphere is a much greater challenge. For example, the mean field at 60 AU is only 0.1 nT, which is comparable to the errors tolerated in the measurements of the magnetic field at 1 AU!

Certain fundamental parameters in interplanetary dynamics are derived from the basic fields discussed above. Three particularly important parameters based on the fundamental field quantities are: (1) the Alfvén speed

$$V_A = \frac{B}{(4\pi NM)^{1/2}} \tag{3.1}$$

where M is the proton mass and N is the proton density (neglecting the small contribution of minor ions) (2), the Alfvén Mach number,

$$M_A = \frac{V}{V_A} \tag{3.2}$$

and (3) the ratio of the magnetic pressure to gas pressure

$$\beta = \frac{Nk(T + T_e)}{B^2/8\pi} \qquad (3.3)$$

The most probable value V_A for the Mariner 2 data in 1962 is 55 km/s, and 82% of the values are in the range from 30 km/s to 100 km/s (Neugebauer and Snyder, 1966). For the period 1965–67 at 1 AU, the most probable value of V_A is 38 km/s and the average value is 43 km/s, the range being 18–88 km/s (Ness et al., 1971). The average V_A from December 1968 to December 1969 is 74 km/s, from the Heos data analyzed by Formisano et al. (1974). Thus, on average the Alfvén speed is approximately 50 or 60 km/s, and it can vary by about a factor of 2 during the solar cycle.

The average Alfvén Mach number derived by NHB and FMA is 10.7 and 6.5, respectively. The rms of the distributions of Alfvén Mach numbers are 4.8 and 5.9, respectively. The important point is that the solar wind is almost always super-Alfvénic. A sub-Alfvénic solar wind was observed by Gosling et al. (1982) for portions of a 5-hour period on November 22, 1979, when the Alfvén speed was extraordinarily high ($V_A = 540$ km/s) owing to an abnormally low density (0.07 cm^{-3}) and the solar wind speed was only 320 km/s, but this observation is unique. The basic state of the solar wind is that of a super-Alfvénic, supersonic, supermagnetoacoustic plasma. Thus, the motion of the solar wind relative to the sun or a planet is like that of a supersonic gas rather than a subsonic fluid.

The value of β measures the relative importance of the magnetic pressure and the thermal pressure. The average value for the proton beta at 1 AU reported by NHB and FMA is 0.95 and 0.34, respectively. The rms of the beta distribution is 0.74 and 0.22, respectively. While these early measurements might not be precise, they demonstrate that the magnetic pressure is comparable to or slightly greater than the proton thermal pressure at 1 AU. Thus, a second fundamental characteristic of the interplanetary medium is that magnetic pressure force is comparable to the thermal pressure force. This means that one must use magnetohydrodynamics to explain the interplanetary dynamical processes, rather than gas dynamics.

3.2 Solar Cycle Variations at 1 AU

The average solar wind speed during the years 1962 through 1969, which included the minimum of solar activity in 1964 and the maximum in 1969, was near 400 km/s. The speed did not change appreciably from year to year (Gosling et al., 1971, 1976a). The difference between the highest yearly average speed and the lowest average speed was only about 25%, so that the variability of the speed with solar activity was small. The largest yearly average speed (490 km/s) occurred in 1962, during the declining phase of the solar cycle (Gosling et al., 1971).

The temporal variation of the solar wind speed at 1 AU from 1973 to 1988 is discussed by Lazarus and Belcher (1988). The speed was relatively high (>500 km/s) in 1973 and 1974, during the declining phase of solar cycle 21, approximately one solar cycle after Gosling et al. (1971) observed relatively high speeds during 1962. Diodato et al. (1974), Feldman et al. (1978a), and Schwenn (1990) all noted the high speeds during 1973 and 1974. Schwenn showed that the speed was again high during the declining phase of the next solar cycle in 1985. Thus, for three successive solar cycles, the highest speeds occurred during the declining phase of solar activity.

A minimum speed, 400 km/s, occurred at solar maximum in 1980, according to Schwenn (1990). On the other hand, Gosling et al. (1971) found that the minimum speed occurred at solar minimum in 1965. Similar diversity concerning the phase of the solar cycle at which V is minimum was reported by other observers. Thus, there is no evidence that the minimum speed occurs at a particular part of the solar cycle.

The solar cycle variations of the density and temperature from 1973 to 1978 are discussed by Feldman et al. (1978) and Lazarus and Belcher (1988). Just after solar minimum in 1977, when the speed was exceptionally low, the density was maximum and the temperature was low (Schwenn, 1990). Quite generally, the density is high, the temperature is low, and the speed is low near the heliospheric current sheet, which is near the solar equator close to the time of minimum solar activity. During the declining phase of the solar cycle, in 1974, the density was low and the temperature was high, corresponding to high speeds associated with corotating streams.

Long-term variations in the electron temperature were studied by Feldman et al. (1978) for the period from 1971 to 1978 using the electron data from IMP-6, IMP-7, and IMP-8. From 1971 to the beginning of 1975, T_e was approximately constant, the core temperature being about 1.2×10^5 K and the total temperature being 1.4×10^5 K. From 1971 to 1975 the electron temperature increased approximately 30%, and then it declined somewhat during 1978. Thus, T_e began to increase in the year prior to solar minimum and it continued to increase at solar minimum and as the solar activity increased. The reason for this behavior is not clear; it is possible that the behavior is not real. Obviously, it is important to extend the studies of $T_e(t)$.

3.3 Relations Among Fields at 1 AU

A quantitative relation between the proton temperature T and the speed V was found by Burlaga and Ogilvie (1970a), namely

$$T^{1/2} = AV + B \qquad (3.4)$$

where T is in units of 10^3 K, V is in km/s, $A = 0.033 \pm 0.001$, and

$B = -4.8 \pm 0.4$. Burlaga and Ogilvie (1973) showed that this relation did not vary over a substantial portion of the solar cycle (from 1966 to 1971) and was not influenced by stream interactions by more than approximately 15%. Hundhausen et al. (1970) confirmed the existence of a general relation between T and V, although they found a slightly different quantitative relationship from that of Burlaga and Ogilvie. Lopez (1987), and Lopez and Freeman (1986) presented additional evidence in support of equation (3.4) and its invariance with solar activity for $V < 500$ km/s using Helios 1 data for the period 1974–80. Lopez (1987) showed that equation (3.4) is also valid for the data from August 25, 1984, to April 17, 1985. Lopez and Freeman (1986) and Lopez (1987) found that the relation $T = AV + B$ provided a better fit to the data with $V > 500$ km/s.

An inverse relation between the density and the solar wind speed was demonstrated by Burlaga and Ogilvie (1970b) using data from several spacecraft. The data from Explorer 34 are described by the relation

$$N = \text{constant} \times V^{-1.5} \tag{3.5}$$

The densities from IMP-1, Vela 3, Pioneer 11, and Mariner show a similar variation with V, but they tend to be higher than the densities measured by Explorer 34.

Something equivalent to an energy relation is needed in order to close the MHD equations, as discussed in Chapter 1. Since the laws of transport of thermal energy in a collisionless, inhomogeneous plasma are not known, it is customary to assume a polytropic relation between the proton pressure and the density and a polytropic relation between the electron pressure and density. For the electrons, Sittler and Scudder (1980) proposed the relation

$$p_e = C \times N^{(1.175 \pm 0.03)} \tag{3.6}$$

based on measurements over a wide variety of conditions. If the electrons cool adiabatically while they move from the sun to 5 AU, one would expect an exponent of $5/3 = 1.667$, and if they remained isothermal the exponent would be 1. Thus the polytropic law for electrons on a large scale and for a variety of flows is more nearly isothermal than adiabatic, indicating that thermal energy is transferred rather effectively from the sun through the interplanetary medium.

3.4 Latitude Variations

Latitude variations in the solar wind speed were first reported by Hewish and Dennison (1967), based on interplanetary scintillation observations from a single radio source 3C 48 during 1966. The speed at 64°N was 490 km/s while that in the ecliptic was 300 km/s, giving a latitude gradient in the speed equal to 3 km/s/deg. The same measurements showed that the solar wind velocity was radial within ±15°. Further evidence for the existence of latitudinal gradients in 1967 was given by Hewish and Symonds (1969) and in 1972 by Coles and Maagoe (1972). Systematic observations of the solar wind speed between latitudes of 80°S and 80°N from 1971 to 1975 (Coles and Rickett, 1986) consistently showed a minimum speed near the ecliptic and a latitude gradient in the speed of 2.1 km/s/deg.

In situ evidence for latitudinal variations of the solar wind speed within ±7.25° of the ecliptic was presented by Hundhausen et al. (1971) four years after the discovery of latitude gradients in the speed by Hewish and Dennison. Their results are based on the fact that the earth's latitude varies with respect to the solar equator by ±7.25°, reaching a maximum solar latitude on approximately September 7 and a minimum solar latitude on approximately March 6. Hundhausen et al. observed low speeds near the solar equator and higher speeds at higher heliographic latitudes between July 1965 and June 1968. They observed a gradient of 6.5 km/s/deg in the northern hemisphere and 1.7 km/s/deg in the southern hemisphere.

Evidence of latitudinal variations of the speed based on simultaneous measurements from two or more spacecraft from June to December 1967 was presented by Rhodes and Smith (1975) for speeds observed near sector boundaries. Their results are consistent with latitude gradients of approximately 10 km/s/deg, but they could not exclude the possibility that the observations were due to radial variations of the solar wind speed, because of the spacecraft they used was Mariner 5, which moved from 1 AU to Venus at 0.6 AU. Rhodes and Smith (1976a,b) observed a speed gradient of 15 km/s/deg when the observations were not restricted to the vicinity of sector boundaries. Rhodes and Smith (1981) observed a very large wind shear of 60 km/s/deg for a single stream during 1967. Mitchel et al. (1981) observed latitude gradients in the speed associated with streams exceeding 20 km/s/deg in a boundary layer whose width was approximately 5° during the period from March 1972 to April 1976. Large latitudinal gradients in the speed were observed in the outer heliosphere during 1986 and 1987 near solar minimum (Barnes *et al.*, 1992).

A model of the latitude dependence of the solar wind speed, based on the coronal magnetic field geometry, was presented by Pneuman (1976). He assumed that the solar magnetic field has the form of a dipole at solar minimum, which is extended by the solar wind as described above to give a neutral line at the source surface and near-equatorial heliospheric current sheet in the interplanetary medium. He made the important suggestion that the speed is inversely related to the field divergence from pole to equator. The latitude gradient in speed would be greatest near solar minimum. His model implies that the speed would be minimum at the neutral line and the HCS. Whang and Sheeley (1990, 1992) arrived at a similar conclusion.

The observations referenced above describe the latitude variations of the speed relative to the solar equator. It was known for many years that the minimum in solar wind speed at 1 AU occurs at sector boundaries (Wilcox and Ness, 1965; Ness et al., 1971; Sawyer, 1976), and it was shown above that the heliospheric current sheet (HCS and neutral line) is not close to the solar equator, except near solar minimum. Following the ideas of Pneuman (1976), Hakamada and Akasofu (1981) and Zhao and Hundhausen (1981) suggested that during a significant part of the solar cycle the speed gradient should be measured with respect to the "magnetic latitude" λ relative to the heliospheric current sheet, with the minimum speed at the HCS. Hakamada and Akasofu (1981) proposed the equation

$$V \text{ (km/s)} = 700\left[\frac{1-1}{\cosh(0.06\,|\lambda|)}\right] + 300 \tag{3.7}$$

Support for this idea was provided by Zhao and Hundhausen (1983) based on Helios data, coronal data, and interplanetary scintillation measurements obtained in 1976. They obtained somewhat different equations

$$V \text{ (km/s)} = 350 + 800 \sin^2 \lambda \quad \text{for} \quad |\lambda| < 35°$$
$$V \text{ (km/s)} = 600 \text{ km/s} \quad \text{for} \quad |\lambda| > 35° \tag{3.8}$$

Hakamada and Munakata (1984) proposed a similar result for the period from May 1976 to August 1977:

$$V \text{ (km/s)} = 408 + 473 \sin^2 \lambda \tag{3.9}$$

The observations described above and the model of Pneuman (1976) provide the basis for a global model of the solar wind speed throughout the heliosphere and throughout the solar cycle. The minimum speeds should be observed at the HCS and should evolve with solar activity in the same way as the neutral line. The speed increases with angular distance from the HCS. At solar minimum, when the HCS is near the ecliptic, large latitude gradients in speed should be observed by spacecraft near the ecliptic, Near solar maximum, when the HCS makes large angles with respect to the ecliptic near the equator, latitude gradients in speed should not be observed. Empirical support for this conceptual model can be found in numerous papers. Kojima and Kakinuma (1987) showed that the major minimum speed regions are along the neutral line throughout the whole solar cycle from 1973 to 1985. They also noted that the width of the low speed region is smallest near solar minimum, hence the velocity gradients are largest at that time.

During 1986 and 1987 there were large latitudinal gradients in the solar wind speed (Gazis et al., 1989; Barnes, 1990), which disappeared during 1988 and 1989, when solar activity was increasing. Voyager 2 was in the ecliptic (within 7.25° of the heliographic equator) throughout the period, and its radial position increased from 18.9 AU at the beginning of 1986 to 30.2 AU at the end of 1989. Voyager 1 was north of the heliographic equator, its heliographic latitude increasing from 26.8° at the beginning of 1986 to 30.8° at the end of 1989. The radial position of Voyager 1 increased from 25.4 AU to 39.9 AU during this period.

A remarkable global view of the latitudinal variations of the solar wind speed from 1977 to 1991 was derived by Sheeley (1992, Fig. 3) from solar magnetic field measurements. The map was derived using the inverse correlation between the solar wind speed at 1 AU, and the flux tube divergence in the corona found by Pneuman (1976), Sheeley et al. (1991), and Whang and Sheeley (1990, 1992). The speed is low at all latitudes at the solar maxima of 1980 and 1990, and the speed is low only near the equator near solar minimum in 1986 when the HCS was near the equator. High speeds occur in the polar regions during the declining phase of solar activity from 1982 through 1983. High speeds are also observed away from the equator during the later years in the declining phase of solar activity and during the ascending phase.

Let us now discuss the large-scale latitudinal variations in the density. A latitudinal variation in the density between ±7.25° heliographic latitude at 1 AU was observed by the Vela 3 and Vela 4 spacecraft from July 1965 to July 1968 (Hundhausen et al., 1971). The latitudinal variation in density was anticorrelated with the latitudinal variation in speed. The density was highest near the solar equator and lowest at ±7.25°. Based on a comparison of interplanetary scintillation measurements with measurements of the average brightness of the UV-E corona, Watanabe et al. (1974) suggested that the latitudinal distribution of the solar wind speed is inversely related to the brightness distribution of the solar UV-E corona. This result implies an inverse relation between speed and density at various latitudes. Hence, the result implies that the density decreased with increasing latitude up to 64°N in 1972 and 1973. A similar result was reported by Sime and Rickett (1978), who observed an inverse relation between the speed inferred from interplanetary scintillation observations and the intensity of the white light coronameter data taken at $1.5R_S$ from 1973 to 1977. This work was extended by Sime and Rickett (1981), who showed that an anticorrelation between the speed and the coronal density to which it maps extends to all latitudes up to 60°N on time scales from 2 days to 1 year from 1972 to 1975. Borrini et al. (1981) and Gosling et al. (1971) found an anticorrelation between speed and density in their plasma measurements at sector boundaries, corresponding to the HCS. They interpreted the high density region in the neigborhood of the HCS as the signal of a coronal streamer at 1 AU.

3.5 Radial Variations

The radial variations of the speed in the inner heliosphere were studied using the data from Mariner 2, Pioneer 9, and Helios 1 and 2. No statistically significant dependence of the average solar wind speed on distances between 0.7 AU and 1 AU was observed in a study of Pioneer 9 data and Ogo 5 data (at 1 AU) for five solar rotations during 1968 and 1969 (Intriligator and Neugebauer, 1975). An analysis of interplanetary scintillation observations from 1971 to 1975 (Coles and Rickett, 1976) showed no significant variation of the speed with radial distance from the sun between 0.4 and 1.1 AU. Other papers on the radial variation of V in the inner heliosphere, particularly those concerning the Helios observations, were reviewed by Schwenn (1990).

Our knowledge of the radial variations of the solar wind speed in the outer heliosphere is based primarily on the data from Pioneers 10 and 11 and Voyagers 1 and 2. Using simultaneous data from Voyagers 1 and 2 and IMP-8, Gazis (1984) showed there is no significant radial variation of the solar wind speed between 1 and 10 AU. Very little variation of the speed with distance between 1 and 15 AU was found by Collard et al. (1982) based on simultaneous observations from Pioneers 10 and 11 and IMP-6, -7, and -8.

More recent observations also show the independence of the solar wind speed and distance from the sun out to larger distances under most cricumstances. Burlaga and Ness (1993b) analyzed the radial variation of the relative difference $(V - V_1)/V_1$ of the yearly averages of the speed V measured by Voyagers 1 and 2 in the outer heliosphere and the speed V_1 measured at 1 AU as a function of distance R from the sun. A linear least squares fit to the data gives a slope 0.0027 ± 0.0026, consistent with zero, indicating no radial variation of the speed between 1 and 20 AU from 1977 through 1984. These results do not support the early prediction of Goldstein and Jokipii (1977) that the solar wind speed should decrease by 7–25 km/s between 1 and 6 AU.

The radial variation of the density is consistently found to be $N \propto R^{-2}$, which is expected as a consequence of the conservation of mass in a spherically expanding solar wind with a speed that is independent of the distance R from the sun. Lazarus and McNutt (1990) show a plot of NR^2 versus time (which is related to the distance from the sun) from 1977 to 1989 (1–30 Au) for 200-day averages of Voyager 2 data. In their data, NR^2 is nearly constant, indicating that the density does vary approximately as R^{-2}.

The radial variation of the proton temperature T measured by Pioneer 11 and Pioneer 10 between 1 AU and 20 AU was discussed by Kayser et al. (1984), who found $T(10^3 \text{ K}) = 73 R^{-0.57}$. Other early measurements of $T(R)$ were reported by Collard and Wolfe (1974), Smith and Wolfe (1979), Mihalov and Wolfe (1978, 1979), Collard et al. (1982), Smith and Barnes (1983), Gazis and Lazarus (1982), Gazis (1984), and Gazis et al. (1988, 1992). These early measurements all showed that $T \propto R^{-\alpha}$ where α is between 0.5 and 0.7. The protons cool more slowly than adiabatically. Since the proton heat conduction is rather ineffective, this implies some heating mechanism. Corotating shocks are important in heating the solar wind in certain regions in the outer heliosphere during the declining phase of the solar cycle (Whang et al., 1991), but other heating mechanisms might also be effective.

The temperature measurements from the Voyager and Pioneer spacecraft depend on time as well as distance. The distance increases with time on a scale of the solar cycle. In particular, the proton temperature varies with solar activity as discussed above. Thus one should consider the radial variation of T normalized by the proton temperature at 1 AU. Gazis et al. (1992) and Barnes et al. (1992) present values of $T(R)$ measured by Voyager and Pioneer normalized by measurements of $T(t)$ by IMP at 1 AU. The Pioneer 10 data extended from 1 AU in 1972 to 53 AU near the end of 1990, while the Voyager 2 data extend from 1 AU in 1977 to 36 AU at the end of 1990. There are significant differences between the temperatures measured by Voyager 2 and Pioneer 10 from 1977 through 1984.

The radial variation of the electron temperature in the inner heliosphere was studied using data from Mariner 10 and Helios 1 and 2. The electron temperature between 0.47 AU and 1 AU was determined by Ogilvie and Scudder (1978) using measurements by Mariner 10 from January 9 to March 30, 1974. They found that the data could be described by a single power law $T_e = \text{constant} \times R^{-0.3}$ for the core electrons. Using the same Mariner 10 data, together with simultaneous IMP-6, IMP-7, and IMP-8 data at 1 AU, Feldman et al. (1978, 1979) arrived at a very different conclusion. They observed a minimum in T_e at 0.6 AU, and they described the radial variation of T_e by two power laws. Feldman et al. concluded that between 0.47 AU and 0.62 AU, T_e was decreasing with increasing R as to $R^{(-1.14\pm0.24)}$ and that between 0.62 AU and 1 AU, T_e was increasing with R as $R^{(+0.28\pm0.13)}$. An increase of T_e with increasing R has not been confirmed.

Combining electron temperature observations from Mariner 10 with similar data from Voyager 2 obtained between 1 AU and 4.76 AU, Sittler and Scudder (1980) found a polytropic relation between the electron pressure p_e and the density of the form $p_e = C \times N^{(1.175\pm0.03)}$. Assuming that N varies as R^{-2} gives $T_e = (1.01 \times 10^4 \text{ K}) \times R^{-(0.35\pm0.03)}$. The upper limits on T_e derived by Burlaga et al. (1990c) from pressure balanced structures out to 24 AU are consistent with this result. Sittler et al. (1981) observed $T_e \propto R^{-(0.34\pm0.16)}$ between 1.36 AU and 2.25 AU. Preliminary electron data from Ulysses obtained between 1.15 AU and 2.76 AU (Bame et al., 1992) show that the electron core temperature varied as $T_e \propto R^{-0.7}$. There are no direct measurements of T_e beyond 5 AU, but Burlaga et al. (1990c) obtained upper limits for T_e between 18.9 AU and 21.0 AU from 11 pressure balanced structures using a technique similar to that of Burlaga (1968). On the assumption of a negligible contribution from the pickup ions to the pressure balanced structures, they found $T_e \leq (5.2 \pm 2.3) \times 10^4 \text{ K}$ at 20 ± 1 AU. This upper limit is to be compared with the result $1.5 \times 10^4 \text{ K}$ extrapolated from the Ulysses data and with the value $3.5 \times 10^4 \text{ K}$ derived from an extrapolation of the results of Sittler and Scudder (1980).

4

Pressure Balanced Structures

4.1 Basic Concept

4.1.1 Theory of Pressure Balanced Structures

Pressure balanced structures are structures across which the total pressure is a constant (Ferraro and Plumpton, 1966; Burlaga, 1971a). Assuming that there is no motion of a pressure balanced structure relative to the solar wind frame, the equation of motion (1.1) is simply $-\nabla p' + (\mathbf{J} \times \mathbf{B}) = 0$. Using equation (1.4) for $\mathbf{J} \times \mathbf{B}$ and neglecting the magnetic curvature force $(\mathbf{B} \cdot \nabla)\mathbf{B}/4\pi$, a solution of the equation of motion is $P = p' + B^2/8\pi$, where P is a constant, the total pressure. Substituting equation (1.3) for p' gives the basic equation of a pressure balanced structure

$$P := Nk(T + T_e) + \frac{B^2}{8\pi} = \text{constant} \tag{4.1}$$

In some cases it is necessary to add a small correction term to include the pressure of the alpha particles (see Burlaga, 1968, and Burlaga et al., 1990c). Beyond 35 AU the pressure of pickup protons can be greater than the pressure of solar wind protons and electrons (Burlaga et al., 1994), which implies an important change in the MHD dynamics in the distance heliosphere in response to the influence of interstellar material.

4.1.2 Observations of Pressure Balanced Structures

The existence of pressure balanced structures in the solar wind at 1 AU was demonstrated by Burlaga (1968) using plasma and magnetic field data from Explorer 34. The panels in Fig. 4.1(a) show one example of a pressure balanced structure (PBS) from that paper. The magnitude of $\mathbf{B}$, B, decreases from 8 nT to about 2 nT in 6 hours. The change in B consists of three discontinuities, some abrupt but not discontinuous changes, and some slower variations. The proton density $n = N$ is anticorrelated with B, and it increases from 12 cm^{-3} to 29 cm^{-3}. The proton temperature T, measured by the thermal speed V_T (which is related to T by $T = MV_T^2/2k$, where k is

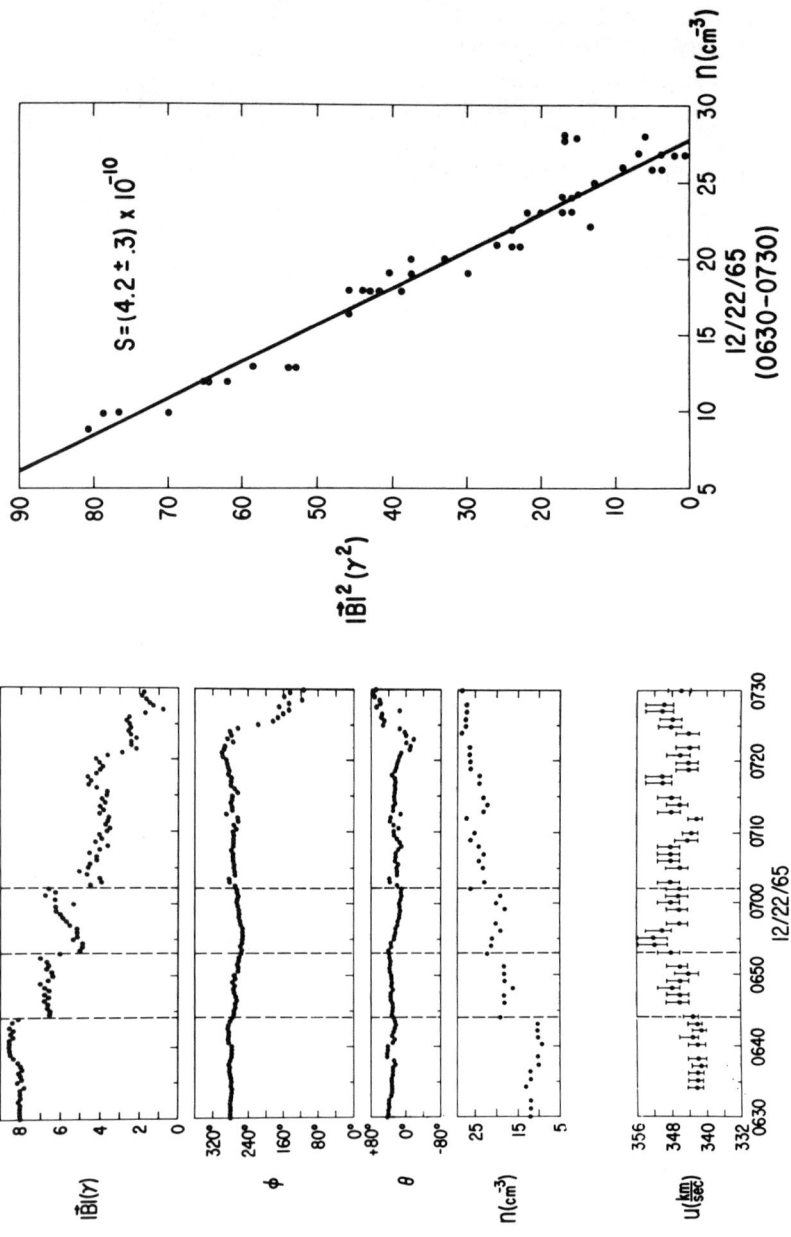

Fig. 4.1. (a) A pressure balanced structure with tangential discontinuities. (b) A linear relation between B^2 and density across the pressure balanced structure in (a). (After L.F. Burlaga, *Solar Phys.*, **4**, 67, 1968.)

Boltzmann's constant), is relatively constant and equal to $T = 2.4 \times 10^4$ K in this event. The solar wind speed $U = V$ is nearly constant, changing slightly between 342 km/s and 350 km/s. The changes probably represent velocity shears across the PBS, which do not affect the pressure balance condition, as we discuss below. Since the PBS is convected passively with the wind past a fixed spacecraft at a speed of 345 km/s in one hour, its radial extent is approximately 0.01 AU, which is the order of magnitude of the size of many pressure balanced structures in the solar wind at 1 AU.

If the electron temperature was constant across the PBS in Fig. 4.1, then equation (4.1) implies that

$$B^2 = -Sn + a \qquad (4.2)$$

where $a = 8\pi P$ and $S = 8\pi k(T + T_e) =$ constant; hence, a plot of B^2 versus n should be a straight line with slope $-S$. Figure 4.1(b) shows that a plot of B^2 versus n for the data in the PBS of Fig. 4.1(a) is indeed a straight line, with a slope $S = (4.2 \pm 0.3) \times 10^{-10}$. Given S and the measured value of $T = 2.4 \times 10^4$ K, one can calculate the electron temperature $T_e = 10^5$ K, which is consistent with the directly measured electron temperatures discussed earlier (Section 3.1). From the consistency of these results with equation (4.1) Burlaga (1968) concluded that the MHD structure in Fig. 4.1(a) is a pressure balanced structure.

Another example of a PBS is shown in Fig. 4.2. The data were obtained from Voyager 2 at 2.1 AU over an interval of 5 hours, corresponding to a radial extent of approximately 0.05 AU. In this case the total pressure P shown at the bottom of the figure is clearly constant across the PBS. Note the detailed anticorrelation between N and B, despite the variability in T that is also present. Both the proton and electron temperatures were measured directly in this case, so that the evidence for a pressure balanced structure is direct and conclusive. The electron temperature calculated from N, T, and B as that value which gives the smallest deviation from $P =$ constant across a PBS is $(5.25 \pm 0.2) \times 10^4$ K, which agrees well with the directly measured moment temperature $T_e = (5.2 \pm 0.05) \times 10^4$ K. Another PBS based on a complete set of measurements, incuding the electron temperature, observed by Voyager at 2 AU was discussed earlier by Burlaga et al. (1981b).

Pressure balanced structures are observed at all distances between 1 AU and 24 AU. An example of a PBS at 23.7 AU is shown in Fig. 4.3. In this case there are large changes in B, N, and T, but again the total pressure P shown in the bottom panel is a constant. Direct measurements of T_e are not available beyond approximately 5 AU, but the electron temperature that gives the smallest deviation from $P =$ constant for this PBS at 23.7 AU is $T_e = (6.3 \pm 0.2) \times 10^4$ K.

The magnetic pressure is anticorrelated with the thermal pressure when equation (4.1) is valid. Burlaga and Ogilvie (1970b) found that at 1 AU the magnetic pressure is anticorrelated with the thermal pressure on a scale of 1

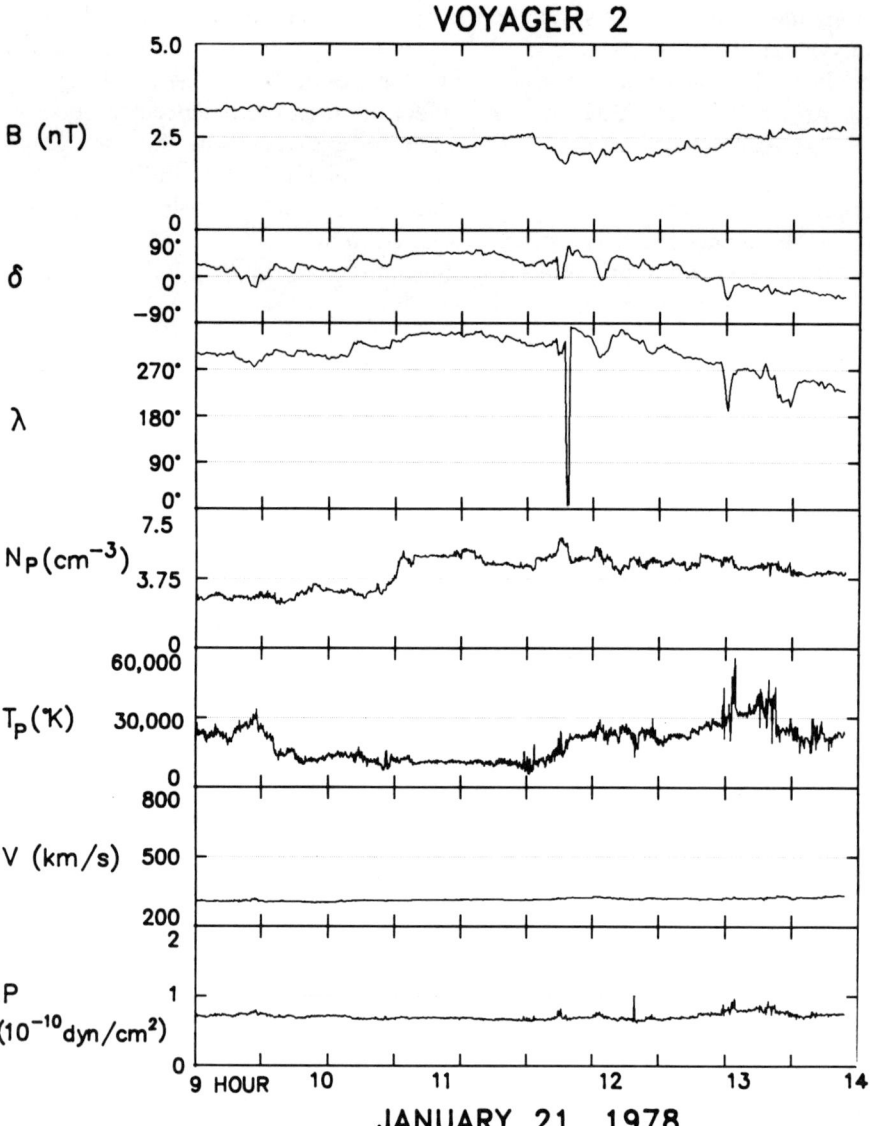

Fig. 4.2. A pressure balanced structure at 2.1 AU. (L.F. Burlaga, J.D. Scudder, L.W. Klein, and P.A. Isenberg, *J. Geophys. Res.*, **95**, 2229, 1990c, copyright by the American Geophysical Union.)

hour (0.01 AU), but these pressures are positively correlated on a scale of greater than 2 days. The tendency to be in equilibrium on a scale of the order of 0.01 AU (the "microscale," Burlaga and Ness, 1968) is a general property of the interplanetary medium at 1 AU and in the outer heliosphere. An anticorrelation between the magnetic pressure and the ion

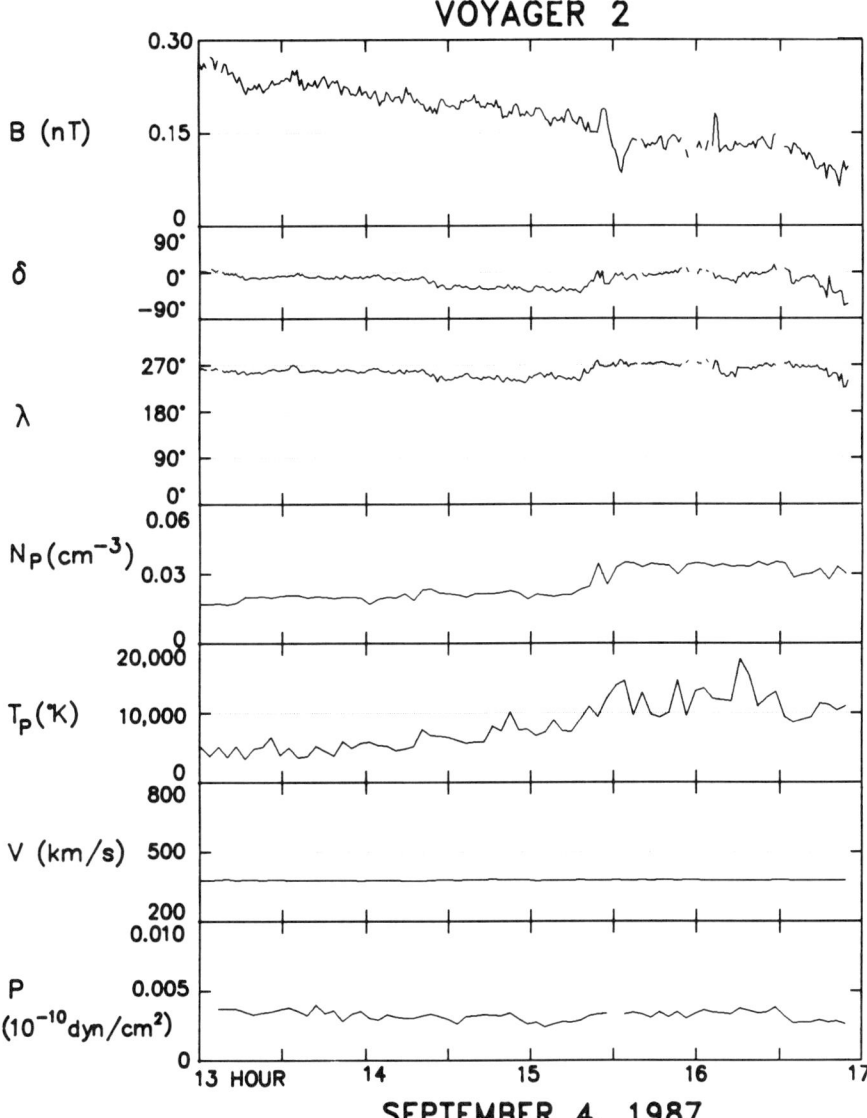

Fig. 4.3. A pressure balanced structure at 23.7 AU. (L.F. Burlaga, J.D. Scudder, L.W. Klein, and P.A. Isenberg, *J. Geophys. Res.*, **95**, 2229, 1990c, copyright by the American Geophysical Union.)

thermal pressure was commonly observed between 1 AU and 10 AU in the Voyager data (Vellante and Lazarus, 1987).

The density generally changes more across a pressure balanced structure than the temperature, as observed in the examples discussed above, so that an anticorrelation between the magnetic pressure and the thermal pressure is largely (but not entirely and not always) the result of an anticorrelation between B and N. Exceptions to this are observed in the inner heliosphere (Roberts, 1990). An anticorrelation between the magnetic field strength and the density appears to be a general feature of the solar wind in the outer heliosphere (Roberts, 1990).

4.2 Tangential and Rotational Discontinuities

4.2.1 Tangential Discontinuities

A tangential discontinuity (TD) is a special case of a pressure balanced structure, viz., a surface across which the pressure normal to the surface is constant. The magnetic field direction can change across a tangential discontinuity, but there can be no component of **B** normal to the surface in any reference frame (see Fig. 4.4). Similarly, there can be a velocity shear across a TD, but there is no component of the velocity normal to the surface in a frame moving with the solar wind. When the proton and electron temperatures are approximately isotropic, equation (4.1) applies to a tangential discontinuity. Otherwise, one substitutes the perpendicular temperatures in this equation.

Three tangential discontinuities are marked by the vertical dashed lines in Fig. 4.1(a). There there are large changes in B and N. There is also a change in the bulk speed $V = U$, which is probably a consequence of a shear across the discontinuity surface, as we discuss below.

Any combination of values of N, T, and B is allowed across a tangential discontinuity, provided equation (4.1) is satisfied. The magnetic field direction can also change across a tangential discontinuity in general. A

TANGENTIAL DISCONTINUITY

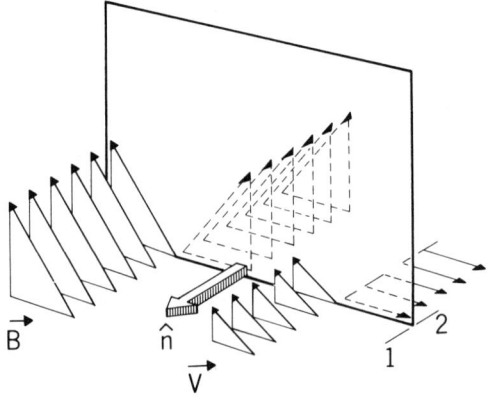

Fig. 4.4. Sketch of a tangential discontinuity.

particularly common case in the solar wind is that for which the magnetic field strength is constant, but the magnetic field direction changes across a TD. In this case there are two possibilities: (1) the density and temperature can both change, such that the thermal pressure p' is the same on both sides of the surface of discontinuity, and (2) the density and temperatures can be the same on both sides of the surface of discontinuity. The latter type of TD cannot be distinguished from a rotational discontinuity (see Section 4.2.2) on the basis of the profiles on N, T, and B, because each of these quantities is constant across the discontinuity in both cases. Some authors erroneously assume that if the magnetic field strength does not change across a discontinuity, then it is not a tangential discontinuity. An example of a tangential discontinuity across which the magnetic field strength does not change is shown in Burlaga (1968, Fig. 8). Other evidence for tangential discontinuities in the solar wind was presented by Siscoe et al. (1969).

The normal to a tangential discontinuity across the magnetic field direction changes is in the direction $\mathbf{B}_1 \times \mathbf{B}_2$, as indicated by Fig. 4.4, since $\mathbf{B}_1$ and $\mathbf{B}_2$ are parallel to the discontinuity surface. The orientation of a tangential discontinuity across which the magnetic field direction changes can also be obtained by a "minimum variance analysis" (Sonnerup and Cahill, 1967; Siscoe et al., 1968). The uncertainties in this method can be estimated by the methods of Siscoe and Suey (1972) and Lepping and Behannon (1980).

Given observations of a tangential discontinuity by two or more spacecraft, one can estimate the local shape of the discontinuity surface by computing the local normals. The existence of one discontinuity that was nearly planar over a distance of 0.01 AU was identified by Ness et al. (1966), using data from Pioneer 6 and IMP-3. Other directional discontinuities are curved on a scale of 0.01 AU, as demonstrated by Burlaga and Ness (1969), who studied six surfaces using three spacecraft (Explorers 33, 34, and 35) (see Fig. 4.5). Since the scale of 0.01 AU corresponds to the autocorrelation length of the magnetic field (Sari and Ness, 1969), it is not surprising that the discontinuity surfaces are curved on this scale. The curvature might result from initial conditions as well as from various microscale physical processes in the solar wind such as waves and turbulence.

A relation between the change in $\mathbf{V}$ and the change in $\mathbf{B}$ across a tangential discontinuity is neither required nor expected. Yet, Denskat and Burlaga (1977) found a tangential discontinuity for which the change in $\mathbf{V}$, $[\mathbf{V}]$, was parallel to the change in $\mathbf{B}$, $[\mathbf{B}]$ and across which there was no change in the magnitude of $\mathbf{B}$. This is a necessary but not a sufficient condition for a rotational discontinuity, as discussed in the next section. Additional evidence that $[\mathbf{V}]$ can be parallel to $[\mathbf{B}]$ across a tangential discontinuity was provided by Neugebauer et al. (1984). Using complete 3-D plasma measurements and magnetic field data Neugebauer (1985) found that $[\mathbf{V}]$ and $[\mathbf{B}]/\rho^{1/2}$ are closely aligned across TDs in the sense associated with the propagation of Alfvén waves outward from the sun at all distances between 1 AU and 2.2 AU independent of the stream structure.

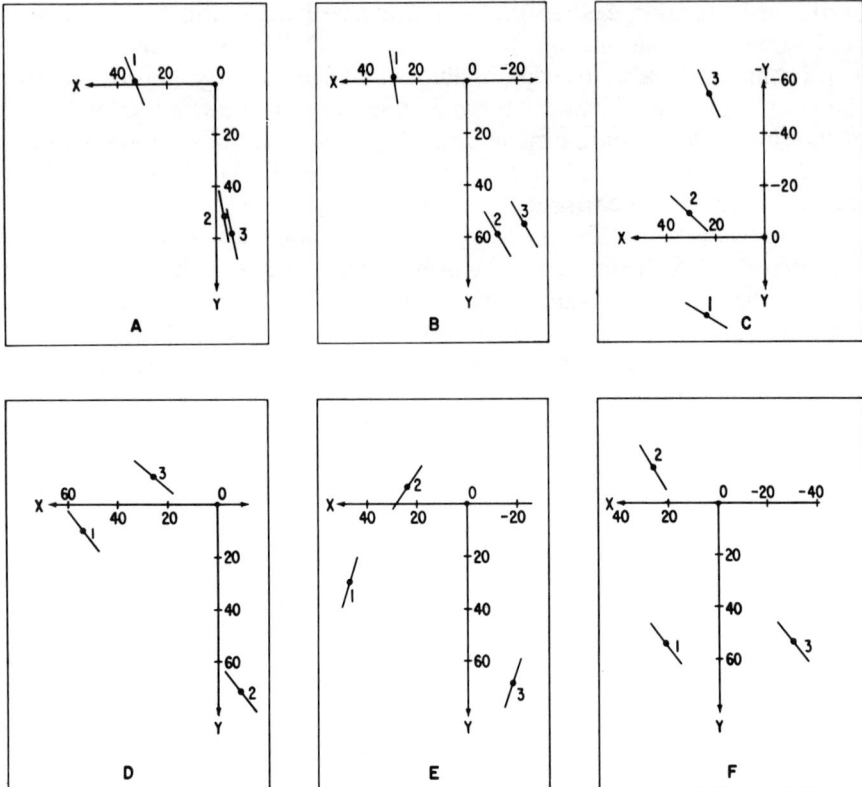

Fig. 4.5. Multispacecraft observations of tangential discontinuities. (L.F. Burlaga and N.F Ness, *Solar Phys.*, **9**, 467, 1969.) Reprinted by permission of Kluwer Academic Publishers.

Tangential discontinuities can coexist with Alfvénic fluctuations, so that the relation between [**V**] and [**B**] across a TD might be a property of the flow in which the TDs are passively embedded (Burlaga et al., 1977). The correlation between [**V**] and [**B**] across TDs might also be a signature of Alfvén waves propagating along TDs (Hollweg, 1982). Another possibility is discussed at the end of Section 4.5.2. The reason for the relation between [**V**] and [**B**] across TDs has not yet been determined definitely.

4.2.2 Rotational Discontinuities

The pressure does not change across a rotational discontinuity, yet an RD is not a static structure. The magnetic field direction must change across a rotational discontinuity, and there must be a nonzero component of the magnetic field normal to the surface of discontinuity. Thus, an RD is equivalent to a large-amplitude Alfvén wave propagating along the large-scale magnetic field. The normal component of **B** can be determined using the method of Sonnerup and Cahill (1967). In the absence of a temperature anisotropy, B, N, and T are all constant across a rotational discontinuity. This necessary condition for an RD in an isotropic medium is erroneously regarded as a sufficient condition for an RD by some authors, who use it for identifying RDs. However, B, N, and T can all be constant across a TD. Changes in B, N, and T across a rotational discontinuity are allowed if there is a thermal anisotropy (Hudson, 1970, 1971).

Rotational discontinuities propagate along the normal to the surface of discontinuity at the Alfvén speed corresponding to the normal component of the magnetic field,

$$V_{An} = \frac{B_n}{(4\pi\rho)^{1/2}} \tag{4.3}$$

where B_n is the component of **B** along the normal to the surface. Rotational discontinuities resemble shocks insofar as they are propagating surfaces of discontinuity through which there is a nonzero mass flux, but there is no increase in entropy across and RD. A necessary condition for an RD is the Alfvén wave condition

$$\mathbf{V}_2 - \mathbf{V}_1 = \frac{\pm(\mathbf{B}_1 - \mathbf{B}_2)}{4\pi\rho} \tag{4.4}$$

The change in **B** is a change in the direction of the magnetic field, since B is a constant across an RD. Equation (4.4) implies that the velocity also changes by rotating across the RD. The ± sign refers to the ability of RDs to propagate either away from the sun or toward it. The structure of a rotational discontinuity is illustrated in Fig. 4.6. Note that the magnetic field

ALFVEN SHOCK
ROTATIONAL DISCONTINUITY

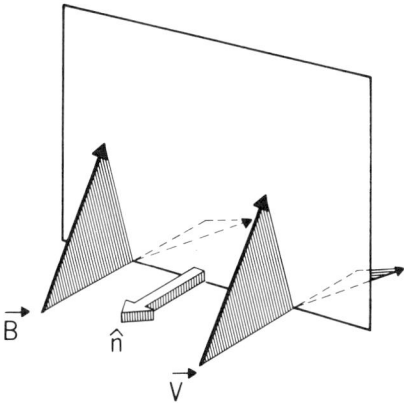

Fig. 4.6. Sketch of an Alfvén shock/rotational discontinuity.

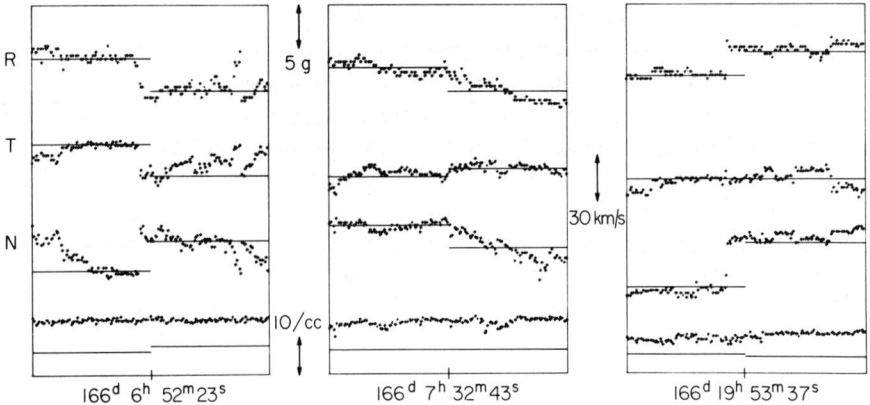

Fig. 4.7. Observation of rotational discontinuities. (J.W. Belcher and L. Davis, Jr., *J. Geophys. Res.*, **76**, 3534, 1971, copyright by the American Geophysical Union.)

vector **B** and the velocity vector **V** rotate about the normal to the surface of discontinuity.

The existence of rotational discontinuities in the interplanetary medium was demonstrated by Belcher and Davis (1971). Figure 4.7, which is taken from their paper, shows the components of the magnetic field and the velocity normalized using equation (4.4) for three RDs. Since equation (4.4) is satisfied and since B, N, and T are constant across the discontinuities, it was concluded that they are rotational discontinuities. Additional observations of rotational discontinuities are discussed in Belcher and Solodyna (1975).

The rotation of **V** across an RD (Fig. 4.6) implies that rotational discontinuities transport vorticity and angular momentum (Jeffrey, 1966; Smith, 1973a). Since rotational discontinuities propagate along the normal to the surface of discontinuity at the speed $V_{An} = B_n/(4\pi\rho)^{1/2}$, the RD propagates at any angle $\theta = \cos^{-1}(B_n/B)$ relative to the magnetic field **B**. Equal numbers of right-handed and left-handed polarizations were found in the set of rotational discontinuities studied by Smith (1973a).

4.3 Directional Discontinuities

4.3.1 Early Observations

The existence of "discontinuous" changes in the interplanetary magnetic field direction was established by Ness et al. (1966). The term "directional discontinuities" was introduced by Burlaga and Ness (1968) and Burlaga (1969a) to describe structures across which the change in the magnetic field direction exceeds 30° in less than 30 seconds. In view of the discussion above, one needs more information to be able to determine whether a directional discontinuity is a tangential discontinuity or a rotational discontinuity. Thus, the subject of directional discontinuities is largely phenomenological, but it raises many questions of fundamental importance to MHD. Lacking plasma data, Siscoe et al. (1968) offered the hypothesis that all interplanetary discontinuities they considered were tangential discontinuities, whereas Smith (1973a,b) and Belcher and Davis (1971) suggested that most interplanetary discontinuities are rotational discontinuities.

Seventy-five percent of the directional discontinuities observed by Pioneer 6 in the period from December 18 to December 25, 1965, did not satisfy equation (4.4) within the experimental uncertainties, hence they were not rotational discontinuities (Burlaga, 1971b). Since directional discontinuities are a mixture of rotational and tangential discontinuities, Burlaga concluded that most of the discontinuities in the interval were tangential discontinuities. The speed during this period was relatively low, ranging from 340 to 470 km/s. Smith (1973b) reported that most of the directional discontinuities he was able to classify were rotational discontinuities. Mariani et al. (1973) found that half the discontinuities observed from February to October 1968 in the Pioneer 8 data were tangential discontinuities. Mariani et al. (1973, 1985) found more RDs than TDs in the Helios 1 and 2 data but fewer RDs than TDs in the Pioneer 8 data in 1967–68; they concluded that different physical situations such as different levels of solar activity lead to different relative contents of RDs and TDs. Tangential discontinuities tend to dominate in the lower speed wind, and rotational discontinuities tend to dominate in the high speed wind (Martin et al., 1973; Solodyna et al., 1977). The important point is that both RDs and TDs are abundant in the solar wind at 1 AU, occurring at a rate of the order of 1 hour. Sometimes RDs dominate while at other times TDs dominate.

4.3.2 Properties of Directional Discontinuities

The magnitude of the magnetic field tends not to change across a directional discontinuity in **B**. This was demonstrated by Siscoe et al. (1968), Burlaga (1969a, 1971b), and Solodyna et al. (1977), among others. The distribution of the relative change of magnitude of **B** was found to be approximately exponential by Tsurutani and Smith (1979), the number of directional discontinuities per day being

$$n_d = 25 \times \exp\left(\frac{-\Delta B/B}{0.4}\right) \qquad \text{at 1 AU} \qquad (4.5)$$

and

$$n_d = 11.4 \times \exp\left(\frac{-\Delta B/B}{0.9}\right) \qquad \text{at 5 AU} \qquad (4.6)$$

The most probable change in the density across a directional discontinuity is likewise zero, and it is small in any case (Solodyna et al., 1977).

The change in the magnetic field direction across directional discontinuities, ω, for the interval considered by Burlaga (1971a) had a distribution

$$P(\omega) \propto \exp\left(\frac{-\omega}{75°}\right)^2 \tag{4.7}$$

Thus, ω is small for most directional discontinuities, and there are few discontinuities with $\omega = 180°$. Siscoe et al. (1968) had noted earlier a relative paucity of discontinuities with large changes in the magnetic field direction, but they also showed a decrease in the fraction of discontinuities between 30° and 60°, which is probably an artifact related to their selection criterion. Mariani et al. (1973) observed a distribution of ω very similar to that of Burlaga (1971a).

The distribution of the time intervals between successive directional discontinuities within a certain band of ω values is an exponential, with a slope depending on ω, as illustrated in Fig. 4.8. This is expected if the separation of discontinuities follows a Poisson distribution. The directional discontinuities have a tendency to cluster, as illustrated in Burlaga (1969a, Fig. 8), but the clusters are not related to the speed gradients in streams (Burlaga, 1972, Burlaga et al., 1977). The Poisson distribution for the separation times and the clustering of discontinuities was confirmed by Tsurutani and Smith (1979). The logarithm of the mean separation of discontinuities depends linearly on ω (Burlaga, 1969a), increasing from approximately 0.02 AU for $30° < \omega < 60°$ to approximately 0.1 AU for $120° < \omega < 150°$.

The radial variation in the rate of occurrence of directional discontinuities was studied by several authors. Burlaga (1971a), using Pioneer 6 data, concluded that most discontinuities originate within 0.8 AU and do not evolve appreciably between 0.8 AU and 1 AU. On the other hand, Mariani et al. (1973) found a 60% decrease in the rate of occurrence of discontinuities from 1 AU to 1.06 AU in the Pioneer 8 data; they could not exclude the possibility that the change was due to latitudinal or temporal variations. Once again we see the difficulty of determining the radial gradient of a quantity from measurements over a small range of distances. Lepping and Behannon (1986) measured n_d, the number of directional discontinuities per day, as a function of distance from 0.46 AU to 1 AU using Mariner 4 data and found

$$n_d \propto R^{-(1.3 \pm 0.4)} \tag{4.8}$$

Assuming 41 discontinuities per day at 1 AU, this equation implies that

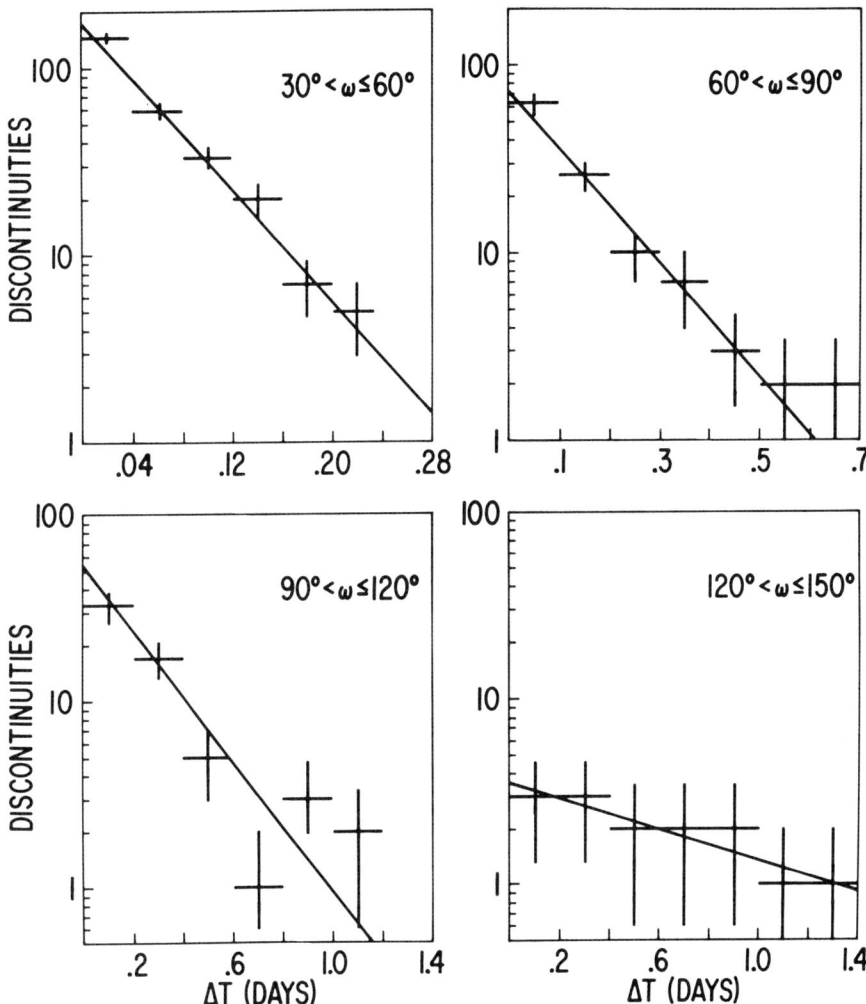

Fig. 4.8. The distribution of time intervals between directional discontinuities. (L.F. Burlaga, *Solar Phys.*, **7**, 54, 1969a.) Reprinted by permission of Kluwer Academic Publishers.

directional discontinuities would be observed at the rate of 0.5/day at 30 AU. Using Pioneer 10 and 11 data, Tsurutani and Smith (1979) found that the number of discontinuities per day decreases from 1 AU to 8.5 AU as

$$n_d = 41 \times \exp\left(-\frac{R-1}{4}\right) \qquad (4.9)$$

where R is in AU. This implies a radial gradient of 25%/AU, which is smaller than that measured by Mariani et al. (1973) but larger than the gradient measured by Lepping and Behannon (1986). The formula of Tsurutani and Smith (i.e., equation 4.9) implies that directional discontinuities should be observed at the rate of only 1 per month at 30 AU. However, Tsurutani and Smith caution that the decreased rate of occurrence given by equation (4.9) might be the result of their selection criterion as a consequence of the increasing thickness of directional discontinuities with increasing R. The thickness of directional discontinuities increased by 56% between 0.46 AU and 1 AU, such that the thickness was 36 ± 5 proton Larmor radii at all distances (Lepping and Behannon, 1986). On the other hand, Burlaga et al. (1977) found that the average thickness of the directional discontinuities at 1 AU was 12 proton Larmor radii.

4.4 Stream Interfaces

A broad transition region between a hot, low density, fast corotating stream (Chapter 7) and the cool, high density, slow solar wind ahead of it was observed by Neugebauer and Snyder (1966). This transition region is often very thin, with a thickness less than 10^6 km (Burlaga, 1974), as shown by Fig. 4.9. These thin boundaries were called "interplanetary stream interfaces" by Burlaga (1974). The existence of stream interfaces was confirmed by Gosling et al. (1978). Stream interfaces at the trailing ends of corotating streams were identified by Burlaga et al. (1990b, Fig. 5). These authors suggested that a pair of stream interfaces can form boundaries of "a heliospheric plasma sheet" between two neighboring corotating streams.

The signature of an interplanetary stream interface is an abrupt decrease in density by about a factor of 2, a similar increase in the proton temperature, and an increase in the bulk speed. The magnetic field strength usually does not change abruptly across a stream interface, and the pressure is often nearly constant across a stream interface. The increase in V across a stream interface is probably a manifestation of a velocity shear across the surface of discontinuity (Burlaga, 1974), because the direction of $\mathbf{V}$ changes as well as the magnitude. The stream interface shown in Fig. 4.9 is an early example. Exceptionally fine observations of stream interfaces were provided by the plasma experiment on Helios, and are reviewed by Schwenn (1990, p. 110).

The origin of discontinuous stream interfaces has not been determined unambiguously. Burlaga (1974) suggested that a broad transition region near the sun would tend to evolve toward a discontinuity as a result of kinematic steepening. The model of Hundhausen and Burlaga (1975) supports this hypothesis, although the authors were not able to generate

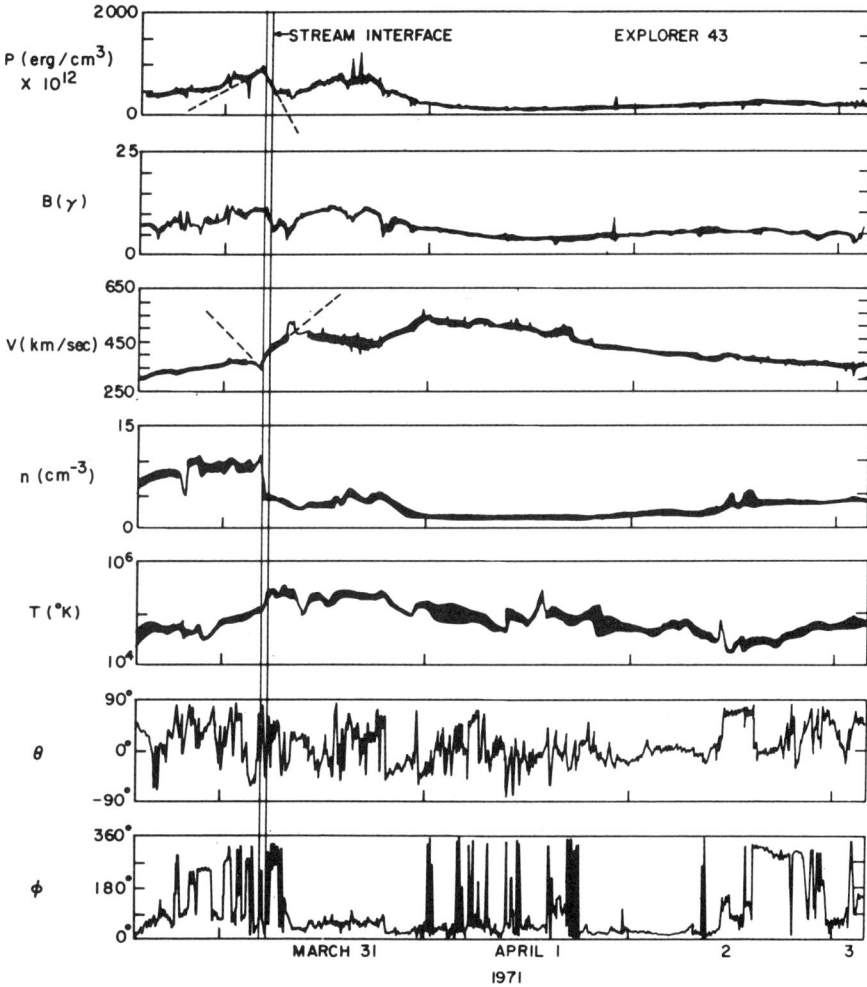

Fig. 4.9. A stream interface. (L.F. Burlaga, *J. Geophys. Res.*, **79**, 3717, 1974, published by the American Geophysical Union.)

real discontinuities with the limited resolution of the numerical model and with the initial condition that was chosen. Gosling et al. (1978) suggested that stream interfaces are discontinuous ($<4 \times 10^4$ km) at the sun and stay this way out to at least 1 AU. Helios observations lend some support to the evolutionary model (Schwenn, 1990). However, it is possible that different interfaces have different origins, so that both models of the origin of stream interfaces might have some validity.

4.5 Kelvin–Helmholtz Instability and Shear Layers

4.5.1 Theory

The Kelvin–Helmholtz instability develops across a tangential discontinuity in a fluid across which there is a sufficiently large velocity change, **V**. Two necessary conditions for the Kelvin–Helmholtz instability in an incompressible MHD flow are (Sen, 1963; Burlaga, 1969b)

$$\frac{B_1^2 + B_2^2}{4\pi} \leq \frac{\rho_1 \rho_2 V^2}{\rho_1 + \rho_2} \qquad (4.10)$$

$$(\mathbf{B}_1 \times \mathbf{B}_2)^2 \leq \frac{4\pi \rho_1 \rho_2 V^2 [\mathbf{B}_{n1}^2 + \mathbf{B}_{n2}^2]}{\rho_1 + \rho_2} \qquad (4.11)$$

where $\mathbf{B}_n$ is the component of **B** perpendicular to the relative velocity **V** (Sen, 1963).

When $B_1 = B_2$ and $\rho_1 = \rho_2$, the conditions for instability given by equations (4.10) and (4.11) reduce to

$$2V_A^2 \leq V^2 \qquad (4.12)$$

$$\sin^2 \omega \leq \left(\frac{V}{2V_A}\right)^2 \frac{\mathbf{B}_{n1}^2 + \mathbf{B}_{n2}^2}{B^2} \qquad (4.13)$$

where V_A is the Alfvén speed in km/s, N is the number density in cm^{-3}, and B is the magnetic field strength in nT,

$$V_A = \frac{21.8 B}{N^{1/2}} \qquad (4.14)$$

One condition for instability from equation (4.12) is that the speed change across the discontinuity, V^2, be greater than $2V_A^2$. The second condition for instability from equation (4.13) is related to the presence of a directional discontinuity. The discontinuity is most unstable if the magnetic field direction does not change across the surface of discontinuity, $\omega = 0$. The discontinuity is most likely to be stable when $\sin^2 \omega = 1$ (i.e., when $\omega = 90°$). In other words, the discontinuity is most stable when the magnetic field direction on one side of the surface is orthogonal to that on the other side.

The theory of the Kelvin–Helmholtz (K–H) instability in the solar wind was considered by others. Parker (1963) studied the case of **V** parallel to **B** and concluded that the K–H instability is not important in the interplanetary medium. Korzhov et al. (1984) suggested that the K–H instability can occur at stream interfaces in the interplanetary medium, leading to turbulent viscosity that can cause the width of stream interfaces to tend to grow with increasing distance from the sun. These ideas need to be critically evaluated in the light of recent observations.

4.5.2 Observations

Large, discontinuous changes in the solar wind speed (>60 km/s in <3 min) not associated with shocks were identified by Burlaga (1969b) in the Explorer 34 plasma data of Ogilvie, with the aim of searching for evidence of the Kelvin–Helmholtz instability. Both increases and decreases in the speed were observed. The density was relatively low for these discontinuities, ranging from 1.5 to 4.3 cm^{-3}, compared to the average density of ≈ 6 cm^{-3}. The magnetic field strength was relatively high, ranging from 3.8 to 12.3 nT. Taking $N = 3$ cm^{-3} and $B = 8$ nT as representative values gives a representative Alfvén speed $V_A \approx 100$ km/s. The Explorer 34 plasma analyzer did not measure velocities, but the change in the bulk speed across the discontinuities exceeded 60 km/s according to the selection criterion. Thus across the discontinuities $2V_A^2$ was probably greater than or comparable to the magnitude of the change of velocity squared, so that the condition of equation (4.12) for instability was probably not satisfied, even though the changes in bulk speed were unusually large.

The pressure was constant across these discontinuities, so that they could be tangential discontinuities, as required for the K–H instability. All the discontinuities in speed were associated with directional discontinuities. However, the distribution of ω, the change in the magnetic field across a discontinuity, was not that observed for directional discontinuities in general, viz., equation (4.7). Rather, the mean value of ω was 93°, the minimum value was 46°, and the largest value was 161°. Recall that $\omega = 90°$ is the condition for greatest stability according to equation (4.13).

The existence of the Kelvin–Helmholtz instability in the solar wind is suggested by the unusual conditions associated with the large velocity discontinuities described above (Burlaga, 1969a). Most directional discontinuities, for which ω is smaller than 90°, would be unstable under normal conditions in the solar wind for velocity changes more than 60 km/s, which accounts for their absence in the set of discontinuities selected by this criterion. Similarly, the unusual large Alfvén speeds for the large velocity discontinuities that were selected by Burlaga (1969b) might be a sign of survival against the tendency to exhibit the Kelvin–Helmholtz instability. Discontinuities with a change of speed greater than 60 km/s having average or smaller than average Alfvén speeds would probably have been unstable.

Using vector measurements of the velocity from Helios 1 and 2, Neugebauer et al. (1986) selected large velocity discontinuities that were probably tangential discontinuities. They required at least a 20% change in the magnitude of **B** across the discontinuity in order to exclude rotational discontinuities. Using Sen's equations for stability (equations 4.10 and 4.11), they showed that tangential discontinuities with velocity shears and changes in the magnitude of B tend to be stable when the change in **V** across the discontinuity is aligned with the change in **B**. They found such alignment in the set of discontinuities selected. They suggested that the alignment of changes in **V** and **B** commonly observed across tangential discontinuities in the solar wind (see Section 4.2.1) might be the result of the destruction of the tangential discontinuities without such alignment by the K–H instability.

The Kelvin–Helmholtz instability was invoked as a possible explanation for the rapid decrease in the number of directional discontinuities with increasing distance from the sun (Mariani et al., 1973). Neugebauer et al. (1986) also suggested that the decrease in the number of discontinuities with increasing distance from the sun observed by Mariani et al. (1973), Tsurutani and Smith (1979), and Lepping and Behannon (1986) is the result of the K–H instability, caused by a decreasing Alfvén speed with increasing distance from the sun.

4.6 Magnetic Holes

4.6.1 Observations

Regions in which the magnetic field strength is relatively small and close to zero are of special interest because they might represent the neighborhoods of singularities in the magnetic field. Such regions were identified by Burlaga and Ness (1968) and discussed in relation to plasma data by Burlaga (1968). Such regions are of fundamental interest insofar as they might identify singularities in the magnetic vector field (points were $B = 0$) or at least regions related to singularities. Examples of such low field regions, called "magnetic holes" by Turner et al. (1977), are shown in Fig. 4.10. Note that the magnetic field strength falls to less than 20% of the value before the magnetic hole. This particular class of magnetic hole is of special interest because it is associated with a large abrupt change in the magnetic field direction. In fact, the magnitude of the minimum magnetic field is equal to that expected if the magnetic field was annihilated in the magnetic hole (e.g., by reconnection). For this reason, Burlaga and Ness (1968) singled out the class of magnetic holes with discontinuous changes in the direction of **B** by the name of "D-sheets." Note that **B**, N, T, and V all change across the D-sheets illustrated in Fig. 4.10, and the temperature was high inside the D-sheets. There are other D-sheets across which the magnetic field direction and plasma parameters do not change.

The occurrence rate of magnetic holes studied by Turner et al. (1977) was intermittent, with an average rate of 1.5/day. The radial extent of the

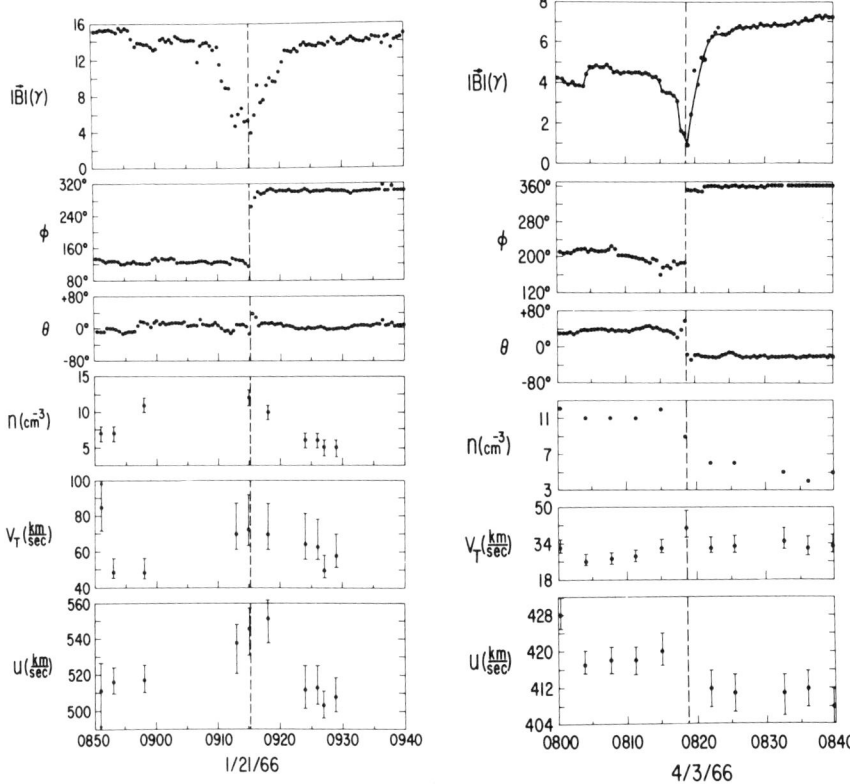

Fig. 4.10. Magnetic holes. (After L.F. Burlaga, *Solar Phys.*, **4**, 67, 1968.)

magnetic holes that they observed was relative small. The time to move past the spacecraft ranged from 2 to 130 seconds, the median time being 50 seconds. Assuming a convection speed of 400 km/s, the radial section had a dimension of the order of 2×10^4 km, of the order of $200R_L$ (Larmor radii) based on the magnetic field strength outside the magnetic hole.

4.6.2 Theory of Magnetic Holes

A particularly interesting example of a magnetic hole, originally discussed by Turner et al. (1977), is shown in Fig. 4.11, which is taken from Fitzenreiter and Burlaga (1978). This magnetic hole is remarkable in that only one component of the magnetic field changed. There was no change in the direction of the magnetic field associated with the structure, so that the decrease in the magneitc field strength could not have been produced by reconnection. The density, temperature, and speed were the same on both sides of the magnetic hole. Turner et al. called such an event a "linear magnetic hole" to distinguish it from the magnetic holes across which the magnetic field direction does change.

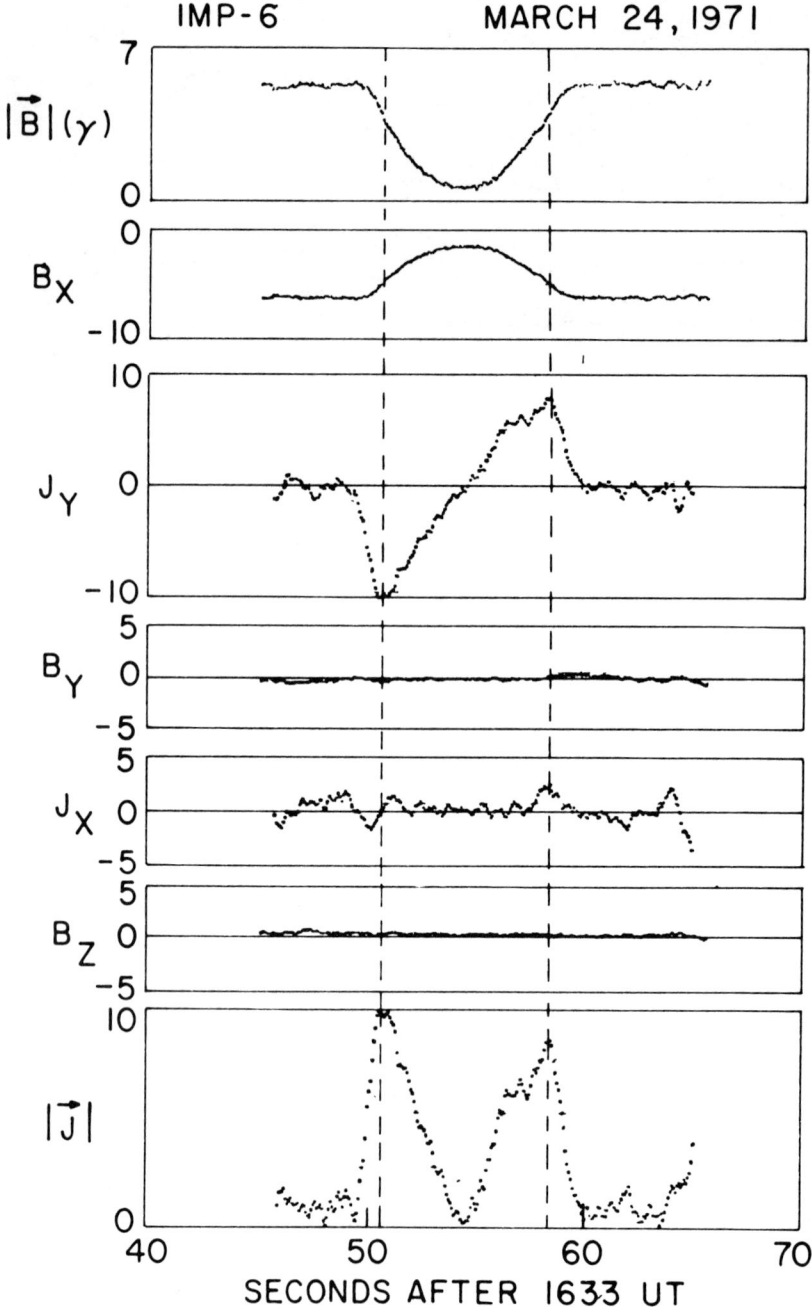

Fig. 4.11. Another magnetic hole. (R.J. Fitzenreiter and L.F. Burlaga, *J. Geophys. Res.*, **83**, 5579, 1978, published by the American Geophysical Union.)

Assuming that a magnetic hole is a planar structure, one can compute the current from $\mathbf{J} = \text{curl } \mathbf{B}$. Figure 4.11 shows the three components of the current and the magnitude of the current in the linear magnetic hole discussed above. This magnetic hole is composed of two current sheets, with the currents flowing in opposite directions in the two sheets, but having essentially the same magnitude in both. Other magnetic holes have more complex structures corresponding to multiple current sheets.

It is possible that magnetic holes cannot all be described by a single physical theory. For example, some might represent sites of reconnection while others might be static structures. Linear magnetic holes cannot be caused by reconnection, so it is reasonable to begin by considering the simpler physical model. Magnetic holes were modeled as equilibrium structures by Burlaga and Lemaire (1978), using the method developed by Lemaire and Burlaga (1976) to explain the structure of individual current sheets. The basic idea is to obtain a self-consistent solution of Vlasov's equation and Maxwell's equations. The proton currents are produced by gradient drifts in the nonuniform magnetic field and by the changing magnetic field direction, if present. The equation of motion of the electrons is $dp_e/dt = -NeE_z$, where e is the magnitude of the unit electric charge and E_z is a charge separation electric field, which is proportional to the magnitude of $(d \ln N)/dz$ when $dT_e/dz \approx 0$ as the model assumes. Strictly speaking, owing to their relatively small size, magnetic holes are found not to be MHD structures upon detailed examination. However, the total pressure P given by equation (4.1) is constant across the magnetic holes, which are thus static pressure balanced structures in this model.

A relatively complex magnetic hole observed at 4.6 AU by Voyager 2 is shown in Fig. 4.12. This is a pressure balanced structure, as shown by the plot of total pressure versus time at the bottom of the figure. This event would consist of at least three current sheets associated with the large gradients in B at the beginning and end of the magnetic hole and with the large directional discontinuity within the hole.

Magnetic holes are observed out to at least 20 AU (Burlaga et al., 1990c). Their size probably increases with distance as the Larmor radius increases as $1/B$ (Lepping and Behannon, 1986). Extrapolating this result to 20 AU implies that a simple magnetic hole should be ≈ 20 times as large as one at 1 AU.

4.6.3 Magnetic Holes, Sector Boundaries, and Stream Interfaces

One might expect magnetic holes to be associated with sector boundaries, where large changes in the magnetic field direction occur. Regions of low magnetic field strength at sector boundaries were identified by Burlaga (1968) and by Bavassano et al. (1976) in the 1–2 second resolution Pioneer 8 data. The size of the structures they observed was in the range

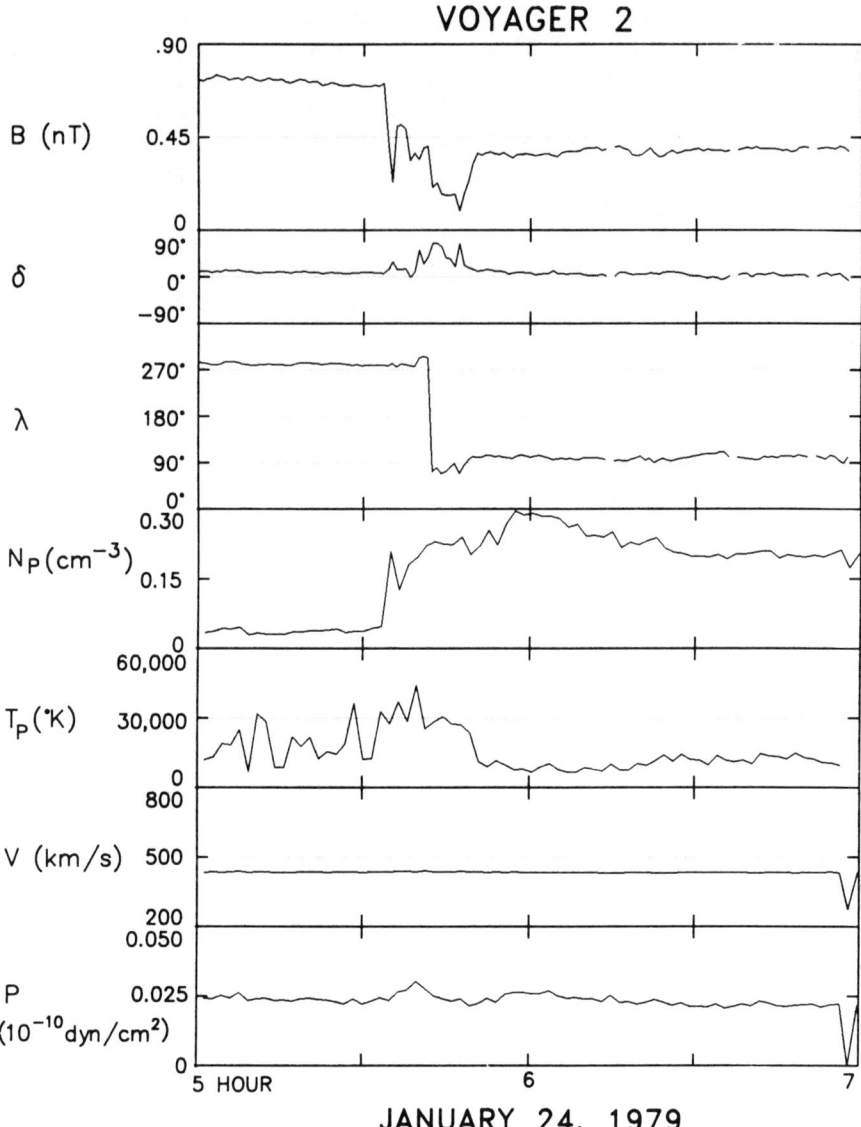

Fig. 4.12. A complex magnetic hole at 4.6 AU. (L.F. Burlaga, J.D. Scudder, L.W. Klein, and P.A. Isenberg, *J. Geophys. Res.*, **95**, 2229, 1990c, copyright by the American Geophysical Union.)

$(0.1-2) \times 10^4$ km, which is comparable to that of magnetic holes. Bavassano et al. interpreted their events as the product of magnetic reconnection, possibly produced by the resistive tearing instability. Magnetic holes near sector boundaries were also observed by Neubauer (1976), Behannon et al. (1981), and Klein and Burlaga (1980).

A sector boundary usually occurs 0–12 hours ahead of a stream interface, when a stream interface is present (Klein and Burlaga, 1980). The magnetic holes discussed by Klein and Burlaga were distributed symmetrically about the stream interfaces, whereas they were asymmetrically distributed about sector boundaries such that they extended to the stream interfaces following the sector boundaries. Thus, magnetic holes appear to be more closely related to stream interfaces than to sector boundaries. The reason for this is not understood, but one possibility that should be examined is that the magnetic holes were produced by the Kelvin–Helmholtz instability associated with shears across stream interfaces. Nevertheless, it is known that magnetic holes can occur anywhere (Fitzenreiter and Burlaga, 1978), including within sector boundaries (Bavassano et al., 1976; Klein and Burlaga, 1980; Behannon et al., 1981; Villante and Bruno, 1982).

4.6.4 Filaments, Discontinuities, Magnetic Holes, and Planar Structures

The flow between an interplanetary shock and an ejection that drives it can have a complex microstructure involving all the structures described above and more. This is illustrated by the Pioneer 6 magnetic field data in Fig. 4.13. A number of directional discontinuities are indicated by the vertical lines. Abrupt changes in the magnetic field strength are probably tangential discontinuities. Note the appearance of filamentary structures in the magnetic field strength profile with a scale length of the order of

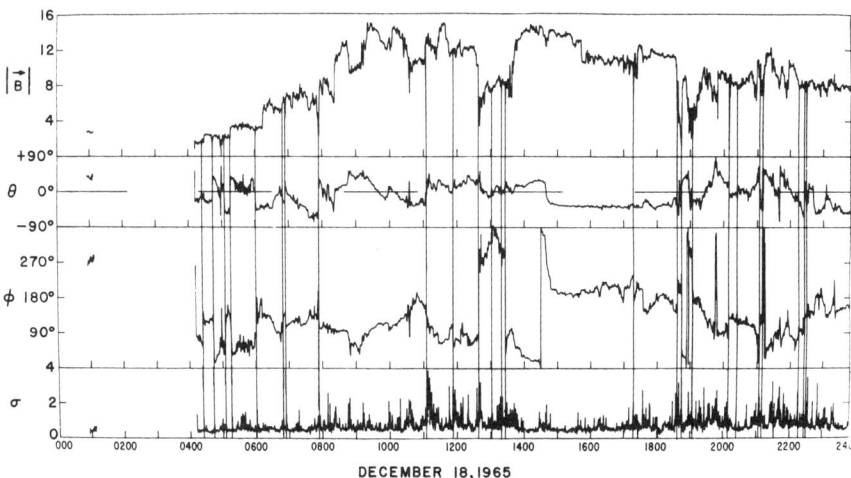

Fig. 4.13. Filaments, directional discontinuities, and planar structures behind a shock. (L.F. Burlaga, in *Solar Wind*, edited by C.P. Sonett, P.J. Coleman, Jr., and J.M. Wilcox, p. 309, NASA Spec. Publ. 308, Washingtion, DC, 1972.)

0.01 AU (1 hour) or less. Filaments are not commonly seen in the solar wind near 1 AU. Two magnetic holes (D-sheets?) associated with large directional discontinuities were observed on hours 18 and 19.

A curious feature of the magnetic field direction in Fig. 4.13 is the tendency for the plots of $\theta(t)$ and $\phi(t)$ to be nearly mirror images of each other. This is most evident from hour 6 to hour 11 and late on December 18. Similar angular variations of the magnetic field in a complex flow on a scale of several hours were analyzed by Nakagawa et al. (1989), using data obtained near 1 AU by the spacecraft Sakagake. Nakagawa et al. found that the magnetic field tended to rotate parallel to a plane throughout the interval; hence they called such regions "planar magnetic structures" (PMSs).

Noting that the plane of the PMSs they observed contained the spiral field direction, Nakagawa et al. suggested that planar magnetic structures are transient ejecta—loops and bottles rooted in the sun and expelled from the sun. A PMS whose plane is orthogonal to the spiral field direction was observed by Farrugia et al. (1990), who questioned the interpretation of Nakagawa et al. (1989). Farrugia et al. suggested that the planar structures might represent magnetic fields "squashed" (draped) around an ejection. Multispacecraft observations of a PMS were analyzed by Farrugia et al. (1991), who showed that the plane of the structure tended to drape around

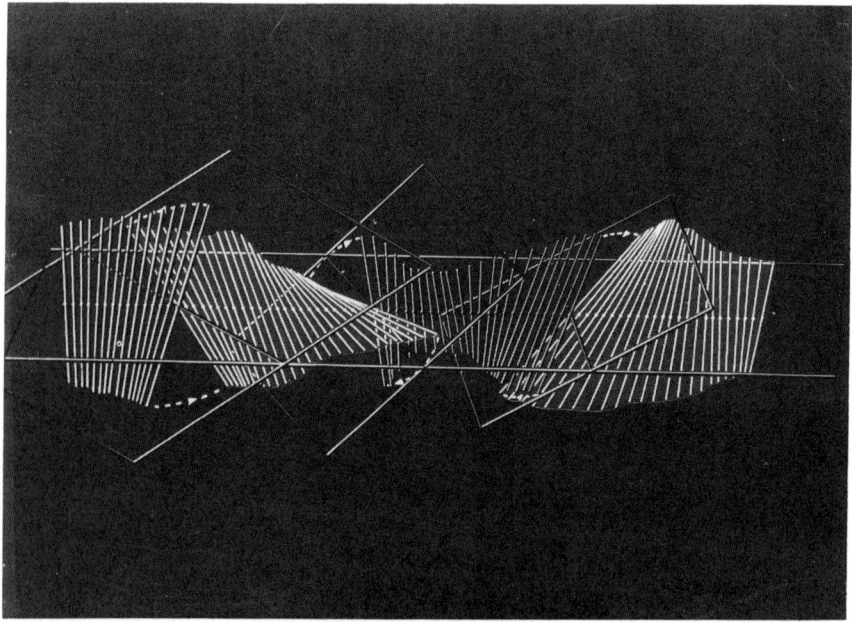

Fig. 4.14. Sketch of several directional discontinuities in a varying magnetic field; a planar structure. (L.F. Burlaga, in *Solar Wind*, edited by C.P. Sonett, P.J. Coleman, Jr., and J.M. Wilcox, p. 309, NASA Spec. Publ. 308, Washington, DC, 1972.)

ambient medium is called a "forward shock." A shock (either fast or slow) moving toward the sun relative to the ambient medium is called a "reverse shock." Since the ambient medium moves supersonically away from the sun, both forward shocks and reverse shocks move away from the sun relative to the sun.

Before observations of the interplanetary medium were available, the existence of fast shocks in the solar wind was inferred by Gold (1955) as the cause of the sudden commencement of geomagnetic storms. A fast MHD shock was first observed in the solar wind by Sonett et al. (1964, 1965). Plasma observations (without magnetic field data) of two interplanetary shocks were presented by Gosling et al. (1968). Other early observations of interplanetary shocks are reviewed by Hundhausen (1972). Most of the early shock observations were semiquantitative and did not emphasize the MHD properties.

Any kind of shock wave, being a surface, has a unit normal **n**, which is assumed to point toward the upstream, lower entropy region. In MHD, the shocks are further classified on the basis of the angle α between **n** and the ambient magnetic field observation $\mathbf{B}_1$. If $\alpha = 90°$, the shock is called a "perpendicular shock." If α is close to $90°$ it is called a "quasi-perpendicular shock." A shock for which $\alpha = 0°$ is called a "parallel shock," and one for which α is close to $0°$ is a "quasi-parallel shock." A shock for which α is neither close to $90°$ nor $0°$ is called an "oblique shock." The jumps in the fields across a shock and the change in the velocity across a shock depend on α as well as on β and a Mach number. The interplanetary magnetic field ahead of the shock and behind a shock is never uniform, as assumed in the basic theory of shocks. The magnetic field contains fluctuations whose nature and structure depend on α. Finally, the internal structure of a shock also depends on α among other things. For all these reasons, α is an important parameter, although it is not the only parameter. Implicitly, then, a fundamental quantity in the quantitative study of shocks is the shock normal **n**.

5.1.2 Rankine–Hugoniot Equations

All types of shocks and indeed all types of surfaces of discontinuity in an MHD fluid must satisfy the fundamental physical relations called the Rankine–Hugoniot equations. These are discussed in many papers, reviews (Colburn and Sonett, 1966), and texts (Landau and Lifshitz, 1960; Jeffrey and Taniuti, 1964; Boyd, 1969; Hundhausen, 1972). We adopt the notation used in the review by Burlaga (1971a). The Rankine–Hugoniot equations are the fundamental MHD equations for the case of a plane surface of discontinuity across which there is a jump in the physical fields from the upstream side (denoted by the subscript 1) to the downstream side (denoted by the subscript 2). It is assumed that the fields on either side of the shock are constant, so that the only field changes occur at the shock. The basic scalar quantities are the density ρ and the total pressure P (assumed to be isotropic). The basic vector quantities are the solar wind velocity **v** and the magnetic field **B**. The components of the vector fields along the shock normal **n** are denoted by the subscript n, and the components perpendicular to the shock normal are denoted by the subscript t.

The Rankine–Hugoniot equations for an MHD discontinuity, relative to a frame with origin at the fixed or moving surface of discontinuity, are as follows.

Conservation of mass:

$$\rho_1 v_{1n} = \rho_2 v_{2n} \tag{5.1}$$

Conservation of normal momentum flux:

$$\rho_1 v_{1n}^2 + P_1 + \frac{B_1^2}{8\pi} = \rho_2 v_{2n}^2 + P_2 + \frac{B_2^2}{8\pi} \tag{5.2}$$

Conservation of tangential momentum flux:

$$\rho_1 v_{1n} v_{1t} + \frac{B_n B_{1t}}{4\pi} = \rho_2 v_{2n} v_{2t} + \frac{B_n B_{2t}}{4\pi} \tag{5.3}$$

Conservation of energy:

$$\left(\frac{\rho_1 v_1^2}{2} + \frac{5P_1}{2} + \frac{B_{1t}^2}{4\pi}\right) v_{1n} - \frac{B_n B_{1t} v_{1t}}{4\pi} = \left(\frac{\rho_2 v_2^2}{2} + \frac{5P_2}{2} + \frac{B_{2t}^2}{4\pi}\right) v_{2n} - \frac{B_n B_{2t} v_{2t}}{4\pi} \tag{5.4}$$

Conservation of magnetic flux:

$$B_{1n} = B_{2n} = B_n \tag{5.5}$$

The frozen-field condition implies that

$$B_n(v_{1t} - v_{2t}) = B_{1t} v_{1n} - B_{2t} v_{2n} \tag{5.6}$$

If a shock is moving radially away from the sun with speed U relative to the sun, and if the upstream and downstream velocities are radial with speeds u_1 and u_2, respectively, then $v_{1n} = u_1 - U$ and $v_{2n} = u_2 - U$. The conservation of mass then gives the shock speed in terms of the density and speed measured by a single spacecraft, viz.,

$$U = \frac{n_2 u_2 - n_1 u_1}{n_2 - n_1} \tag{5.7}$$

This formula is useful for estimating shock speeds, but one must remember that radial motions are assumed in deriving it.

The Rankine–Hugoniot equations are essential for identifying shocks, shock speeds, and the specific kind of shock. However, they require a knowledge of the shock normal and the shock speed, since they are written in a frame moving with the shock speed oriented with the normal along the **n**-direction. To analyze an interplanetary shock, it is necessary to determine the shock normal. Several methods have been derived for this purpose, which are reviewed briefly in the next section.

5.1.3 Calculation of Shock Normals

There are two classes of methods for determining shock normals. The first uses measurements of the physical fields from a single spacecraft. The second method uses measurements of the shock arrival times at two or more spacecraft, together with the Rankine–Hugoniot equations. Let us consider each method in turn.

Theoretically, one can compute the shock normal **n** from the magnetic field observations at a single spacecraft using the coplanarity theorem (Colburn and Sonett, 1966). The coplanarity theorem states that $\mathbf{B}_1, \mathbf{B}_2, \mathbf{n}$ are in a plane. The continuity of the normal component of the magnetic field B_n across the shock implies that $(\mathbf{B}_2 - \mathbf{B}_1)$ is parallel to the shock surface. One can show that $\mathbf{B}_{1t}$ is parallel to $\mathbf{B}_{2t}$, so that $\mathbf{B}_1 \times \mathbf{B}_2$ is also parallel to the shock surface. Thus, $(\mathbf{B}_1 \times \mathbf{B}_2) \times (\mathbf{B}_2 - \mathbf{B}_1)$ is along the shock normal direction. The unit shock normal is therefore

$$\mathbf{n} = \frac{(\mathbf{B}_1 \times \mathbf{B}_2) \times (\mathbf{B}_2 - \mathbf{B}_1)}{|(\mathbf{B}_1 \times \mathbf{B}_2) \times (\mathbf{B}_2 - \mathbf{B}_1)]|} \tag{5.8}$$

The advantage of this method, called the magnetic coplanarity method, is that it requires only measurements of the magnetic field before and after the shock. The disadvantage is that is not very accurate, because the change of the magnetic field direction across the shock is small (of the order of 10°) and because the uncertainties in **B** are relatively large owing to fluctuations that are usually present in the magnetic field both upstream and downstream of the shock. The magnetic coplanarity method does not work for parallel shocks, for which $\mathbf{B}_1 \times \mathbf{B}_2 = 0$, and it is very inaccurate for quasi-parallel shocks.

A similar method for calculating the shock normal, based only on the measurements of the velocity at one spacecraft, was introduced by Abraham-Shrauner (1972) and Abraham-Shrauner and Yun (1976). This method is based on the coplanarity of the shock normal and $\mathbf{v}_1 - \mathbf{v}_2$. Both velocity coplanarity and magnetic coplanarity hold for isotropic and anisotropic plasmas (Hudson, 1970). The "velocity coplanarity method" is rarely used because it is not very accurate. Abraham-Shrauner (1972) and Abraham-Shrauner and Yun (1976) also introduce a method using both plasma and magnetic field data that is superior to both the velocity coplanarity method and the magnetic coplanarity method.

A method of determining the shock normal based on the eleven-dimensional space of unknown variables $\mathbf{B}_1$, $\mathbf{B}_2$, ρ_1, ρ_2, and $(\mathbf{v}_2 - \mathbf{v}_1)$ and the requirements that these variables satisfy eight nonlinear equations derived from the Rankine–Hugoniot equations was introduced by Lepping and Argentiero (1971). This method is accurate for oblique shocks, approximately three times more accurate than the magnetic coplanarity method. However, it does not apply to either parallel or perpendicular shocks, and the solution obtained is not necessarily unique. Since the Lepping–Argentiero method involves specifying eleven non–linearly coupled variables, it is computation intensive and relatively slow. Acuna and Lepping (1984) introduced an algorithm that makes the Lepping–Argentiero method more efficient, but the essential nonlinearity of the method remains. A further extension of the Lepping–Argentiero method was discussed by Hsieh and Richter (1986) and Richter et al. (1985a).

Perhaps the best method for determining shock normals using data from a single spacecraft was introduced by Vinas and Scudder (1986). This method is fast and efficient, because it is based on invariants that reduce the problem to one involving seven linear variables and only four nonlinear variables. It gives a unique solution. The Vinas–Scudder method applies to parallel shocks and perpendicular shocks as well as to oblique shocks. It converges equally rapidly for shocks of all types, rarely taking more than a few seconds of computer time.

The fluctuations ahead of and behind a shock are a major source of uncertainty in the calculation of the shock normal and source speed, especially for quasi-parallel shocks. An analysis on the effect of these fluctuations and the use of various averaging intervals on determining shock normals was carried out by Hsieh and Richter (1986). They introduced a method for dealing with these fluctuations and applied it to 105 fast forward shocks observed by Helios.

An accurate method for determining the shock normal using the arrival times of the shock at three spacecraft was introduced by Ogilvie and Burlaga (1969). The method assumes that the shock surface is planar and that the speed is constant between the spacecraft. A similar method using data from four spacecraft was introduced by Russell et al. (1983a).

5.1.4 Perpendicular Shocks

A perpendicular fast MHD shock is one for which the magnetic field is perpendicular to the shock normal both before and after the shock; thus $B_n = 0$. Figure 5.1 illustrates the magnetic field and velocity in relation to the shock surface. The only wave that propagates perpendicular to **B** is the magnetoacoustic wave. The flow enters at a speed greater than the magnetoacoustic speed and leaves at a speed less than the magnetoacoustic speed. Thus, the shock propagates supermagnetoacoustically.

For a perpendicular shock, the mass flux and frozen-field equations (5.1) and (5.6) give

$$\frac{\mathbf{B}_1}{\rho_1} = \frac{\mathbf{B}_2}{\rho_2} \tag{5.9}$$

which states that the magnetic field direction does not change across a perpendicular shock, and also that the following relation holds.

$$\frac{B_2}{B_1} = \frac{\rho_2}{\rho_1} \tag{5.10}$$

These two necessary conditions are very useful in identifying perpendicular shocks in the data. For a perpendicular shock, the equations for the normal momentum flux (5.2) and the energy flux normal to the surface (5.4) are the same as the corresponding equations for a gas dynamic shock, except that a term $B^2/8\pi$ is added to the pressure in the MHD shock (Burlaga, 1971a, p.

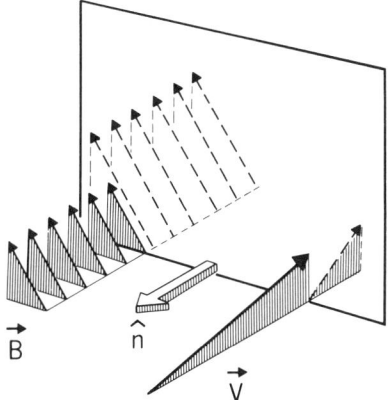

Fig. 5.1. Sketch of a perpendicular fast shock.

640). A perpendicular MHD shock is like a gas dynamic shock, except that the sum of the gas pressure and the magnetic pressure appears rather than the gas pressure, and the shock speed is supermagnetoacoustic rather than supersonic relative to the ambient medium.

A perpendicular fast forward shock was observed in the solar wind by Ogilvie and Burlaga (1969). The shock was observed on August 29, 1967, by three spacecraft (Explorers 33, 34, and 35), so that an accurate shock normal could be determined by the three-spacecraft method introduced by Ogilvie and Burlaga (1969). The magnetic field direction ahead of the shock ($\theta = 51°$, $\phi = 295°$) was essentially the same as that behind the shock ($\theta = 51°$, $\phi = 296°$), in agreement with the parallelism indicated by equation (5.9) above. The density, speed, and magnetic field strength all increased across the shock, but the magnetic field direction did not change. The ratio $B_2/B_1 = 1.3$ was equal to the ratio $\rho_2/\rho_1 = 1.4$ within the measurement errors, in agreement with equation (5.10). Finally the direction of the magnetic field before and after the shock was perpendicular to the shock normal within the errors. The temperature increased across the shock, as it must across any shock, from 6.5×10^4 to 12×10^4 K. The solar wind speed relative to the spacecraft, V, increased from 418 km/s to 452 km/s, corresponding to a decrease in the speed relative to the shock. The change in the momentum flux $[\rho(V_n - U)^2] = 1.5 \times 10^{-10}$ dyn/cm^3 was equal to the change in the sum of the thermal pressure and magnetic pressure $[P] = 1.4 \times 10^{-10}$ dyn/cm^2 within the experimental uncertainties. This calculation is based on the value $U = 496$ km/s determined from the mass conservation equation (5.1), and the assumed electron temperature $T_e = 1.5 \times 10^5$ K.

A reverse perpendicular fast shock in the solar wind was first identified by Burlaga (1970). The shock was observed in the Explorer 34 data on September 28, 1967 (see Fig. 5.2). The signature of a reverse fast shock (a decrease in the density, proton temperature, magnetic field strength, and total pressure) is clear in Fig. 5.2. Physically, forward shocks and reverse shocks are identical; they differ only in the direction of propagation relative to the ambient solar wind. Note that the bulk speed relative to the spacecraft increases across a reverse shock, just as it does across a forward shock. This is a kinematic effect. In both cases, the speed decreases from the low entropy side to the high entropy side relative to the shock.

The reverse perpendicular shock of September 28, 1967, was observed by three spacecraft (Explorers 33, 34, and 35), and the three-spacecraft method was used to obtained an accurate normal. The angle between the shock normal and the magnetic field was 87°, consistent with 90° within the experimental errors, as expected for a perpendicular shock. The magnetic field direction ahead of the shock ($\theta = 60°$, $\phi = 130°$) was nearly parallel to that behind the shock ($\theta = 43°$, $\phi = 125°$), as one expects for a quasi-perpendicular shock from equation (5.9). The ratio $B_2/\rho_2 = 1.5 \pm 0.1$ was approximately equal to the ratio $B_1/\rho_1 = 1.2 \pm 0.2$ within the measurement errors, in agreement with equation (5.9). The change in the momentum flux $[\rho(V_n - U)^2] = 8.6 \times 10^{-10}$ dyne/cm^3 was equal to the change in the sum of

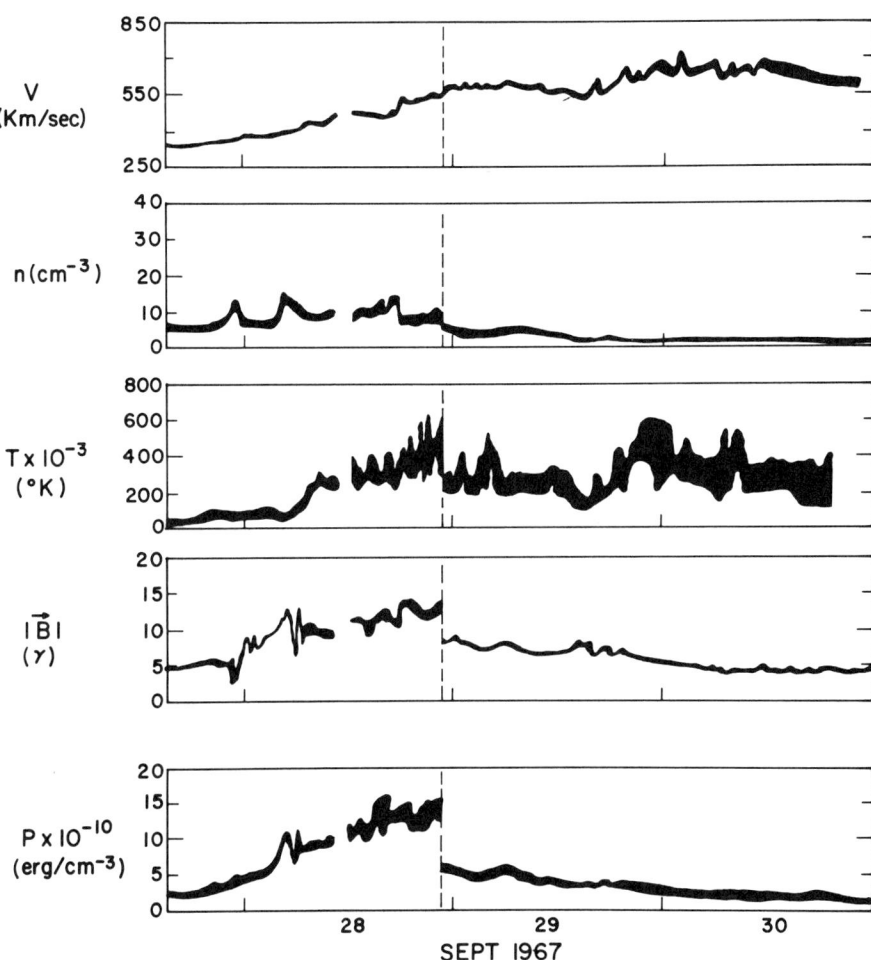

Fig. 5.2. Observation of a reverse perpendicular fast shock. (L.F. Burlaga, *Cosmic Electrodyn.*, **1**, 233, 1970.)

the thermal pressure and magnetic pressure $[P] = 8.04 \times 10^{-10}$ dyne/cm^2 within the experimental uncertainties. This calculation is based on the value $U = 424$ km/s determined from the mass conservation equation (5.1). The speed relative to the shock on the sunward (low entropy) side was 141 km/s relative to the shock, which is greater than the magnetoacoustic speed, 95 km/s. The speed of the solar wind relative to the shock on the opposite side was 76 km/s, which is lower than the magnetoacoustic speed there, 110 km/s. Thus, the shock was supermagnetoacoustic relative to the upstream flow and submagnetoacoustic relative to the downstream flow, as required for a fast MHD shock.

A quasi-perpendicular forward fast shock observed by Voyager 1 on

November 27, 1977, was analyzed in detail by Vinas and Scudder (1986). The data for the shock (except the temperature) are shown in Fig. 5.3. Note that the shock jump is very sharp, and the fluctuations before and after the shock are small, as is typical for perpendicular shocks. Using their method for calculating shock normals and speeds, Vinas and Scudder obtained an angle between the upstream magnetic field and the shock normal equal to $84.2 \pm 9°$ and a shock speed $U = 305 \pm 19$ km/s. The magnetic coplanarity method and the Lepping–Argentiero method did not give an accurate normal for this shock.

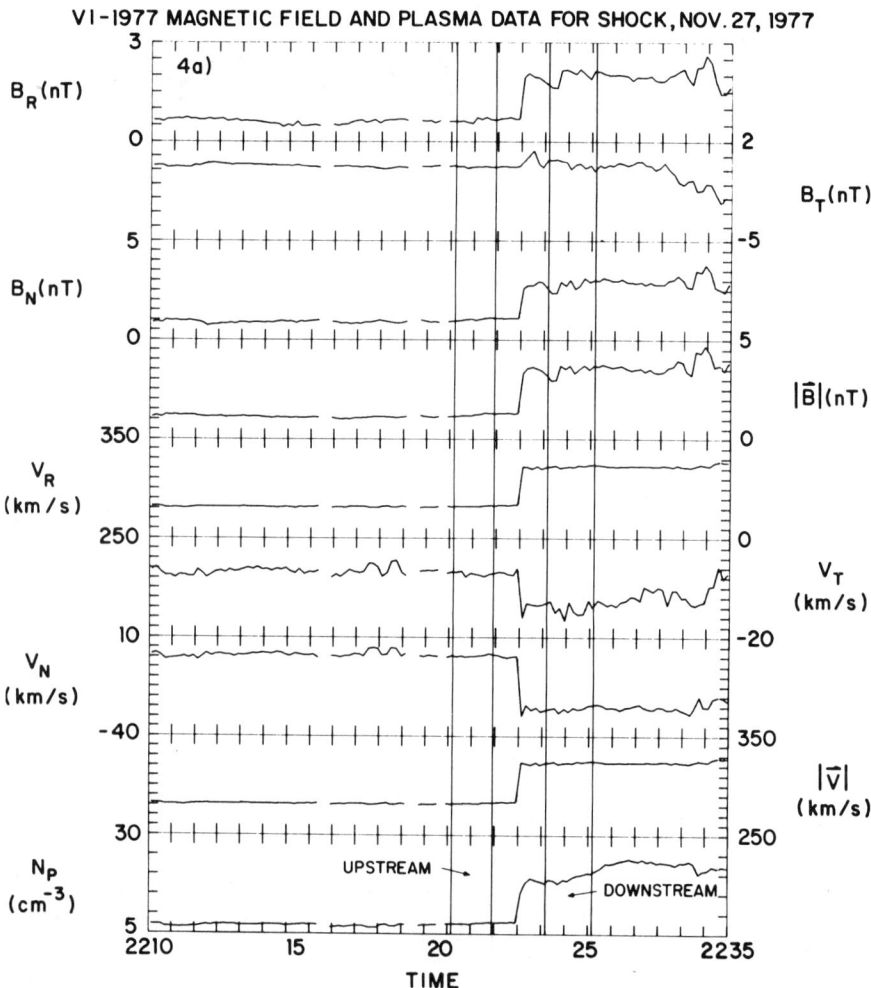

Fig. 5.3. Observations of a forward perpendicular fast shock. (A.F. Vinas and J.D. Scudder, *J. Geophys. Res.*, **91,** 39, 1986, published by the American Geophysical Union.)

5.1.5 Parallel Shocks

A parallel shock is a fast MHD shock whose normal is parallel to **B**. Since **B** = B_n**n**, which is the same on both sides of the shock, neither the magnetic field nor the magnetic field intensity changes across a parallel shock (see Fig. 5.4). Unlike the perpendicular shock, for which there is only one characteristic speed upstream (the magnetoacoustic speed), there are two characteristic speeds (the sound speed and the Alfvén speed) for a parallel shock.

If the sound speed is greater than the Alfvén speed ahead of the shock, then the sound speed is dominant and the shock is an ordinary gas dynamic shock, with gas entering faster than the sound speed and leaving slower than the sound speed. The flow is super-Alfvénic on both sides of the shock, and the sound speed is greater than the Alfvén speed on both sides.

If the Alfvén speed is greater than the sound speed ahead of the shock, there are three possibilities. A parallel shock with gas entering super-Alfvénically and leaving sub-Alfvénically probably does not occur in nature (Jeffrey and Taniuti, 1964). A parallel shock with gas entering supersonically and super-Alfvénically and gas leaving subsonically and still super-Alfvénically is allowed. Finally, the gas might enter supersonically and sub-Alfvénically and leave subsonically and either sub-Alfvénically or super-Alfvénically.

For a parallel shock, $\mathbf{v}_{1t} = \mathbf{v}_{2t}$, so that there is no velocity coplanarity. The magnetic field drops out of the Rankine–Hugoniot equations, which become identical to those for gas dynamic shocks, except that the pressure can be anisotropic.

The existence of quasi-parallel fast MHD shocks has been demonstrated

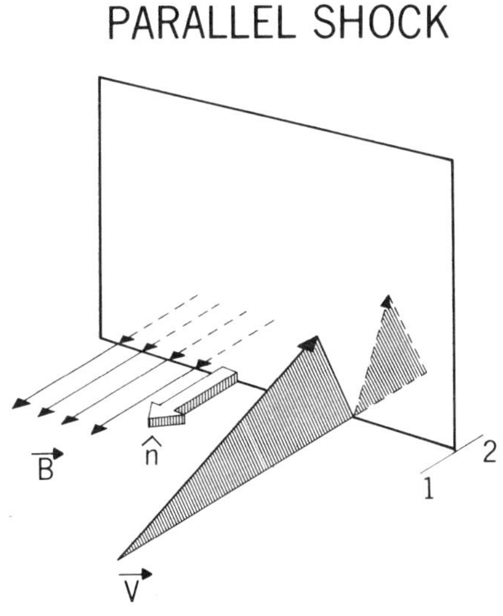

PARALLEL SHOCK

Fig. 5.4. Sketch of a parallel fast shock.

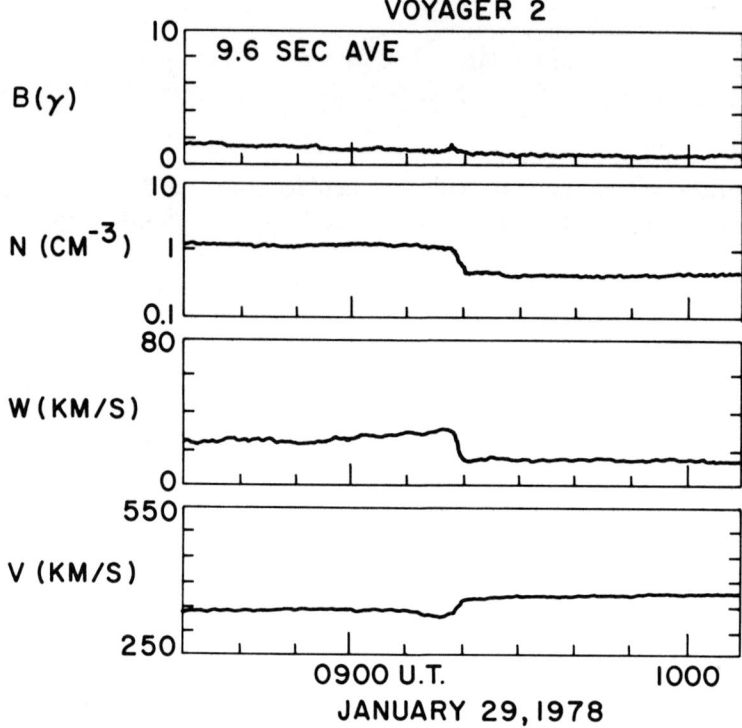

Fig. 5.5. Observations of a reverse parallel fast shock. (M.H. Acuna, L.F. Burlaga, R.P. Lepping, and N.F. Ness, in *Solar Wind Four,* edited by H. Rosenbauer, p., 143, Rep. MPAE-100-81-31, Max-Planck Institute, Lindau, Germany, 1981.)

at <1 AU (Neubauer et al., 1977; Burlaga et al., 1980; Richter et al., 1984; Hseih and Richter, 1986) at 1 AU (Kennel et al., 1982; Scudder et al., 1984), and beyond 1 AU (Acuna et al., 1981).

A reverse parallel fast MHD shock was observed by Voyager 2 on January 29, 1978, at about 2 AU (Acuna et al., 1981). The magnetic field strength **B**, density N, and proton thermal speed (proportional to the square root of the proton temperature and the bulk speed) are shown in Fig. 5.5. The shock signature is a decrease in density and thermal speed and an increase in speed (characteristic of a reverse shock) with no change in the magnetic field strength. High resolution measurements of the magnetic field strength and direction during hour 9 (Fig. 5.6) confirm that there was no jump in either the magnetic field strength or direction across the shock at 0920 UT, but there was intense wave activity near the shock and extending a considerable distance upstream and downstream of the shock.

The parallel shock of January 29, 1978, was studied further by Scudder et al. (1984) and Vinas et al. (1984). An accurate analysis of the shock was made by Vinas and Scudder (1986) using their method for determining shock normals. The angle between the shock normal and the upstream

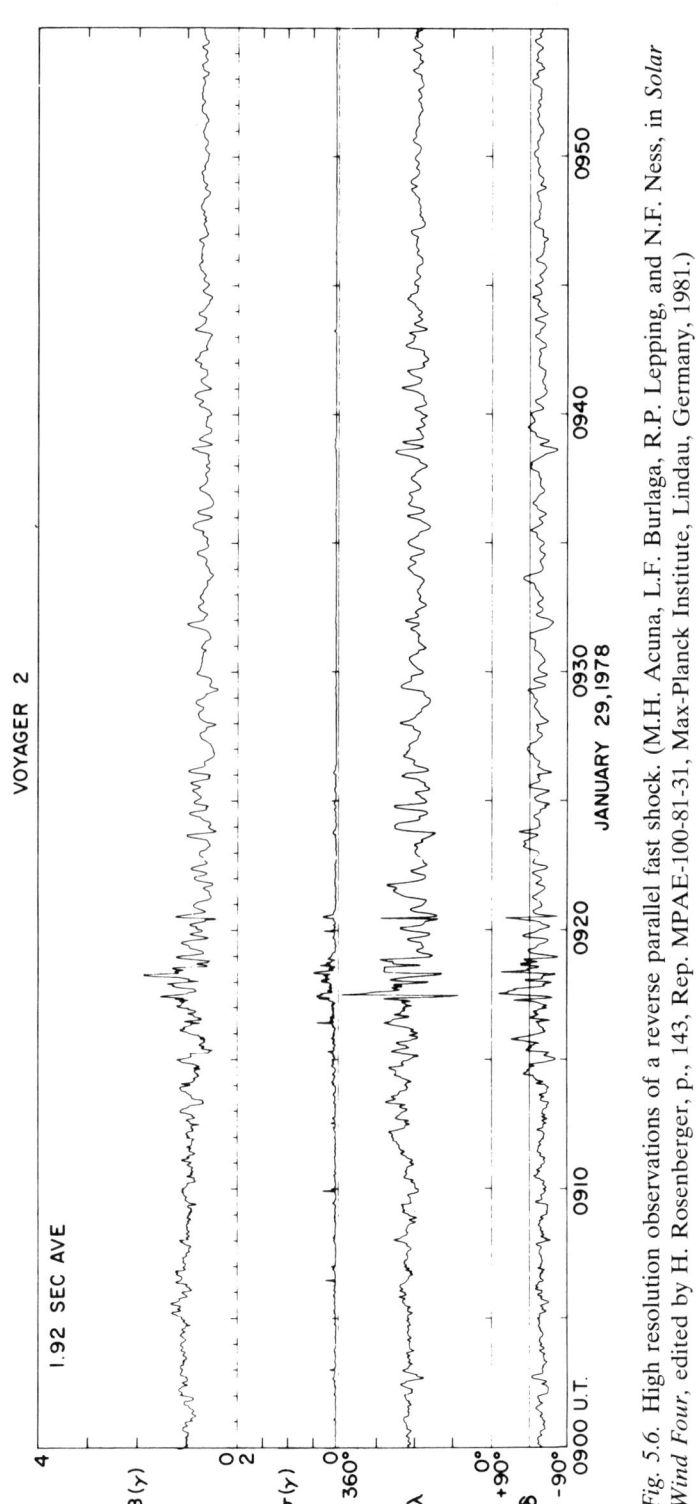

Fig. 5.6. High resolution observations of a reverse parallel fast shock. (M.H. Acuna, L.F. Burlaga, R.P. Lepping, and N.F. Ness, in *Solar Wind Four*, edited by H. Rosenberger, p., 143, Rep. MPAE-100-81-31, Max-Planck Institute, Lindau, Germany, 1981.)

magnetic field was $\alpha = 20 \pm 18.0°$, and the shock speed was 261 ± 39 km/s. The shock is thin as far as the plasma observations are concerned, but very thick as far as the magnetic field is concerned (Acuna et al., 1981; Scudder et al., 1984).

Waves near quasi-parallel shocks were discussed by Tsurutani et al. (1983) and others. A comparison of the ISEE-3 observations of waves upstream of quasi-parallel shocks with the results of a simulation was made by Mandt et al. (1986). Since the subject of upstream waves is properly a subject of plasma physics rather than MHD, it is not discussed further. The reader is referred to review papers in Tsurutani and Stone (1985) for more extensive and detailed results on this vast subject.

In a survey of 140 fast forward shocks between 0.29 AU and 1 AU, Richter et al. (1986) identified between 2 and 17 parallel shocks, depending on the criteria used. This result demonstrates the rarity of parallel shocks, even within 1 AU. They found that the jump conditions across parallel shocks are similar to those for shocks in general, except for the magnitude of **B**. They also found that parallel shocks are observed more frequently closer to the sun, as one might expect because the spiral magnetic field is more radial as one approaches the sun.

5.1.6 Oblique Shocks

Most shocks (e.g., Volkmer and Neubauer, 1985; Richter et al., 1986) in the solar wind are neither perpendicular nor parallel. They are oblique, meaning that the angle between the upstream magnetic field direction is neither close to 90° nor close to 0°. For an oblique shock both $\mathbf{B}_n$ and $\mathbf{B}_t$ are non-zero, $\mathbf{B}_{1t}$ is parallel to $\mathbf{B}_{2t}$, $\mathbf{v}_{1t} - \mathbf{v}_{2t}$ is parallel to $\mathbf{B}_{1t}$ and $\mathbf{B}_{2t}$, and $\mathbf{n}$, $\mathbf{B}_1$, $\mathbf{B}_2$, $\mathbf{v}_1 - \mathbf{v}_2$ are coplanar (see Fig. 5.7). The jump conditions for oblique shocks can be written in a number of forms, all of which are too complex to reproduce here. They are given in many references, including Burlaga (1971a), Landau and Lifshitz (1960), Jeffrey and Taniuti (1964), and Boyd (1969).

Hundreds of oblique fast shocks have been observed in the solar wind (e.g., see Borrini et al., 1982; Bavassano-Cattaneo et al., 1986; Richter et al., 1986) but relatively few have been analyzed thoroughly. Here we demonstrate the existence of oblique fast forward shocks in the solar wind with the results in Fig. 5.8. The same shocks were analyzed by Lepping and Argentiero (1971) using their method for determining shock normals, with more accurate results and better fits. The signature of an oblique fast forward shock is an increase in density, temperature, speed, and magnetic field strength, and a change in all three components of the magnetic field and velocity vectors. The predicted magnitudes of the jumps in these quantities, shown by the dashed lines in Fig. 5.8 are in good agreement with the observed values.

There are several problems associated with quantitatively describing oblique fast shocks. One is the determination of the shock normal, as discussed above. Another is the choice of averaging intervals before and

FAST SHOCK

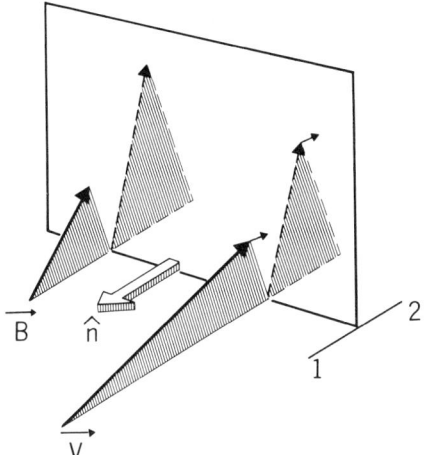

Fig. 5.7. Sketch of an oblique fast shock.

after the shock, which is necessary to minimize the effect of the fluctuations in the direction of the velocity and magnetic field. Finally, there are experimental uncertainties, particularly in the measurement of the density, temperature, and the flow direction. The uncertainties are smaller in more modern instruments, but the modern measurements introduce new problems such as temperature anisotropies and nonthermal distribution functions.

A reverse oblique fast shock in the solar wind away from terrestrial influences was first identified by Chao et al. (1974). The earth's bow shock is also a reverse fast shock, which is generally an oblique shock. The supersonic solar wind is expected to terminate somewhere between 100 AU and 200 AU in a strong reverse fast shock. The termination shock is likely to be quasi-perpendicular at many locations along the shock, but it will be oblique in general owing to the ever-present fluctuations in the interplanetary magnetic field.

5.1.7 Shape of Fast Shocks

One can identify two classes of shocks on the basis of their origin: shocks driven by the ejecta from solar eruptions ("transient shocks") and shocks associated with corotating streams ("corotating shocks"). To first approximation, the transient shocks within 1 AU are spherical (Parker, 1963), whereas the corotating shocks have a spiral form. The early observations of oblique shocks discussed above were probably observations of transient shocks. A corotating (reverse) shock was first observed at 1 AU by Burlaga (1970). The first evidence of corotating shocks beyond 1 AU was presented

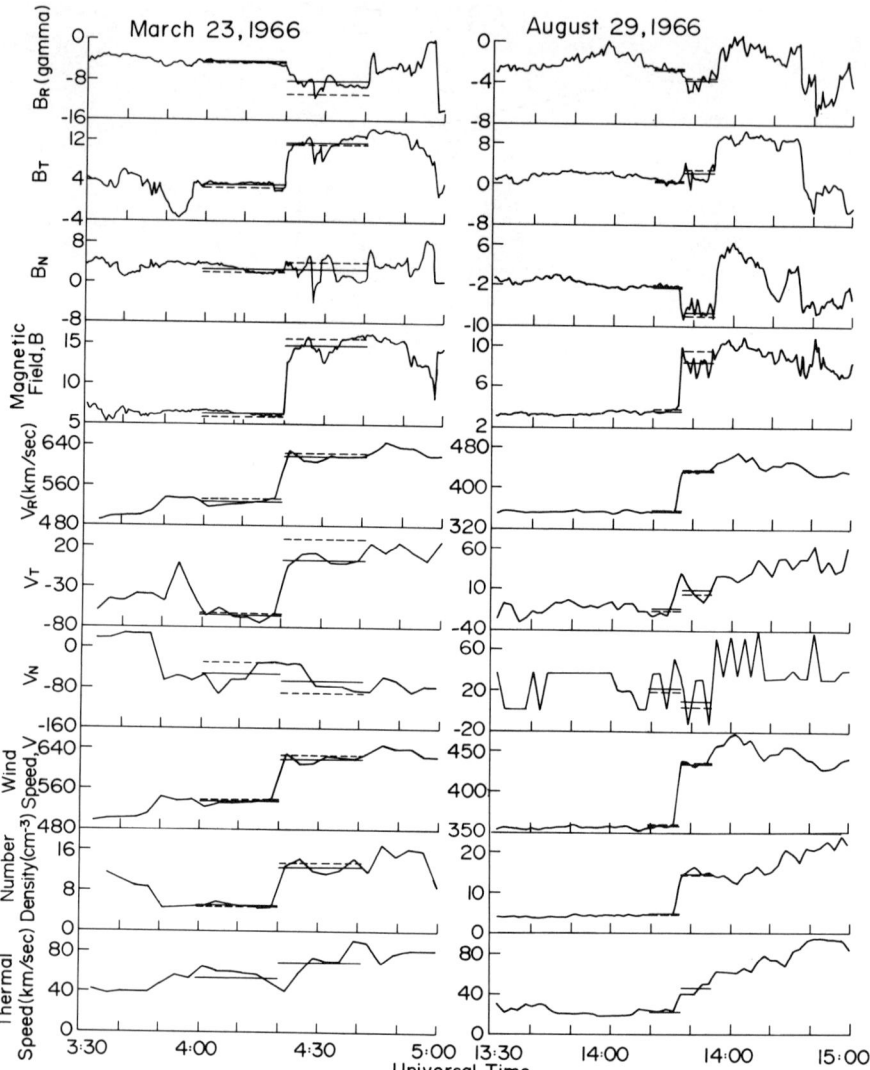

Fig. 5.8. Observations of an oblique shock. (J.K. Chao and S. Olbert, *J. Geophys. Res.*, **75**, 639, 1970, copyright by the American Geophysical Union.)

by Smith and Wolfe (1976, 1977), and other such phenomena were analyzed by Smith (1983, 1985).

There are many studies of the shape of transient shocks. Most of these results are based on the orientation of the shock normal and on the position of the assumed source. Most of the early papers assume that transient shocks are caused by solar flares (see Hundhausen, 1972). The identification of a solar flare as a source of a given shock is always difficult. More importantly, recent results suggest that solar flares are not the primary cause of transient shocks. Thus, the shock shapes deduced from flare associations must be assessed very carefully.

Deviations from a spherical shape for transient shocks have been considered on three scales: a scale of 1 AU or more; a scale of a fraction of an AU, corresponding to the width of corotating interaction regions; and a small scale of hundredths of an AU. Among the early results are those of Hirshberg (1968), Ogilvie and Burlaga (1969), Taylor (1969), and Chao and Lepping (1974). All these papers suggest a quasi-spherical shape, with the shock extending over at least 180° in longitude. An exception is the result of Cane (1985), who suggested that the shock is nearly spherical everywhere except possibly near the east limb of the sun.

Mesoscale distortions of the shape of a transient shock shape from a spherical form can be caused by the interaction of shocks with streams (Ogilvie and Burlaga, 1978). Specific models of the interaction of transient shocks with corotating streams and interaction regions were introduced by Heineman and Siscoe (1974), Burlaga and Scudder (1975), and Hirshberg et al. (1974).

Small-scale distortions of shock shape and fluctuations in shock speed were observed in a transient shock using data from Helios 1 and 2, IMP-7, and Voyagers 1 and 2 (Burlaga et al., 1980). A shock moved from 0.6 AU to 1.6 AU with apparently constant speed on a scale of 1 AU, but the local measurements showed that the shock speed fluctuated up to $\pm(100 \pm 20)$ km/s, and the local directions of the shock fluctuated by $\pm(40 \pm 20°)$. Thus, the shock shape can be rippled on a small scale. Small-scale fluctuations in corotating shocks between 6.5 AU and 9.4 AU were observed by Gazis et al. (1985), who found fluctuations in the shock shape with an amplitude of approximately 0.1 AU on a scale of 0.1 AU. Multispacecraft observations of small-scale shock shapes were also made by Russell et al. (1983a,b, 1984). Evidence for ripples on transient and corotating shocks near 1 AU was presented by Russell and Alexander (1984); they found deviations of the normals up to 20° on a scale of the order of 150 Earth radii. Since a shock surface often propagates through a turbulent medium with a power spectrum of $f^{-5/3}$, the shock surface should have a fractal dimension. Ripples on a shock surface can also be produced by the interaction between the shock and inhomogeneities such as tangential discontinuities (Gazis et al., 1985). Shock surface ripples can play an important role in the acceleration of particles by shocks (Decker, 1990).

5.2 Slow MHD Shocks

5.2.1 Theory

Slow MHD shocks owe their existence to the presence of slow mode waves in an MHD fluid. Gas enters a slow shock at a speed greater than the upstream slow mode speed and it leaves at a speed less than the slow mode

SLOW SHOCK

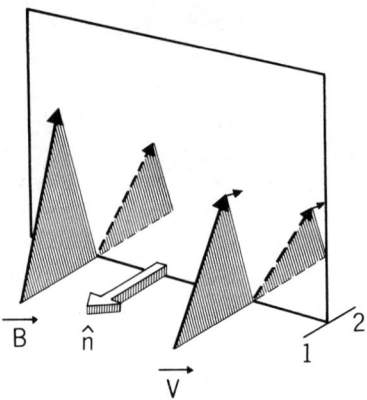

Fig. 5.9. Sketch of a slow shock.

speed. The flow is sub-Alfvénic on both sides of the shock. The relation between the shock normal, the magnetic field vectors, and the velocity vectors across a slow shock is illustrated in Fig. 5.9. The Rankine–Hugoniot equations are the same for slow shocks as for fast shocks. Across a slow shock observed by a spacecraft in the solar wind, the temperature, density, and speed increase, while the magnetic field strength decreases. Thus slow shocks and fast shocks have the same signature in all quantities except the magnetic field strength, which decreases across slow shocks and increases across fast shocks. Magnetic energy increases across a fast shock at the expense of kinetic energy, but magnetic energy decreases across a slow shock, presumably by conversion to thermal energy. Recall that the increase in speed relative to an inertial frame is a kinematic effect; the flow speed decreases across a shock in the frame moving with the shock for slow shocks as well as for fast shocks.

5.2.2 Interplanetary Observations of Slow Shocks

Forward slow shocks in the solar wind were first identified by Chao and Olbert (1970) using Mariner 5 plasma and magnetic field data. Their observations and fits to two slow mode shocks are shown in Fig. 5.10. The slow mode Mach number was 1.5 ahead of the shock and 0.7 behind the shock of July 20, 1967. The slow mode Mach number was 1.8 ahead of the shock and 0.6 behind the shock on August 30, 1967. The flow was sub-Alfvénic on both sides of each of the shocks. The existence of slow mode shocks was confirmed by Burlaga and Chao (1971) using data from Pioneer 6 near 1 AU. An observation of a slow shock at 0.31 AU was

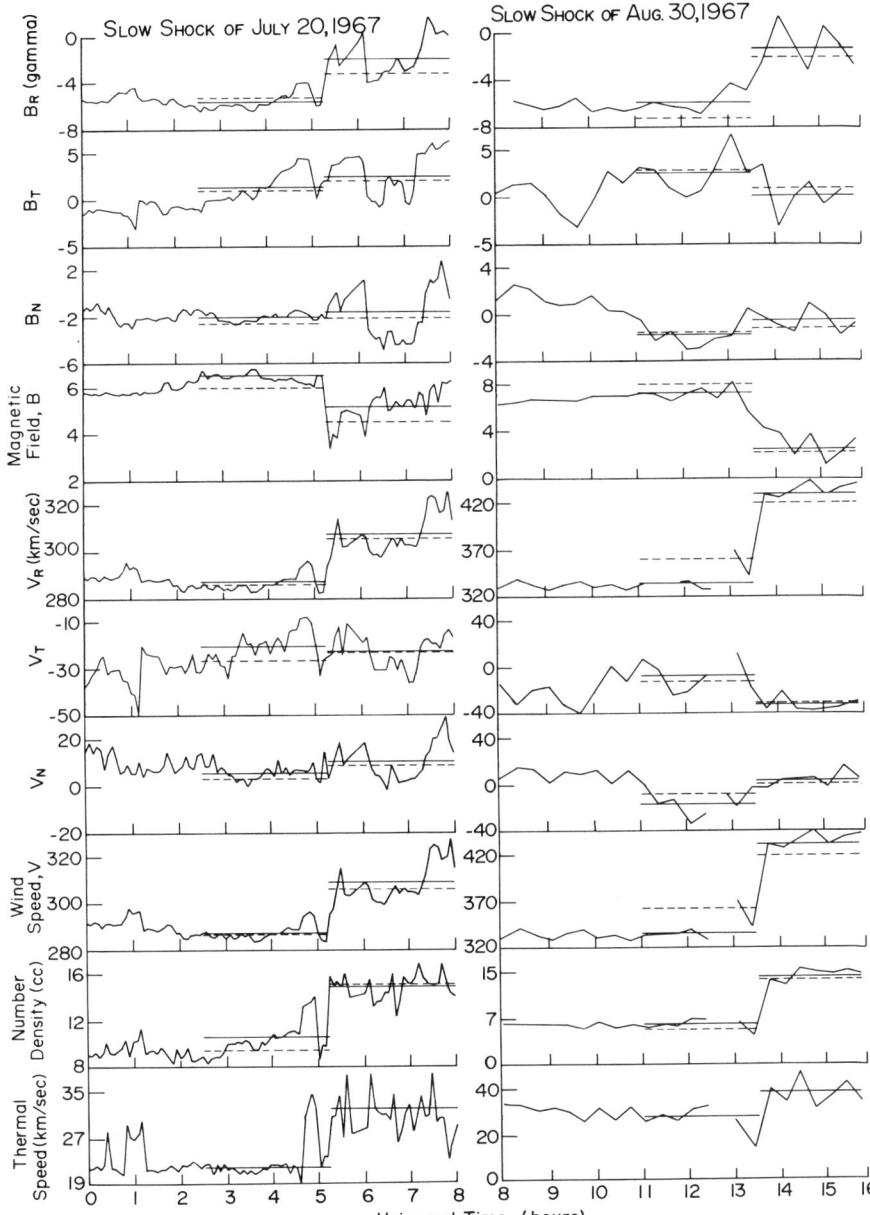

Fig. 5.10. Observations of a forward slow shock. (J.K. Chao and S. Olbert, *J. Geophys. Res.,* **75,** 639, 1970, copyright by the American Geophysical Union.)

reported by Richter et al. (1985b). Additional observations of slow shocks by Helios are discussed by Richter and Marsch (1988) and by Richter (1991), who also considers their nature and origin.

The existence of a reverse slow shock in the solar wind was demonstrated by Burlaga and Chao (1971). This shock was detected by the instruments on Pioneer 6. The slow mode Mach number was 1.2 ahead of the shock and 0.8 behind it. The Alfvén Mach number was 0.9 ahead of the shock and 0.8 behind it (i.e., less than unity both upstream and downstream, as required for a slow shock).

5.2.3 Formation and Destruction of Slow Shocks

Interplanetary slow shocks occur near stream interfaces (Burlaga, 1974; Richter, 1991). Since stream interfaces are barriers to the flow, the magnetic field and plasma tend to pile up there, because the normal component of the speed relative to the stream interface goes to zero (the velocity field is singular there). Such a flow condition is evidently conducive to the formation of slow shocks. Mechanisms for the formation of slow shocks were discussed by Neubauer (1976), Rosenau and Suess (1977), and Hada and Kennel (1985).

Slow shocks might also form at coronal mass ejections in the corona, where the plasma beta is very low (Hundhausen et al., 1987; Whang, 1987, 1988; Steinolfson and Hundhausen, 1989, 1990). A slow shock can transform to a fast shock between the sun and 1 AU, so that the slow shocks might not be observed at 1 AU (Whang, 1987, 1988). On the other hand, Hu and Habbal (1993) suggest that a forward–reverse slow shock pair might form in a depression of the magnetic field strength between the shocks in a forward–reverse fast shock pair, and that such shock pairs might be observed at 1 AU. A standing slow shock might also exist in the solar corona (Whang, 1982, 1986). Since these shocks have not been observed in situ, they are not considered further here. They might be observed by a future mission such as the Solar Probe.

6

Magnetic Clouds and Force-Free Magnetic Fields

6.1 Interplanetary Ejecta

The concept of ejecta ("plasma clouds") emitted from the sun, propagating through interplanetary space, and producing geomagnetic disturbances by their interaction with earth is very old (Lindeman, 1919; Chapman and Ferraro, 1929). The early work suggested that the plasma clouds are ejected by major solar flares with a speed of the order of 1000 km/s determined from the time between the flare and the corresponding solar disturbance (Chapman and Bartels, 1940). Initially, it was thought that the plasma clouds propagate through a vacuum. After Parker introduced the idea of a solar wind moving at a speed of the order of 400 km/s, it was realized by Gold (1955, 1959) that a plasma cloud can move supersonically and should drive a shock wave. Gold proposed that the sudden commencement of a geomagnetic storm is the effect of such a shock interacting with the earth's magnetic field. Gold (1962) introduced the term "magnetosphere" for the region controlled by the earth's magnetic field.

From the observation of an association between flares and ejecta for certain spectacular events, there evolved the paradigm that flares are the cause of all interplanetary ejecta at 1 AU. From the observation of an association of certain shocks at 1 AU with solar flares, there evolved the idea that all transient shocks, even those at 50 AU, can be traced to solar flares. The early associations have validity, but the subsequent generalizations and extrapolations are not correct. It is not true that solar flares are the cause of all transient interplanetary ejecta and shocks (e.g., Kahler, 1992). Likewise, the complex interactions that occur between the sun and 50 AU make it unlikely that all shocks can maintain their identity out to 50 AU and can be traced to particular solar flares. Unfortunately, this important fact is not widely appreciated.

Plasma clouds should carry along solar magnetic fields, by virtue of the high electrical conductivity of the plasma (Alfvén, 1950). In Alfvén's picture of a magnetized plasma cloud, the magnetic field lines are ordered like elastic bands stretched out from the sun, being anchored at both ends.

Coconni et al. (1958) and Gold (1955, 1962) developed this concept further, suggesting that the magnetic field in the ejecta from flares is strong and highly ordered, extending as planar curves from the sun into interplanetary space and filling a large volume in the form of a "magnetic tongue" or bottle. The magnetic field lines in a magnetic tongue are not helical, and they have a radius of curvature of the order of 1 AU. Gold proposed that the magnetic tongue is a strong barrier to cosmic rays because of its strong ordered magnetic fields, and he suggested that the magnetic tongue causes Forbush decreases by deflecting particles by gradient drifts.

In Gold's picture of a magnetic tongue, the magnetic field lines remain connected to the sun for some days. Piddington (1958) suggested that the magnetic field lines might become disconnected from the sun, so that the magnetized plasma cloud might become detached from the sun. The idea that magnetized plasma clouds become disconnected after days or weeks, as a result of reconnection on a magnetic neutral plane in the tongue, was also advocated by Gold (1962). This view has been espoused most recently by McComas et al. (1989, 1992). In a brief section of their book, Akasofu and Chapman (1972) suggested that the magnetic field lines in all interplanetary ejecta are detached from the sun, and they proposed that the ejecta are spherical, with a magnetic field line geometry given by the force-free solution of Friere (1966). However, (1) the solution of Friere is incorrect (Vandas et al., 1991), (2) the magnetic field lines are connected to the sun in the cases permitting the testing of magnetic field line topology, and (3) most ejecta do not have smooth force-free magnetic fields.

A different concept of a magnetized plasma cloud was introduced by Morrison (1954, 1956). He too imagined that a magnetized plasma cloud fills a volume like a tongue or ellipsoid (as opposed to a geometry such as a flux rope). However, Morrison suggested that the magnetic field inside the ejection is highly disordered and the plasma very turbulent. He proposed that the ejecta cause Forbush decreases, but he suggested that particles are excluded from the ejecta by diffusion in the turbulent fields, rather than by drifts in ordered fields. The early studies of in situ observations of transient ejecta behind shocks suggested that the magnetic fields in the ejecta are disordered as Morrison suggested (see Hundhausen, 1972).

In the 1960s and 1970s, it was thought that ejecta could be modeled by solar explosions, analogous to bomb explosions with a near-instantaneous energy input (e.g., Parker, 1963; Hundhausen, 1972; Dryer, 1975, 1982; Dryer et al., 1978a, 1984, 1986). In some cases the initial condition was assumed to be a fast stream, while in other models the initial disturbance was assumed to be a temperature perturbation. The earliest models were gas dynamic models based on codes written to describe bomb explosions (Hundhausen, 1972). Dryer and his colleagues pioneered in the development of multidimensional MHD codes.

Although none of the early models describes all ejecta, each model contains some feature that is relevant to the study of interplanetary ejecta. The literature on interplanetary ejecta is vast and poorly organized, with numerous conflicting ideas and imaginative extrapolations based on very incomplete observations. These interplanetary ejecta are frequently called CMEs in the current literature, but it seems more appropriate to use the term CME as it was originally intended, for "coronal mass ejections," referring to mass ejections observed in the solar corona (e.g., Hundhausen, 1977, 1979).

It is probably futile to search for a single model that applies to all ejecta. Some ejecta might be ordered, others turbulent. Some ejecta might be connected to the sun, others disconnected. Magnetic clouds, which are the subject of the remainder of this chapter, are a subset of interplanetary ejecta, with a well-defined signature in the in situ observations and unique physical properties. Magnetic clouds are very different from the concepts of interplanetary ejecta discussed above. They provide a fresh approach to the study of interplanetary ejecta that is amenable to analytic treatment.

6.2 Magnetic Clouds

6.2.1 Existence of Magnetic Clouds

The existence of magnetic clouds was established using data from IMP, Helios, and Voyager (Burlaga et al., 1981b). The essential features of a magnetic cloud are (1) strong magnetic fields, (2) a smooth rotation of the magnetic field direction through a large angle, close to 180°, as the magnetic cloud moves past a spacecraft, and (3) a low proton temperature and low proton β. At 1 AU, a magnetic cloud moves past the spacecraft in about 24 hours, corresponding to a radial cross section of the order of 0.25 AU. For subsequent considerations it is important to keep in mind that the duration of passage over a spacecraft at 1 AU, of the order of 1–2 days, is comparable to the travel time of the magnetic cloud from the sun to 1 AU, of the order of 3–4 days. Thus the in situ observations contain important information on a substantial part of the history of the magnetic cloud. An example of the local observations of a magnetic cloud as it moves past a spacecraft is shown in Fig. 6.1.

It is essential that all three of the features listed above be used in the identification of a magnetic cloud. It is incorrect to say that since magnetic clouds have low temperatures, events with low temperatures are magnetic clouds. For example, the events identified by Geranios (1978, 1981, 1982, 1987) on the basis of low temperatures are not magnetic clouds. Similarly, there are events (e.g., certain planar structures and some sector boundary crossings), in which the magnetic field rotates smoothly, but such events do not satisfy the other two criteria for magnetic clouds. The requirement of high magnetic field intensity is also important, especially at 1 AU. The field intensity might drop to the ambient value as a magnetic cloud evolves to large distances from the sun (Osherovich et al., 1993b). In short, a

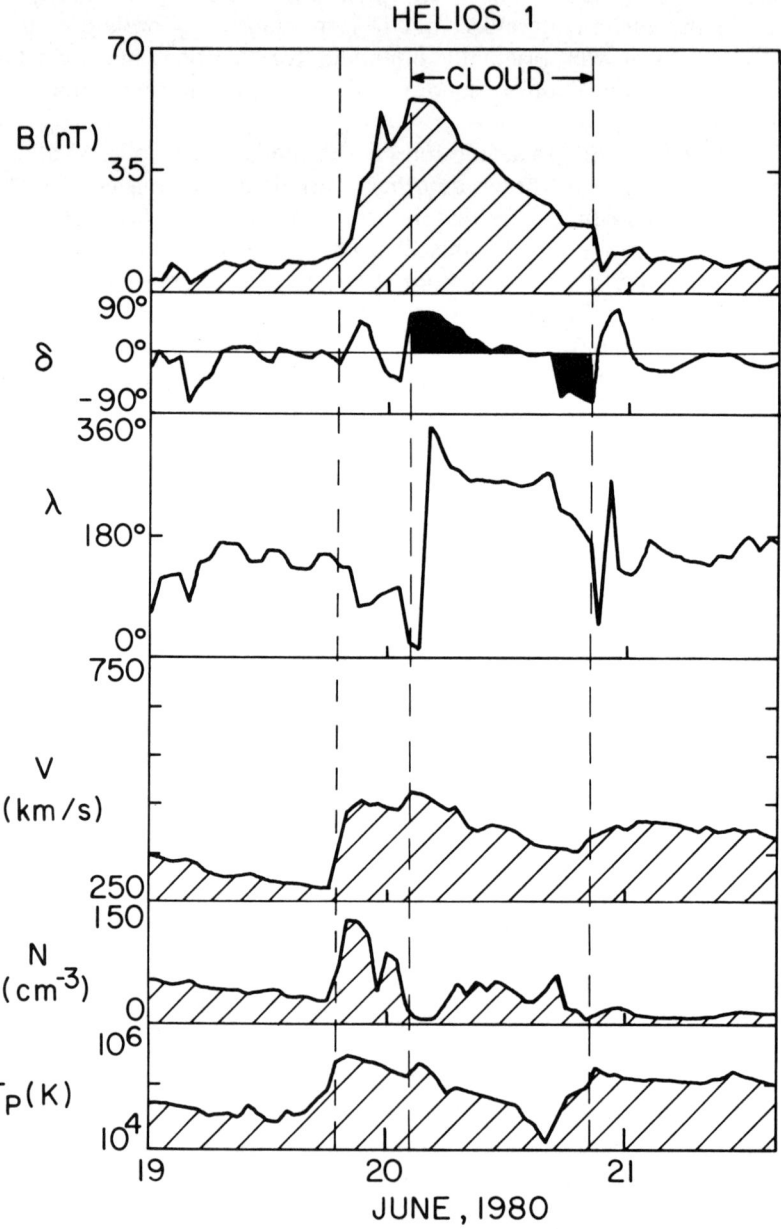

Fig. 6.1. Observation of a magnetic cloud. (L.F. Burlaga, L.W. Klein, N.R. Sheeley, Jr., D.J. Michels, R.A. Howard, M.J. Koomen, R. Schwenn, and H. Rosenbauer, *Geophys. Res. Lett.,* **9,** 1317, 1982a, copyright by the American Geophysical Union.)

condition that is necessary for a magnetic cloud is not sufficient for the identification of a magnetic cloud. Unfortunately, some authors use the term "magnetic cloud" in reference to transient ejecta that are not magnetic clouds as defined above.

Several tabulations of magnetic clouds have appeared in the literature. The earliest list, by Klein and Burlaga (1982), has been widely used in correlative studies, but it contains some events that would not at present be considered to be magnetic clouds. Another list by Zhang and Burlaga (1988) identifies magnetic clouds observed later. The most recent list of magnetic clouds was published by Lepping et al. (1990). In all these lists, the determination of the beginning and end times of the magnetic clouds is subjective, because there is no single signature of the boundary of a magnetic cloud. Other lists of magnetic clouds have appeared in the literature, but many of the events do not meet all three of the criteria set forth in the definition above. The reader must take care not to confuse magnetic clouds with interplanetary ejecta of other types.

6.2.2 Origin of Magnetic Clouds

Magnetic clouds are a subset of interplanetary ejecta. Several studies have shown that interplanetary ejecta are associated with solar flares and disappearing filaments. For example, see the reviews by Schwenn (1983), Hundhausen (1988), and Gosling (1990). The solar flare association might not be causal, since mass ejections sometimes precede flares (e.g., Kahler, 1992). Disappearing filaments are eruptive prominences, seen on the disk of the sun.

A relation between a magnetic cloud observed over the west limb of the sun by Helios 1 at 0.4 AU and a coronal mass ejection observed in the corona was presented by Burlaga et al. (1982a). The ejection was also detected by the zodiacal light photometer on Helios 1 between 0.2 AU and 0.4 AU (Jackson et al., 1985). The time interval from the passage of the ejection through the corona to its arrival at Helios 1 is consistent with the speed of the magnetic cloud measured at Helios 1. Statistical evidence of an association between magnetic clouds and coronal mass ejections was reported by Wilson and Hildner (1984). The solar data are consistent with the hypothesis that all magnetic clouds are associated with coronal mass ejections.

Some magnetic clouds are associated with solar flares (Burlaga et al., 1981b, 1987a). Other magnetic clouds are associated with disappearing filaments (Burlaga et al., 1982a; Wilson and Hildner, 1986). Five magnetic clouds observed near 1 AU from 1977 through 1979 were identified by Tsurutani et al. (1989); three of the magnetic clouds were associated with solar flares and two were associated with disappearing filaments. Rust (1993) found that of the eighteen magnetic clouds in a list published by Lepping et al. (1990), six were associated with solar filament eruptions and ten with eruptive solar flares. This suggests very strongly that magnetic clouds are extensions of the flux ropes in eruptive filaments into interplanetary space.

6.3 Magnetic Flux Tube Model of Magnetic Clouds

6.3.1 Geometry of Magnetic Clouds

The large-scale configuration of a magnetic cloud was estimated from multispacecraft observations by Burlaga et al. (1981b). They suggested that the cross section of a magnetic cloud is approximately circular or possibly elliptical with irregularities superimposed. A more complicated bubblelike cross section was suggested by Crooker et al. (1990). Burlaga et al. (1981b, 1990a) showed that the axis of the magnetic cloud is curved on a scale of 0.5 AU at 1 AU. Thus, they suggested that a magnetic cloud has the form of a cylinder locally, but on a scale of 0.5 AU it has the form of a loop, such as a curved flux tube, with the ends extending toward the sun.

A detailed analysis of multispacecraft observations of a magnetic cloud was carried out by Burlaga et al. (1990a). The X's in Fig. 6.2 show the estimated positions of the front boundary of the magnetic cloud at each of the spacecraft at the indicated time. The O's show the corresponding

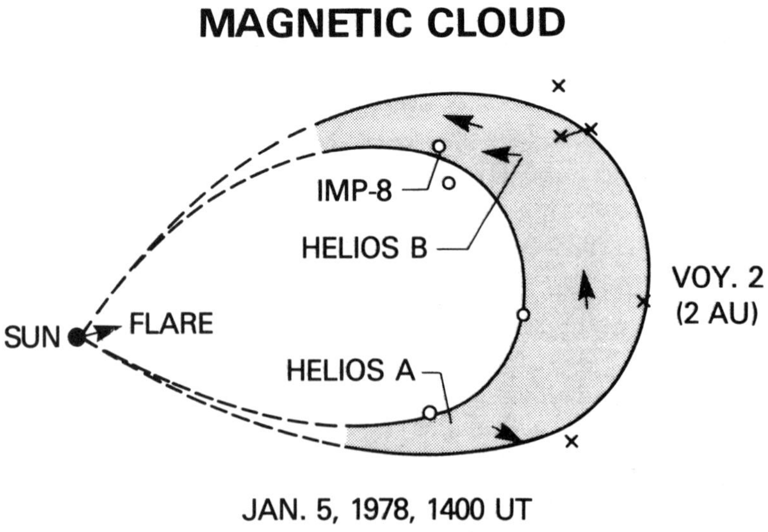

MAGNETIC CLOUD

JAN. 5, 1978, 1400 UT

Fig. 6.2. Geometry of magnetic clouds. (L.F. Burlaga, R. Lepping, and J. Jones, in *Physics of Flux Ropes,* edited by C.T. Russell, E.R. Priest, and L.C. Lee, p. 373, AGU Geophysical Monograph 58, American Geophysical Union, Washington DC, 1990a, copyright by the American Geophysical Union.)

positions of the rear boundary of the magnetic cloud. Smooth curves drawn through the X's and O's indicate the shape of the magnetic cloud. These curves are extended to a hypothetical source on the sun; they are drawn as dashed lines near the sun to indicate that the in situ observations do not prove that the tube is connected to the sun. The curve through the center of the magnetic cloud in Fig. 6.2 is the axis, measured both by a minimum variance technique and by the direction of the magnetic field when its inclination with respect to the ecliptic is zero. Burlaga et al. concluded that the magnetic cloud had the form of a tube extending close to the sun, and possibly connected to the sun at both ends (Fig. 6.2). At 1 AU the radius of the tube was approximately 0.125 AU and its radius of curvature was of the order of 0.3 AU.

6.3.2 Topology of Magnetic Clouds

The basic issue concerning the topology of magnetic clouds is whether their magnetic field lines are connected to the sun. A flux tube can be either connected or disconnected from the sun, whereas both a torus and a sphere must be disconnected from the sun. The connectivity cannot be determined by direct measurements of the magnetic field or by remote sensing observations of high density regions. Some authors have suggested that bidirectional streaming of either suprathermal electrons or low energy protons is a signature of magnetically disconnected ejecta ("plasmoids"), but such a signature could also be associated with flux ropes connected to the sun (e.g., Gosling, 1990). There are, however, observations indicating that magnetic clouds are generally magnetically connected to the sun.

A magnetic cloud was observed on January 14–15, 1988, near 1 AU (Farrugia et al., 1993a). The arrival of the magnetic cloud at IMP-8 during hour 5 of January 14, 1988, was signaled by an abrupt decrease in proton temperature, a magnetic hole, a drop in the intensity of low energy particles accelerated by a shock driven by the magnetic cloud, and bidirectional field-aligned energetic particles. There was also a density spike at the front boundary, but this is not a general feature of magnetic clouds.

Five hours after the passage of the front boundary there was enhancement of ≤ 4 MeV ions streaming away from (not toward) the sun that was characterized by an abrupt increase and a duration of 6 hours. The enhancement has all the characteristics of an impulsive solar particle event (Richardson et al., 1991). The transit time of such particles along spiral field lines is of the order of a few hours. Flares often follow within an hour of the filament eruptions that are associated with interplanetary ejecta (Kahler and Reames, 1991; Kahler, 1992). Thus the solar particle event was probably related to the solar flare that was followed by the ejection of a magnetic cloud. The prompt arrival of the solar flare particles streaming away from the sun implies that the magnetic field lines in the magnetic cloud were connected to the sun near the flare site. Further examples of injections and flow anisotropies of solar energetic particles in magnetic clouds, supporting the magnetic connectedness of these configurations to the sun, have been given by Kahler and Reames (1991) and Farrugia et al. (1993b).

If magnetic clouds were magnetically disconnected from the sun, and if the protons and electrons obeyed a polytropic law, they would cool as the magnetic cloud expands. Chen and Garren (1993), following earlier work of Chen (1989, 1990, 1992), calculated that if the protons in the magnetic cloud were thermally isolated from the sun, their temperature at 1 AU would be of the order of a few degrees Kelvin! Since the temperatures of protons and electrons in magnetic clouds at 1 AU are of the order of 10^4 or 10^5 K, the magnetic field lines in a magnetic cloud must be connected to the sun in order to provide a conduit for the transport of thermal energy.

6.4 Force-Free Field Models of Locally Cylindrical Magnetic Clouds

6.4.1 Early Ideas

For a single spacecraft crossing such a magnetic cloud near the apex of the flux tube, the tube appears to be locally cylindrical to first approximation (Burlaga et al., 1981b; Burlaga and Behannon, 1982; Klein and Burlaga, 1982). Goldstein (1983) proposed that the variation of magnetic field vectors observed during the passage of a magnetic cloud magnetic as sketched by Burlaga and Behannon (1982) is consistent with a force-free magnetic field configuration in a magnetic flux rope with locally cylindrical geometry. Force-free magnetic field configurations are those for which $\mathbf{J} \times \mathbf{B} = 0$. This implies that $\mathbf{J} = \alpha \mathbf{B}$, representing field-aligned currents associated with twisted magnetic field lines. In a force-free field configuration, the magnetic curvature force owing to the helical magnetic field lines is balanced by the force related to the gradient of the magnetic pressure (Ferraro and Plumpton, 1966). Goldstein (1983) considered the general case in which α is a function of position in the magnetic cloud, but he did not obtain specific solutions that could be compared with observations. Marubashi (1986) assumed a special functional form for α and obtained solutions that compared favorably with the observations of a few magnetic clouds.

6.4.2 Constant α Force-Free Magnetic Field Configuration

Constant α force-free magnetic field configurations (Lust and Schluter, 1954; Woltjer, 1958; Chandrasekhar and Kendall, 1957; Buck, 1965; Taylor, 1974, 1986; Miller and Turner, 1981) play a special role in MHD, because they are easy to calculate, the field equations being linear. Constant α fields represent an energy minimum for fixed total magnetic helicity in a specified volume. Force-free tubes on the sun were discussed by Gold and Hoyle

(1960), who suggested that two loops of opposite twist would attract, meet, and annihilate, leading to a solar flare.

A cylindrically symmetric constant α force-free magnetic field is described by Lundquist's solution (Lundquist, 1950):

$$B_R = 0 \tag{6.1}$$

$$B_\varnothing = \pm B_0 J_1(\alpha R) \tag{6.2}$$

$$B_z = B_0 J_0(\alpha R) \tag{6.3}$$

The integral curves describing the field lines of Lundquist's solution are helices wrapped on coaxial cylinders. Thus, this solution is related to the Euclidean group $E(1)$, which is the direct product of the translation group and the rotation group SO(1). The axis of the cylinder is the limiting case of a helix that is a straight line. On the outermost boundary of the cylinder, the helices have the limiting form of circles, if one cuts off the cylinder at a zero of J_0. The $\pm$ sign in equation (6.2) is very important, referring to the existence of both right- and left-handed helices. The handedness of the flux rope is an invariant and should not change between the sun and the arrival of a magnetic flux rope (magnetic cloud) at Earth or beyond. Thus the observation of Rust (1993) that the handedness of the magnetic clouds agrees with the handedness of the solar filaments with which they are associated is particularly significant, and provides strong support for the flux rope model.

Lundquist's solution provides good fits to the variations of the magnetic field direction observed across a variety of magnetic clouds (Burlaga, 1988a). The panel on the right of Fig. 6.3 shows the observations of a magnetic cloud in which the magnetic field direction varied smoothly from north to south; the panel on the left shows the results derived from Lundquist's solution for a magnetic cloud whose axis is inclined only 20° from the ecliptic and 120° from the Earth–Sun line. The theoretical boundary of the magnetic cloud is taken as the point where the longitudinal component of the magnetic field is zero. Clearly, Lundquist's solution provides a good approximation to the magnetic field in the observed magnetic cloud.

A magnetic cloud in which the field is always pointing southward, except at the boundaries where it is in the ecliptic, is shown on the right of Fig. 6.4. This corresponds to a magnetic cloud whose axis is nearly normal to the ecliptic, as demonstrated by Lundquist's solution for a magnetic cloud inclined −80° with respect to the ecliptic (Fig. 6.4, left).

Lundquist's solution implies that the magnetic field strength is maximum at the center of the flux rope, whereas the observations in Figs 6.3 and 6.4 show an asymmetry in the magnetic field strength profile. The magnetic field strength always reaches a maximum before the center of the magnetic cloud interval, unless the magnetic cloud is compressed by a flow or shock

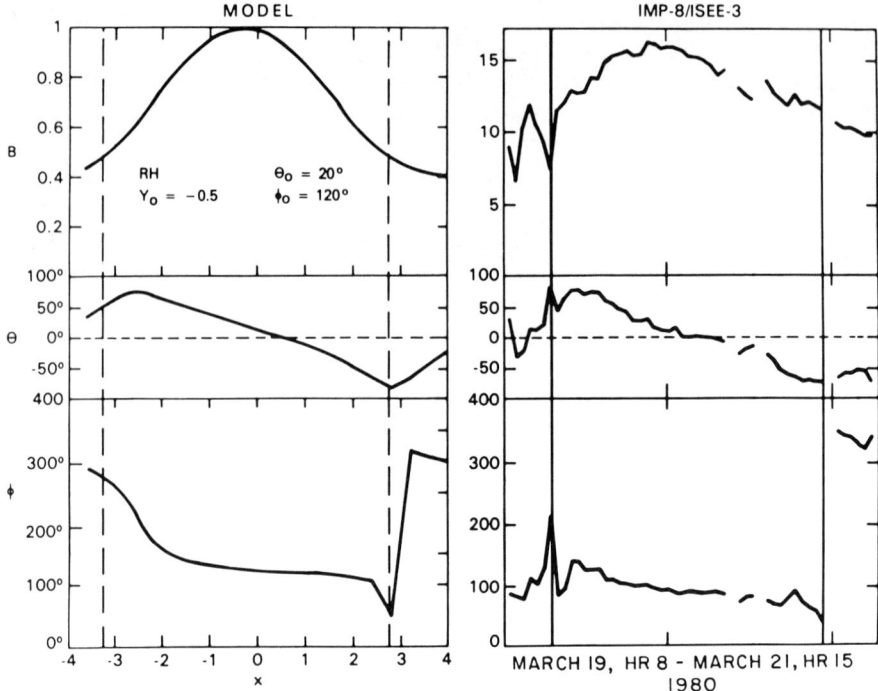

Fig. 6.3. Magnetic cloud and fit to the constant-α force-free field model for a magnetic cloud whose axis is near the ecliptic. (L.F. Burlaga, *J. Geophys. Res.*, **93**, 7217, 1988a, published by the American Geophysical Union.)

overtaking it from behind. We shall show below that this asymmetry can arise as a consequence of the expansion of a magnetic cloud moving past a spacecraft.

Given the large-scale geometry of a magnetic cloud shown in Fig. 6.1 and given that the magnetic field lines are helices described approximately by Lundquist's solution, one can reconstruct the projection of the magnetic field lines in a plane approximately to scale. Figure 6.5 shows such an image of a magnetic cloud. This figure was originally constructed by Burlaga et al. (1990a), but it has been reproduced in several other papers since then.

6.5 Motions of Magnetic Clouds

6.5.1 Local Expansion

The magnetic cloud reported by Burlaga et al. (1981b) was moving faster than the ambient medium and was driving a shock. These authors noted that the speed decreased monotonically as the magnetic cloud moved past the

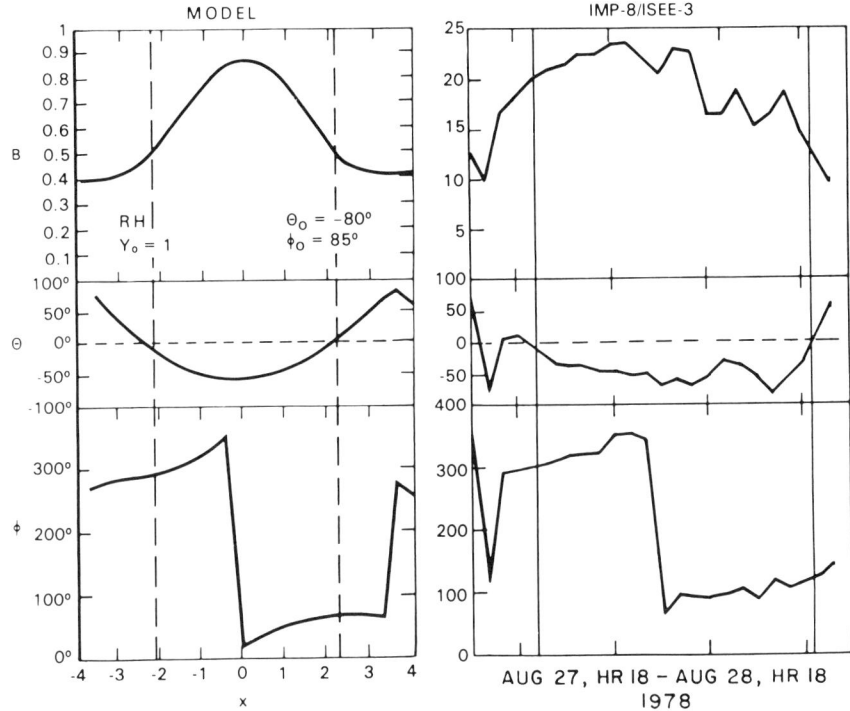

Fig. 6.4. Magnetic cloud and fit to the constant α force-free field model for a magnetic cloud whose axis is nearly perpendicular to the ecliptic. (L.F. Burlaga, *J. Geophys. Res.*, **93**, 7217, 1988a, published by the American Geophysical Union.)

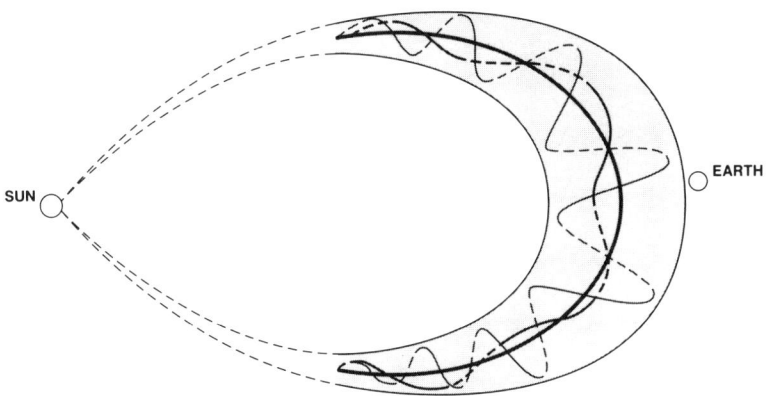

Fig. 6.5. Global configuration of a magnetic cloud and its magnetic field lines. (L.F. Burlaga, R. Lepping, and J. Jones, in *Physics of Flux Ropes,* edited by C.T. Russell, E.R. Priest, and L.C. Lee, p. 373, AGU Geophysical Monograph 58, American Geophysical Union, Washington D.C., 1990a, copyright by the American Geophysical Union.)

spacecraft. Klein and Burlaga (1982) found that the speed typically decreases across magnetic clouds. They interpreted the decreasing speed as evidence of expansion of a magnetic cloud at a speed of the order of half the Alfvén speed. Expansion of a magnetic cloud is clearly necessary if it originates at the sun and has a radial extent of 0.25 AU at 1 AU.

Additional evidence of the expansion of magnetic cloud was presented by Burlaga and Behannon (1982), who studied magnetic clouds observed between 2 AU and 5 AU by the Voyager spacecraft. The magnetic clouds at larger distances are larger on average than the magnetic clouds observed at 1 AU. The largest magnetic cloud observed to date was seen at 11.5 AU (Burlaga et al., 1985a).

Suess (1988) attributed the expansion of the minor radius of magnetic clouds to a kinematic effect. Imagine the cross section of a magnetic cloud to be a circle in a meridian plane in a region above the corona. The particles at the top and bottom of the circle (and elsewhere in the magnetic cloud) are moving predominantly along radial lines passing through the center of the sun. Thus, the separation between the top and bottom of the circle increases as the circle moves away from the sun, giving an increase in the cross section of the magnetic cloud and a corresponding meridional expansion velocity. This model does not directly account for the radial expansion of the magnetic cloud unless it is assumed that forces tend to maintain the circular cross section.

A dynamical model for the radial expansion of a constant α, force-free, zero β flux tube was introduced by Farrugia et al. (1992a,c). A class of exact solutions was found for self-similar expansion with a similarity variable $\xi = r/y(t)$, where $y(t)$ ("the evolution function") is the solution of $(y'/y)' = -(t-t_0)^{-1}$. This gives the relative speed for expansion in an inertial frame

$$v = \frac{U + r_0/t_0}{1 + t/t_0} \qquad (6.4)$$

The solution of equation (6.4) with the average speed $U = 631$ km/s gives a good fit to the speed profile for the magnetic cloud of January 14–15, 1988, as shown by the curve and the observations in Fig. 6.6. The fit also gives the parameters $t_0 = 65.4$ hours and $r_0 = 0.18$ AU, which says that when the magnetic cloud first arrived at the spacecraft it had been expanding freely a little over 2.5 days and had a radius of 0.18 AU. The simplest case of self-similar radial expansion leads to a velocity law $v = r/t$, which gives a good fit to the data in Fig. 6.6. The average expansion speed of the boundary was $r_0/t_0 = 114$ km/s, which is about 0.7 times the Alfvén speed on the axis of the cloud, consistent with the estimate of half the Alfvén speed by Klein and Burlaga (1982). This solution also describes the speed profile in many other clouds.

The expansion of a cylindrical flux rope with a constant α, force-free field solution was analyzed in a somewhat different way by Yang (1990). He

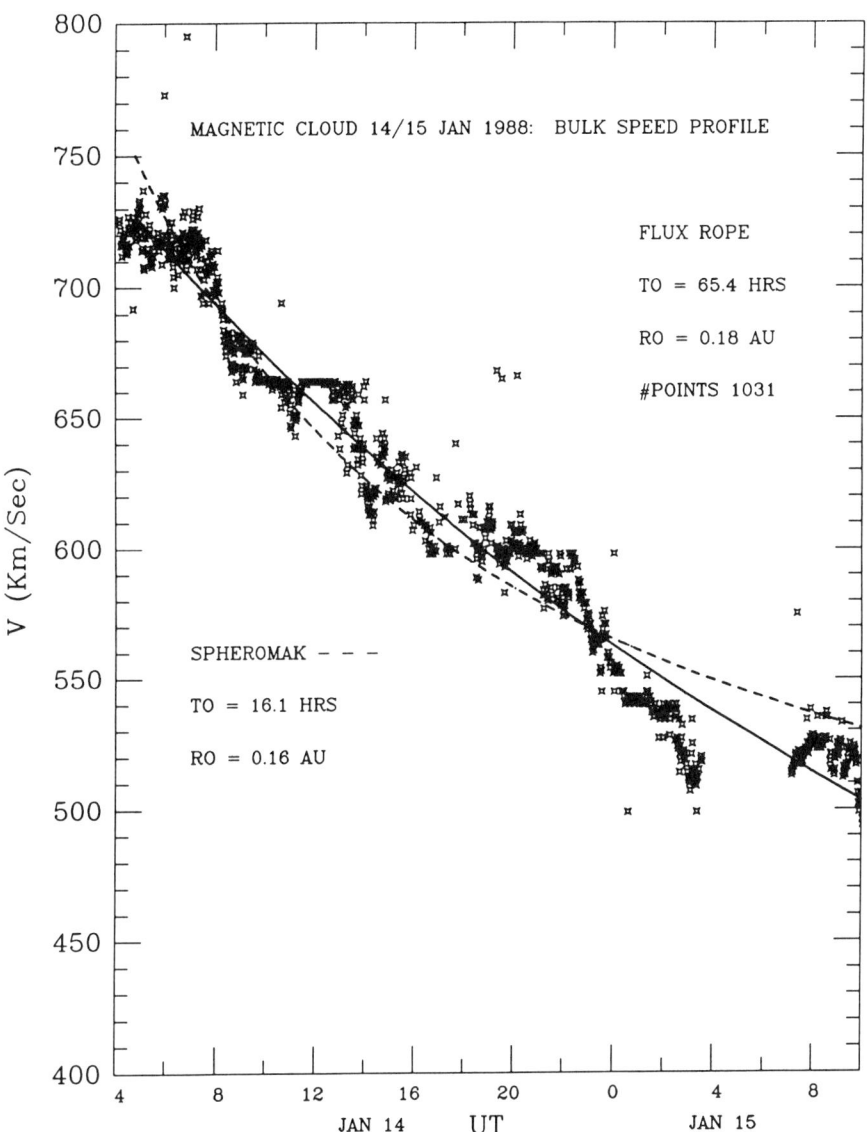

Fig. 6.6. Speed profile of an expanding magnetic cloud. (After C.J. Farrugia, L.F. Burlaga, V.A. Osherovich, I.G. Richardson, M.P. Freeman, R.P. Lepping, and A.J. Lazarus, *J. Geophys. Res.,* **98,** 7621, 1993a, copyright by the American Geophysical Union.)

assumed that the expansion is approximately linear with increasing time, and he showed that the force-free equilibrium can be approximately maintained in this limit. Yang found an expansion speed of the order of half the Alfvén speed, consistent with the results of Klein and Burlaga (1982). When the magnetic cloud expands, it does work on the ambient solar wind.

Yang showed that the work done is equal to the loss in magnetic energy of the magnetic cloud as a result of the expansion.

6.5.2 Asymmetry of the Magnetic Field Strength Profile

The magnetic field strength profile recorded during the motion of a magnetic cloud past a spacecraft is usually asymmetric, with the peak field intensity occurring in the front half of the magnetic cloud. Occasionally, the peak intensity is observed at the rear of a magnetic cloud, but this occurs when a faster flow or shock is overtaking the magnetic cloud and compressing it. Burlaga et al. (1987a) suggested that a peak field intensity at the front of a magnetic cloud might similarly be caused by an interaction of the magnetic cloud with the ambient medium. Such an effect is expected if the magnetic cloud is moving much faster than the ambient medium, just as strong fields are observed in interaction regions ahead of fast flows (see Section 7.1.2).

An asymmetry in the observed magnetic field strength profile is expected even for the Lundquist constant α, force-free field solution, which is symmetric relative to the axis of the magnetic cloud (Farrugia et al., 1993a, 1994a; Osherovich et al., 1993b). This surprising conclusion is a consequence of the expansion of the magnetic cloud while it moves past the spacecraft. The process is illustrated in Fig. 6.7, where the symmetrical curves

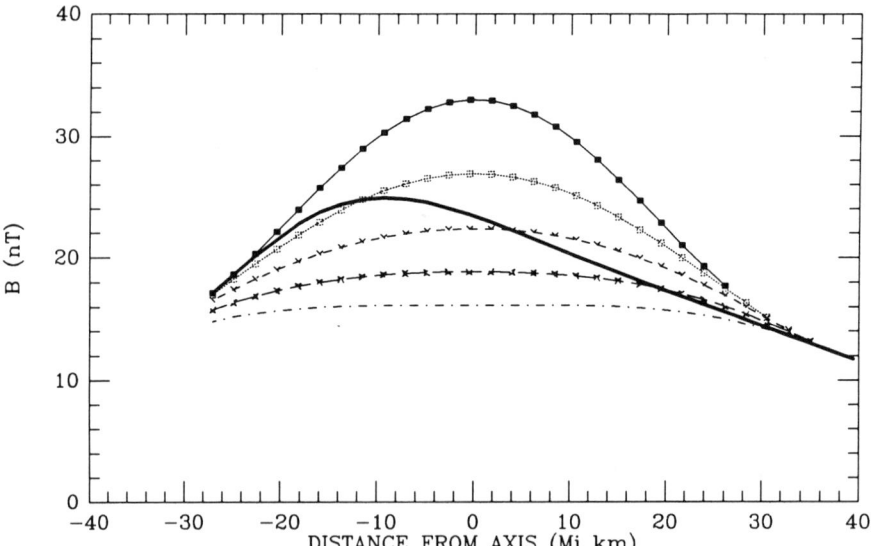

Fig. 6.7. Asymmetry of the magnetic field strength profile in a magnetic cloud. (C.J. Farrugia, L.F. Burlaga, V.A. Osherovich, I.G. Richardson, M.P. Freeman, R.P. Lepping, and A.J. Lazarus, *J. Geophys. Res.*, **98**, 7621, 1993a, copyright by the American Geophysical Union.)

containing points are Lundquist's solutions for a magnetic cloud with increasing size. The initial profile, corresponding to the time of arrival of a magnetic cloud at a spacecraft at 1 AU, is shown by the curve with the strongest fields, marked by the solid squares. The lower symmetrical curves represent snapshots of the magnetic field intensity profile across the magnetic cloud at equal time steps later.

If the magnetic cloud moved at infinite speed, a spacecraft would observe the symmetrical profile at the top of Fig. 6.7. However, the magnetic cloud expands appreciably during its motion past a fixed spacecraft, and the spacecraft samples the weaker magnetic field strength profiles as time goes on. Consequently, one observes the asymmetric magnetic field strength profile shown by the solid curve in Fig. 6.7. The effects of any evolution of the magnetic field and thermodynamic structures should be noticeable in the signatures. In particular, the magnetic field components decrease in time. For the spherical case, the decrease is the same for all the components; but for the flux rope, the axial component decreases faster than the azimuthal component. The fields after the time of maximum magnetic field strength have had a longer time to decrease, because the spacecraft samples them later. This brings in a dual asymmetry, a shift in the time of the maximum field strength and weaker fields at the rear of the magnetic cloud.

Another effect of the expansion of the magnetic cloud is that the magnetic field strength decreases faster with distance than it would in the absence of expansion. This is simply because the magnetic flux is distributed over a larger volume at larger distances in a tube that expands relative to the ambient medium than in a tube that does not expand. The magnetic field strength in a magnetic cloud at 1 AU is stronger than the ambient Parker magnetic field, by definition. However, as a result of expansion, at some distance the magnetic field strength in the cloud will be equal to the ambient magnetic field strength. Osherovich et al. (1993b) calculate that this distance will be between 2 AU and approximately 12 AU, depending on the magnetic field strengths in the magnetic cloud and the ambient medium and the polytropic index of the gas. This is consistent with the observation that the most distant magnetic cloud observed to date was at 11.5 AU (Burlaga et al., 1985a). At large distances, the requirement that a magnetic cloud have higher than average magnetic field strength is no longer satisfied, and magnetic clouds would not be identified by definition.

6.5.3 Local Rotation

Some magnetic clouds rotate, as demonstrated by Fig. 6.8. The south-to-north rotation of the magnetic field direction indicates that the magnetic cloud axis was close to the ecliptic. The decreasing speed profile clearly indicates that the magnetic cloud was expanding. The evidence for rotation is in the north–south flow angle. Near the center of the magnetic cloud, the profile is consistent with that for rigid rotation, but the flow angle returns to

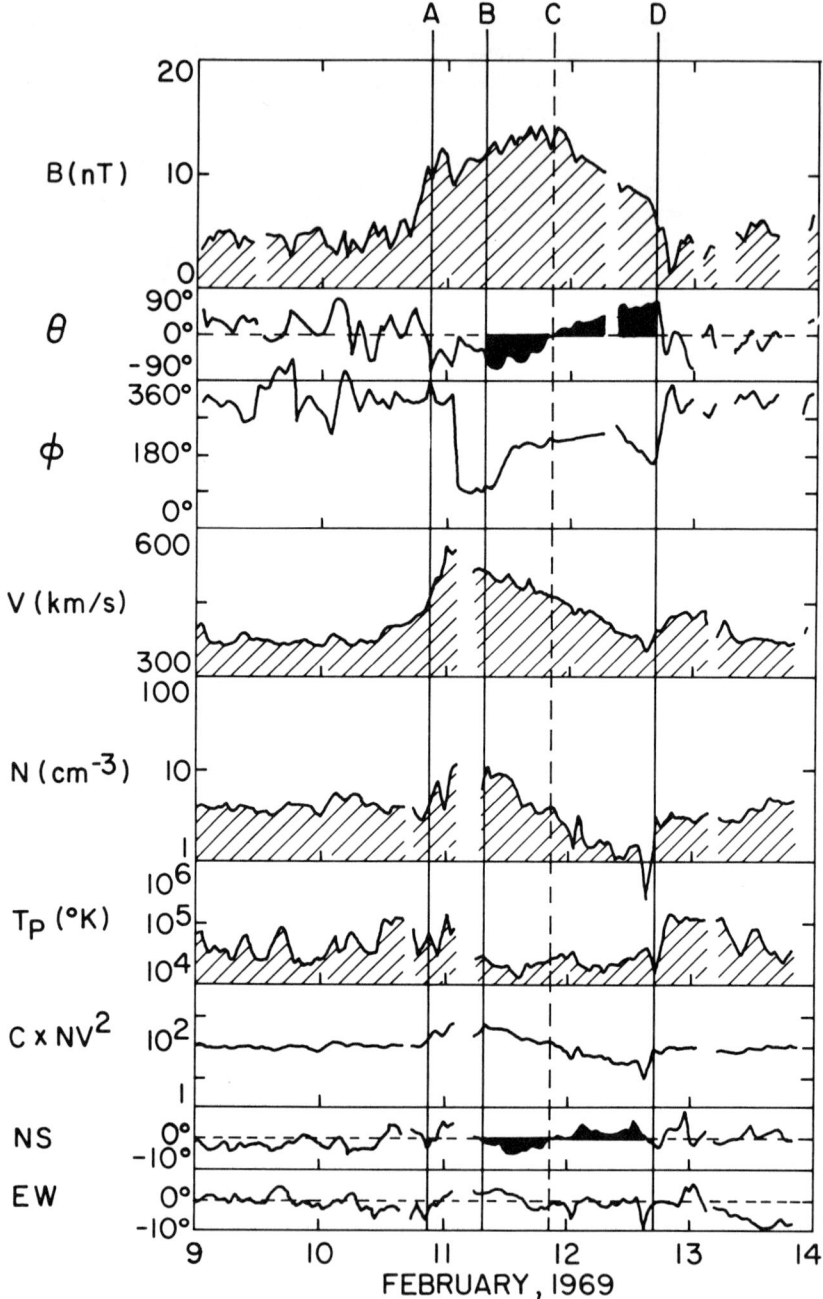

Fig. 6.8. A rotating magnetic cloud., (L.W. Klein and L.F. Burlaga, *J. Geophys. Res.*, **87**, 613, 1982, copyright by the American Geophysical Union.)

0° at the front and rear of the boundary. Overall, the change in direction is much like that in a Rankine vortex. Other rotating clouds were found by Crooker et al. (1990) and Farrugia et al. (1992b).

An ideal MHD solution for a rigidly rotating, cylindrically symmetric magnetic cloud was derived by Farrugia et al. (1992b) for the case of finite beta. The velocity relative to the axis of the cylinder is

$$\mathbf{v} = vr\mathbf{e}_r + v_\varnothing \mathbf{e}_\varnothing = \left(\frac{y'}{y}\right) r \mathbf{e}_r + \Omega(t) r \mathbf{e}_r \qquad (6.5)$$

where $\Omega(t) = \Gamma y^{-2}(t)$ is the angular frequency of rotation, Γ is the angular momentum per unit mass, and $y(t)$ is the evolution function that satisfies the ordinary differential equation

$$y'' = y^{-3} - y^{-1} - \kappa y^{(-2\gamma+1)} \qquad (6.6)$$

where κ is a constant dependent on the polytropic index, the plasma beta, and the ratio of the azimuthal to axial field components. This solution assumes that the evolution is self-similar and the rotation is rigid.

The solution above was used to interpret the rotation of the magnetic cloud observed at 1 AU on January 14, 1988 (Farrugia, 1992b). The angular velocity of the rigidly rotating core at 1 AU corresponds to a rotation period of approximately 3 days. Since the fit to the speed profile for a self-similar radial expansion gave an age of 65 hours, the core of the magnetic cloud rotated approximately once in transit from the sun to earth. The rotation was not rigid across the entire magnetic cloud. Farrugia et al. (1992b) estimated that the average angular speed was 217 km/s, which is at least 0.8 times the Alfvén speed on the axis of the magnetic cloud. In other words, the average angular speed is comparable to the Alfvén speed. A similar result was found for the rotating magnetic cloud of February 1969 identified by Klein and Burlaga (1982).

6.5.4 Oscillations and Stability of Cylindrical Magnetic Clouds

The equilibrium of galactic jets modeled as cylindrical force-free fields was studied by Koenigl and Choudhuri (1985, 1986) (see the erratum note by Koenigl and Choudhuri (1986)). They studied basically a kink instability, and Burlaga (1988a) estimated that the growth rate of this instability is negligible for magnetic clouds in the solar wind.

The nonlinear stability of a cylindrical flux rope with a low plasma beta was studied by Osherovich et al. (1993a). The problem can be reduced to an ordinary differential equation with an effective potential. Osherovich et al.

found that a flux tube with maximum magnetic field strength on the axis is stable. It oscillates, but it does not expand. The restoring force is provided by magnetic tension, and the force driving the oscillation outward is the magnetic pressure gradient. The oscillations are about the force-free state. In the linear limit, the oscillation period is of the order of the radius of the tube divided by the Alfvén speed, which is very large in the solar wind but short in the solar corona, where the Alfvén speed is high and the tube radius is small. The period of the oscillations grows exponentially with increasing energy.

6.5.5 Acceleration of Magnetic Clouds and Propagation to Earth

A model for propelling a flux rope from the sun into interplanetary space at supersonic speeds was introduced by Chen and Garren (1993) based on the ideas developed by Chen (1989, 1990, 1992). In their model, the flux rope is rooted in the sun at both ends. It expands initially in response to magnetic flux supplied from below the photosphere. The magnetic cloud is accelerated in the lower corona primarily by the Lorentz force associated with the poloidal current along the flux tube and the toroidal magnetic field around the flux tube. Thus, the currents transfer stress from the magnetic field to the gas. The model includes both a magnetic pinch force related to the curvature of circular magnetic field lines in the flux rope and a magnetic pressure gradient force owing to the stronger magnetic field in the center of the current carrying tube. These forces are important in the evolution of the minor radius. The magnetic cloud propagates through a solar wind model. It is accelerated in part by the large radial pressure gradients across the magnetic cloud near the sun. The magnetic cloud is also convected by the solar wind. In this model the magnetic field geometry inside the tube is like that of a magnetic pinch rather than a force-free configuration, but that is not essential for the basic mechanism that drives the flux tube away from the sun and causes its minor radius to expand.

For the prescribed input flux profile, the loop reaches 1 AU after 4.8 days with an apex speed of 457 km/s and a minor diameter of 0.12 AU, in the case of a polytropic law with $\gamma = 1.1$. The motion of the apex of the flux loop is not very different for a polytropic exponent of 5/3, but the minor diameter is smaller by a factor of 2 in this case.

6.6 Force-Free Tori and Spheroids

The magnetic flux rope model discussed above has strong support in observations, and it is easy to imagine that a magnetic cloud is a solar flux tube or filament that has been driven into the solar wind by a mechanism such as that of Chen, maintaining its connection to the sun under most conditions. However, several papers advance the idea that magnetic clouds are magnetically disconnected from the sun and have the form of either a torus or a spheroid.

The idea of a toroidal geometry for magnetic clouds was advocated by Ivanov et al. (1989). They used a constant α, toroidal force-free field solution to fit observations of a magnetic cloud made by the Soviet spacecraft Vega 1 and Vega 2 at 1 AU. The agreement between the observations is only qualitative. In any case, the agreement between observations and the local toroidal solution would not prove that magnetic clouds are actually toroidal. It would simply indicate that the toroidal solution is locally a good approximation to a bent flux rope. Indeed, it is reasonable to use a segment of a torus as a local approximation to a bent flux rope, but this tube could be connected to the sun, as shown in Fig. 6.2.

If a magnetic cloud has the global form of a torus, then one should pass through it twice when the axis is in the ecliptic. For example, one might see the magnetic field rotate from north to south as in Fig. 6.1 when the front of the torus moves past the spacecraft; then, some time later, the magnetic field should rotate from south to north as the rear of the torus moves past the spacecraft. Such double crossings are not observed. It is reasonable to use the toroidal solution as a local correction to the cylindrical force-free magnetic cloud solution that allows for the curvature of the axis that must be present, but one must not infer that agreement with local observations implies global toroidal topology.

A spherical or ellipsoidal geometry for magnetic clouds was proposed by Vandas et al. (1991, 1992a,b, 1993). Spherical topology for magnetized interplanetary ejecta in general was also proposed by Ivanov and Harshiladze (1985) following earlier ideas by Parker (1957), although the Soviet workers did not apply the results to magnetic clouds. Of course, spherical topology implies disconnection of a magnetic cloud from the sun.

Spherical constant α, force-free magnetic field solutions have been advocated by Vandas et al. (1991, 1992a,b, 1993) as the appropriate models of magnetic clouds. The same authors have also proposed more complex prolate and oblate spheroidal models. The magnetic field geometries in these solutions are described in figures of Vandas et al. (1993). In the spheroidal case, the magnetic field is essentially that of a toroidal magnetic cloud that is swollen such that there is no gap on the axle (if one pictures the torus as a wheel on an axle), and the whole configuration is inflated such that it fills a sphere. An excellent illustration of this configuration is given by Rosenbluth and Bussac (1979, Fig. 1), who analyzed the MHD stability of such a spheromak in relation to tokamaks. When the axis (the axle) is normal to the ecliptic, and the axis moves past the spacecraft, one should effectively see two cylindrical magnetic clouds side by side. For example, one might see the field rotate from north to south, and then it would immediately rotate back from south to north. When Vandas et al. (1993) reinterpret the magnetic clouds identified by Burlaga and coworkers (e.g., Lepping et al., 1990), they must extend the boundaries such that their magnetic cloud includes both the signature of a cylindrical magnetic cloud with the signature of an adjacent magnetic cloud with the opposite sense of helicity. Thus, the magnetic clouds of Vandas et al. are of the order of twice as large as those modeled by the cylindrical solution. In some cases, Vandas et al. (1993) chose the cloud boundary very close to the shock that precedes a cloud or extends into regions where the temperature is high and therefore did not satisfy the definition of a magnetic cloud. They predict a double peak in the magnetic field strength under certain conditions. The double peak observed in the December 1980 event was described by Vandas et al. (1993) in this way. However, it is more likely that the second peak of the December 1980 event is associated with the compression produced by a second stream that is overtaking the magnetic cloud.

It does frequently happen that the magnetic field is out of the ecliptic before and after a magnetic cloud according to the definition of Burlaga and coworkers. An example of this is in Fig. 6.1. However, it is possible that such perturbations are produced by draping of magnetic field lines about magnetic clouds (e.g., see the discussion of draping by Gosling and McComas, 1987; Gosling, 1990; Detman et al., 1991) or simply by turbulence that is always present in the solar wind.

Observations show that field generally peaks in the forward half of the magnetic cloud. For a spheromak geometry (i.e., the one Vandas uses), there are orbits intersecting the configuration along which asymmetries in the magnetic field strength profile occur even for a static magnetic cloud (i.e., regardless of expansion). Clearly, the peak magnetic field should occur with equal likelihood on the forward half or the rear half of a spherical magnetic cloud, in contrast to the observations. Of course, a spheroidal magnetic cloud is disconnected from the sun, whereas the available evidence favors connection.

The fits of toroidal and spheroidal solutions to observations mentioned above assume static configurations, whereas it is known that magnetic clouds expand and that expansion can significantly affect the observed magnetic field profile. Thus good fits to the data with static toroidal or spheroidal solutions might imply poor fits for an expanding configuration with the same parameters. Farrugia et al. (1992a, 1994a) were the first to construct theoretical expanding spheromak solutions and compare them with data. They note undesirable features of the expanding spheromak fit to the data. When the thermodynamic structure is considered (Farrugia et al., 1994a), a further undesirable feature occurs. When one considers the expansion of magnetic clouds, it is found (Farrugia et al., 1994a) that self-similar, radially expanding spheromak solutions are known for $\gamma = 4/3$, which is in conflict with the observation that γ in magnetic clouds is about 0.5. It is desirable to construct solutions for expanding tori and spheroids and compare them with observations.

At present, however, the observations are consistent with a curved, locally cylindrical flux rope. There is no convincing evidence for toroidal topologies, and the evidence for spheroidal topologies is inconclusive.

6.7 Internal Structure

6.7.1 Temperature and Polytropic Laws

The proton temperature in a magnetic cloud is low by definition. The density in a magnetic cloud is also frequently lower than the ambient density. These results are a natural consequence of the expansion of the magnetic cloud. Expansion would obviously decrease the density, although a decrease might not be observed at certain distances if the initial density in the magnetic cloud were high. Expansion will also decrease the proton temperature, if the proton temperature obeys an adiabatic law. Thermal conduction of heat from the sun would tend to increase the proton temperature, but it is a very inefficient process and cannot offset the decrease caused by expansion of the magnetic cloud.

Electron thermal energy, on the other hand, is transported very efficiently along magnetic field lines from the sun to the magnetic cloud at 1 AU, if the field lines remain connected to the sun. Thus, the electron temperature in magnetic clouds is high compared to the proton temperature, and it is comparable to the average electron temperature of the solar wind. Because the electron temperature remains high while the density decreases as a result of expansion of the magnetic cloud, the polytropic law for electrons in magnetic clouds at 1 AU has an exponent less than one! Observational evidence for this was given by Osherovich et al. (1993c). Figure 6.9 shows an example of this. The electron temperature decreases approximately linearly with increasing density so that the electrons in a magnetic cloud at 1 AU obey a polytropic law of the form

$$P_e = N^{0.41 \pm 0.05} \tag{6.7}$$

Whether this relation is a true polytropic law valid at all distances from the sun, as opposed to the observation of the relation between pressure and density across a section of 1 AU, remains to be determined.

6.7.2 Relation to Plasma Waves

Intense ion acoustic waves in a magnetic cloud were observed in the Helios plasma wave data of Gurnett by Burlaga et al. (1980), as illustrated in Fig. 6.10. This event was not recognized as a magnetic cloud at the time, but it clearly satisfies the definition of a magnetic cloud. Observations not shown in Fig. 6.10 indicate that the magnetic field direction continued to rotate northward after December 1. This is the first evidence for a magnetic hole at the front of a magnetic cloud. There also were large decreases in the density and in the proton temperature at the front of the magnetic cloud, but they did not occur simultaneously. Since the magnetic hole and the change to

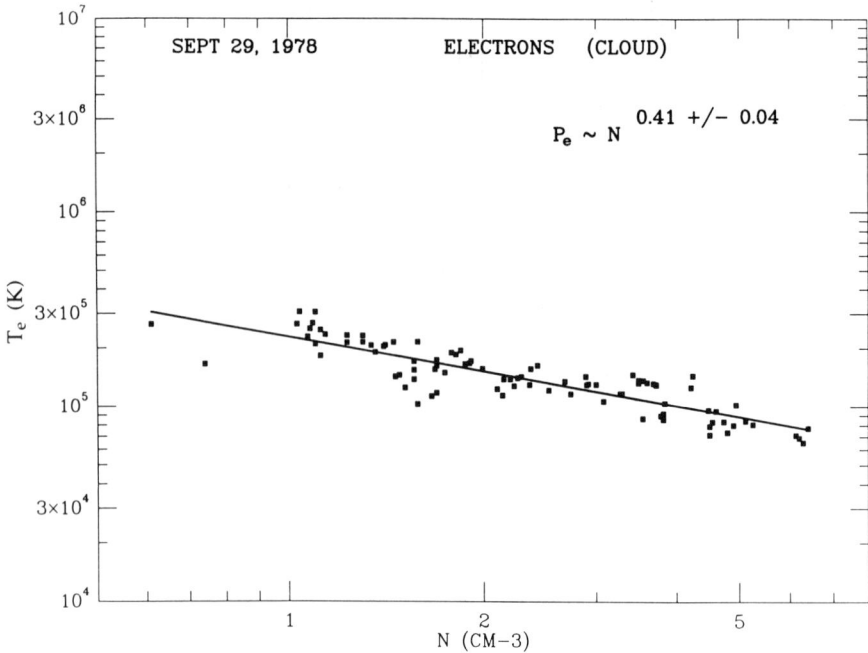

Fig. 6.9. Electron temperature versus density in a magnetic cloud. (V.A. Osherovich, C.J. Farrugia, L.F. Burlaga, R.P. Lepping, J. Fainberg, and R.G. Stone, *J. Geophys. Res.*, **98**, 15331, 1993c, copyright by the American Geophysical Union.)

southward magnetic fields occur nearly simultaneously, one might assume that they mark the forward boundary of the magnetic cloud.

The ion acoustic waves occur in the magnetic cloud of Fig. 6.10 when the proton temperature decreases and remains low. Since the electron temperature is generally high in a magnetic cloud, as discussed in Section 6.7.1, the ratio T_e/T_p was large when the ion acoustic waves occurred. This is consistent with the fact that the ion acoustic instability drives ion acoustic waves that are weakly damped when T_e/T_p is large (Gurnett, 1991). Similar observations of ion acoustic waves in a magnetic cloud with $T_e/T_p \gg 1$ were found recently by Farrugia and Gurnett (private communication, 1992). Farrugia and Burlaga (1994) predict that there should be plenty of ion acoustic emission in the terrestrial magnetosheath when a magnetic cloud moves past Earth.

6.7.3 Turbulence

The variance of the fluctuations of the magnetic field direction is relatively low compared to that in the ambient medium. This was reported by Burlaga et al. (1981b) and confirmed by others (e.g., Marsden et al., 1987). The fluctuations inside of one magnetic cloud were actually turbulent with the

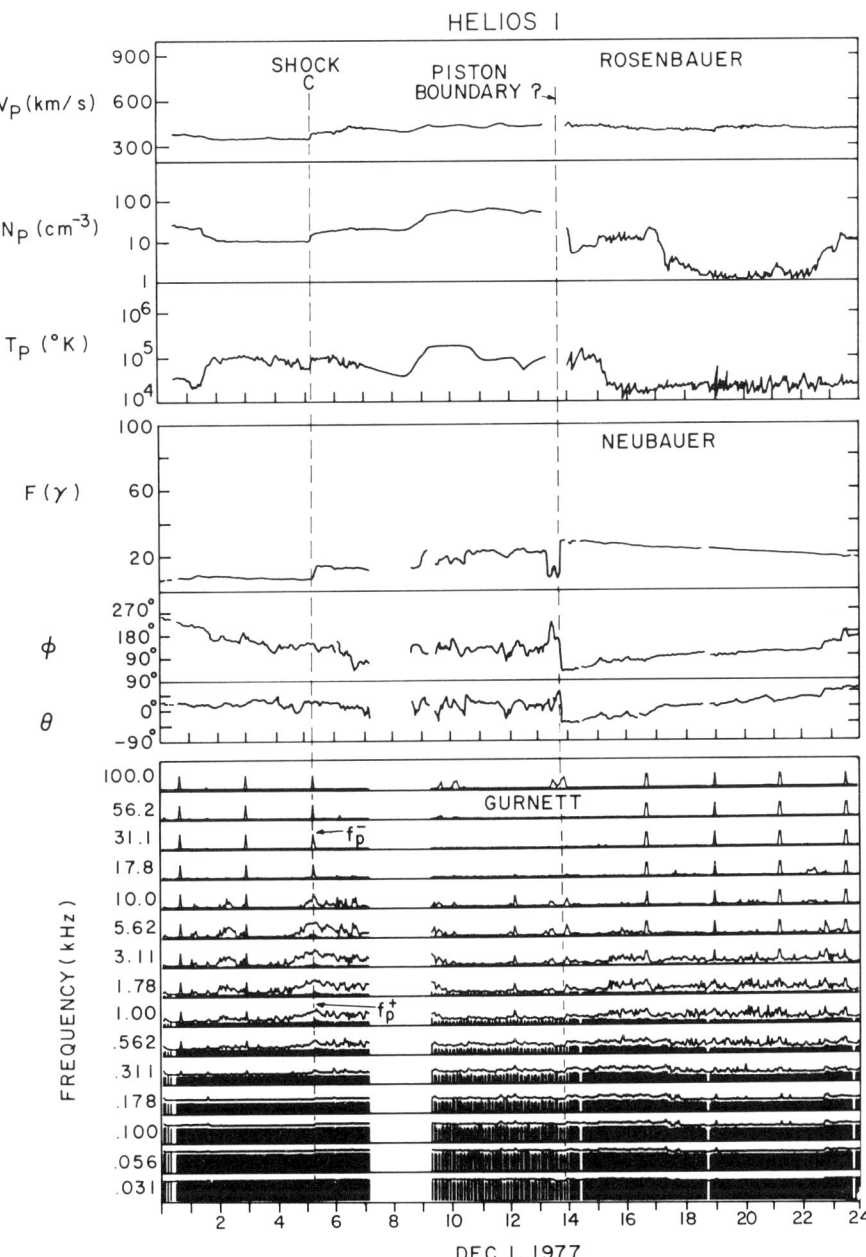

Fig. 6.10. Plasma waves in a magnetic cloud. (L.F. Burlaga, R. Lepping, R. Weber, T. Armstrong, C. Goodrich, J. Sullivan, D. Gurnett, P. Kellogg, E. Keppler, F. Mariani, F. Neubauer, H. Rosenbauer, and R. Schwenn, *J. Geophys. Res.*, **85**, 2227, 1980, copyright by the American Geophysical Union.)

classical 5/3 spectral exponent, just as the fluctuations outside the magnetic cloud (Lepping et al., 1991). However, the level of the turbulence was much lower inside the magnetic cloud than outside it.

It is not surprising that the magnetic fluctuations are small in a magnetic cloud, where the proton beta is unusually low. Quite generally the magnetic fluctuations are low when beta is low and they are high when the proton beta is high (Burlaga et al., 1969). The magnetic field tends to be high and the temperature low in all ejecta, of which magnetic clouds are just a subset. Thus it is to be expected that the magnetic fluctuations are small in ejecta in general (Zwickl et al., 1983).

6.7.4 Filaments and Discontinuities

A remarkable filament at the end of a magnetic cloud was reported by Burlaga et al. (1981b), who showed that it consisted of two tangential discontinuities. Discontinuities in a magnetic cloud were analyzed by Farrugia et al. (1993a), who showed that they produce sudden impulses in the geomagnetic field. Several tangential discontinuities associated with a magnetic cloud were analyzed by Crooker et al. (1990), who inferred from their orientation that the cross section of a magnetic cloud is not necessarily circular.

6.7.5 Boundaries of Magnetic Clouds

One of the most important unanswered questions concerning magnetic clouds is the signature of their boundaries. One frequently observes an abrupt decrease in the proton temperature near the front of a magnetic cloud. This is illustrated in Fig. 6.10, for example. Thus, Vandas et al. (1992a) suggested that these proton signatures represent the boundaries of a magnetic cloud. On the other hand, where necessary to obtain the signature of a spheroidal magnetic cloud, Vandas et al. (1993) put the boundary in a region of high temperature. The density also decreases abruptly near the front of a magnetic cloud, but not always simultaneously with the decrease in proton temperature, as illustrated in Fig. 6.10. It is not clear in such a case whether one should use the temperature decrease, the density decrease, or the directional discontinuity and magnetic hole as the boundary of the magnetic cloud. When a magnetic hole is observed at the front of a magnetic cloud, as in Fig. 6.10, it is tempting to take this as the boundary of the magnetic cloud, but the magnetic hole preceded the drop in temperature by one hour in Fig. 6.10. Such a magnetic hole could be a pressure balanced structure, but the possibility that it could be the signature of magnetic reconnection should also be considered in future studies. The subject of magnetic cloud boundaries requires further research.

Another approach to identifying the boundary of magnetic clouds is to examine the low energy particles. If the boundary of a magnetic cloud is a tangential dicontinuity, then ambient energetic protons such as those associated with a shock driven by a magnetic cloud should not penetrate into the cloud. The magnetic cloud boundary would be marked by an abrupt decrease in the intensity of low energy protons. Such a decrease was reported by Marsden et al. (1987), and Sanderson et al. (1983, 1985, 1990).

One approach to identifying the boundary of a magnetic cloud is to examine the bidirectional streaming of suprathermal electrons, which are regarded as a signature of closed magnetic field lines (see Bame et al., 1981; Gosling et al., 1987, and the review by Gosling, 1990). Not all magnetic clouds show such bidirectional streaming. Moreover, it is possible that the onset of bidirectional streaming occurs after the arrival of the front boundary of a magnetic cloud. A bidirectional stream of low energy protons is also interpreted as evidence for a magnetic loop (Tranquille et al., 1987), but the proton streaming events are not in one-to-one correspondence with the electron streaming events.

Despite the several means of identifying magnetic cloud boundaries discussed above, there is no consistency among the various approaches. Thus the problem of identifying the boundaries of magnetic clouds is unsolved. One possibility is that the boundary of a magnetic cloud is not a single discontinuity. It might be a broad transition region, or a filamentary structure, for example. It is also possible there are several types of structure for magnetic cloud boundaries.

6.8 Relation to Shocks

6.8.1 Magnetic Clouds and Forward Shocks

Approximately one-third of the magnetic clouds observed at 1 AU are preceded by shocks (Klein and Burlaga, 1982). A model for a shock wave in front of a circular cylinder and a sphere was published by Hida (1953). Most magnetic clouds at 1 AU are not moving fast enough to drive a shock. Because of their expansion, the density in a magnetic cloud, hence the momentum flux, decrease rapidly with distance. Thus, it is possible that a magnetic cloud that drives a shock near the sun will decelerate rapidly and the shock will move freely on ahead. Such as shock was probably observed by Voyager 2 near 2 AU (Burlaga et al., 1981b).

The region between a shock and a magnetic cloud, called the "sheath," is turbulent. Magnetic turbulence in the sheath can effectively reduce the galactic cosmic ray intensity (Burlaga et al., 1981b; Zhang and Burlaga, 1988). Thus the sheath between a shock and a magnetic cloud can play the role of a plasma cloud in the model of Morrison. The magnetic cloud can also cause a decrease in the cosmic ray intensity, as in Gold's model of a magnetic tongue, but this decrease is generally smaller than that caused by the shock (Zhang and Burlaga, 1988). However, the magnetic cloud can produce a larger decrease in the cosmic ray intensity than the shock if the shock is weak and the fields in the magnetic cloud are very strong (Cane, 1993).

The Alfvén Mach number just upstream of the shock is low; hence MHD effects on the magnetic field in the sheath of the magnetic cloud should be pronounced. A strong increase of magnetic field strength and magnetic pressure while keeping the total pressure constant was observed in a large part of the sheath region in front of a magnetic cloud (Farrugia et al., 1993c). This is a signature similar in some respects to the magnetic barrier (also called the depletion layer) in the terrestrial magnetosheath, only much thicker in the magnetic cloud sheath.

6.8.2 Magnetic Cloud–Transient Shock Pair

The 1-D gas dynamic "bomb" explosion models of ejecta tend to predict shock pairs at 1 AU, but shock pairs are rarely observed in transient ejecta at 1 AU (Hundhausen 1972; Richter et al., 1985a). This is probably because the models neglect the magnetic field, which produces a higher characteristic speed (Burlaga, 1975) and thus a lower Mach number, and they neglect flow deflections and shears that can effectively relieve the stress that drives the shocks. A forward–reverse shock pair driven by an ejection is rarely observed at 1 AU, because it requires a strong solar event near central meridian (Steinolfsen et al., 1975a,b.)

A forward–reverse shock pair associated with a magnetic cloud was observed on November 1, 1972, by IMP-8. The geometry of the system, based on unpublished observations by Lepping, Ipavich, and Burlaga, is shown in the review by Burlaga (1991c). Ions in the MeV range, are typically accelerated by a forward shock driven by a magnetic cloud, do not penetrate effectively into the magnetic cloud. A forward–reverse shock pair at a magnetic cloud at 1 AU was also reported by Gosling et al. (1988), but the reverse shock might have been produced by a corotating stream that was being overtaken by the magnetic cloud. A forward–reverse shock pair was associated with a magnetic cloud observed at 11 AU (Burlaga et al., 1985a), but in this case the reverse shock was almost certainly associated with a corotating stream that was being overtaken by the magnetic cloud.

7

Corotating Streams and Interaction Regions

7.1 Corotating Streams and Interaction Regions <1 AU

7.1.1 Introduction

In many cases, the existence of an object is inferred from its effects long before it is demonstrated by direct observations. The existence of corotating streams was inferred from their effects on geomagnetic activity (Maunder, 1905; Chapman and Ferraro, 1929; Chapman and Bartels, 1940; Parker, 1963) long before their existence was established by the in situ observations of Neugebauer and Snyder (1966). The inference of the existence of streams served as a hypothesis that guided theoretical work as well as observational studies. The early work of Parker (1958, 1963) is particularly significant.

The existence of corotating stream was demonstrated by the observations of Neugebauer and Snyder (1966). More recent observations of recurrent streams were made by Helios and are reviewed by Schwenn (1990). Several fundamental characteristics of corotating streams are evident in Fig. 7.1. The corotating streams tend to recur with a period of 27 days. The density within a corotating stream tends to be relatively low, while that ahead of the stream and in the rising portion of the stream tends to be relatively high. The transition from high to low densities tends to occur abruptly, coincident with a rapid increase in the speed. The studies of corotating streams show that they tend to be relatively hot compared to the slow solar wind (Neugebauer and Snyder, 1966; Burlaga and Ogilvie, 1973; Schwenn, 1990) and that the magnetic field tends to be enhanced at the leading edge of the corotating stream, except in the inner heliosphere (Schwenn, 1990). The magnetic polarity is always essentially constant in a corotating stream. One or two corotating streams are embedded in a single magnetic sector; hence, the corotating streams presumably originate in a unipolar region at the sun. The early observations and gas dynamic models of corotating streams are reviewed in the book by Hundhausen (1972), by Pizzo (1983a; 1985), and in the recent overview of Schwenn (1990).

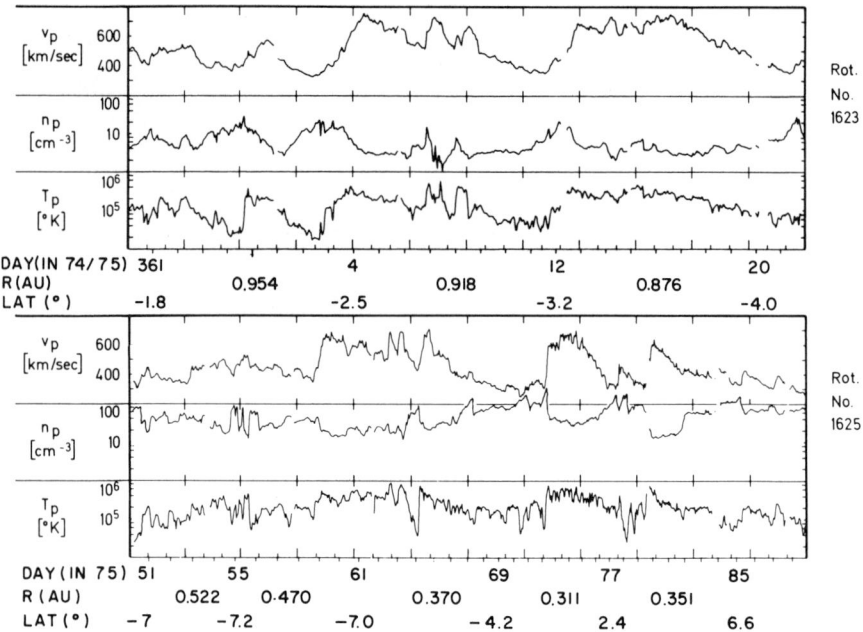

Fig. 7.1. Corotating streams observed by Helios. (After H. Rosenbauer, E. Marsch, B. Meyer, H. Miggenrieder, M. Montgomery, K.H. Mulhauser, W. Pillip, W. Voges, and S.K. Zink, *J. Geophys. Res.*, **42**, 561, 1977, copyright by the American Geophysical Union.)

7.1.2 Interaction Regions

The total pressure $P = p' + B^2/8\pi$ (the sum of the thermal pressure and magnetic pressure) is generally highest at the leading edge of fast streams at 1 AU (Burlaga and Ogilvie, 1970b). This very important observation is not surprising in view of the high densities and magnetic field strengths ahead of corotating streams. The high pressure regions passing a spacecraft at 1 AU in ≈ 36 hours were called "interaction regions" by Burlaga and Ogilvie (1970b), who stressed the dynamical importance of interaction regions. The most probable pressure in the solar wind at 1 AU is $(2.2 \pm 0.3) \times 10^{-10}$ dyn/cm^2 in the Explorer 34 data, and the maximum total pressure in an interaction region is ≈ 2 to ≈ 10 times the most probable value in the solar wind as a whole at 1 AU. A plot of the magnetic pressure and thermal pressure is shown in Fig. 7.2.

The pressure gradients in interaction regions play a major role in the dynamics of mesoscale processes in the solar wind. Neglecting the magnetic curvature term in the Lorentz force (equation 1.4) and neglecting the viscous stress, the equation of motion (1.1) in Lagrangian form becomes $d\mathbf{V}/dt = -\nabla P$, where P is the sum of the magnetic pressure and thermal

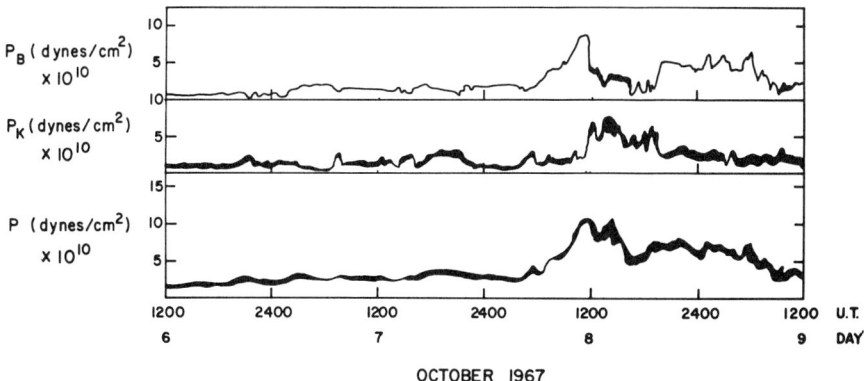

Fig. 7.2 Relation between the magnetic pressure P_B and the thermal pressure P_K. (L.F. Burlaga and K.W. Ogilvie, *Solar Phys.*, **15**, 61, 1970b. Reprinted by permission of Kluwer Academic Publishers.)

pressure, which are comparable at 1 AU. Interaction regions are always present ahead of corotating streams at 1 AU. The interaction regions associated with corotating streams are called "corotating interaction regions" (Smith and Wolfe, 1976) or CIRs. Interaction regions are also associated with transient ejecta and shocks.

7.1.3 Observations of Corotating Streams and Corotating Interaction Regions Near 0.3 AU

Prior to the launch of Helios 1 and 2 it was thought that the "leading edge" of corotating streams (the region in which the speed increases) is broader near the sun than at 1 AU. It was commonly assumed that corotating streams are approximately sinusoidal near the sun. Thus it came as a great surprise to find that the leading edge of a corotating stream near 0.3 AU is steeper than that at 1 AU (Rosenbauer et al., 1977). The broadening of the leading edge of the speed profile as Helios moved from 0.3 AU to 1 AU is evident in Fig. 7.1. The trailing edges of corotating streams near 0.3 AU tend to be steeper than those near 1 AU. The corotating stream profiles near 0.3 AU were described as "mesa-like" by Rosenbauer et al., inspired by the landscape near Los Alamos. Note that the density in the corotating streams has an inverse mesa-like profile, being distinctly lower in the corotating streams than between streams. This is evidence that the low density in corotating streams at 0.3 AU is primarily a signature of the boundary conditions imposed by the source, rather than a rarefaction produced by the expansion of a stream, as we shall discuss below. The temperature near 0.3 AU also has a mesa-like profile, indicating that the high temperatures in corotating streams are imposed by the boundary conditions at the source rather than by dynamical processes. Figure 7.1 also shows a weak, narrow enhancement in the density and temperature at 0.3 AU relative to that at 1 AU. The magnetic field strength in corotating interaction regions near the sun tends to be constant near the sun, increasing relative to the ambient field with increasing distance from the sun (e.g., see the reviews by Pizzo, 1983a, and Whang, 1991).

The interaction regions near 0.3 AU are relatively small compared to those farther from the sun. This is shown in the review by Whang (1991), for example. Pressure waves form in the interplanetary medium and grow in amplitude and width as they move toward 1 AU. The creation and growth of interaction regions is an important dynamical subject, which is discussed in Section 7.1.6.

7.1.4 Sources of Corotating Streams

The sources of corotating streams (the cause of recurrent geomagnetic storms with a period of 27 days) were discussed for more than 80 years (e.g., by Maunder, 1905, and Chapman and Bartels, 1940). The name "M-regions" was given to the source long before the source was identified. Billings and Roberts (1964) suggested that the sources of corotating streams are regions of open magnetic field lines extending from the photosphere to the interplanetary medium. It was demonstrated that the sources of corotating streams are coronal holes, which are associated with open magnetic field lines (Krieger et al., 1973; Noci, 1973; Pneuman, 1973; Neupert and Pizzo, 1974; Nolte et al. 1976; Sheeley et al., 1976). The X-ray observations from Skylab, potential field extrapolations of the photospheric magnetic field, and in the in situ observations of the interplanetary plasma and magnetic field showed definitively that the sources of corotating streams are coronal holes in which the magnetic field lines are open. An excellent review of the subject was written by Hundhausen (1977). Coronal holes appear as dark regions in X-ray images of the sun (hence their name), because they are regions of low density. The relatively low density observed in corotating streams is a consequence of the low density of their sources, the coronal holes.

An instructive example of the relations among corotating streams, coronal holes, and interplanetary magnetic fields is shown in Fig. 7.3. Three corotating streams are shown in Fig. 7.3, labeled 2, 1, and 7. Each of these streams is associated with a coronal hole observed in the X-ray data of Nolte et al. (1976), shown at the bottom of Fig. 7.3. The broad, fast corotating stream 2 was associated with a large equatorial coronal hole, so that the spacecraft, which was in the ecliptic, passed close to the middle of the stream from that coronal hole. It appears that the equatorial coronal

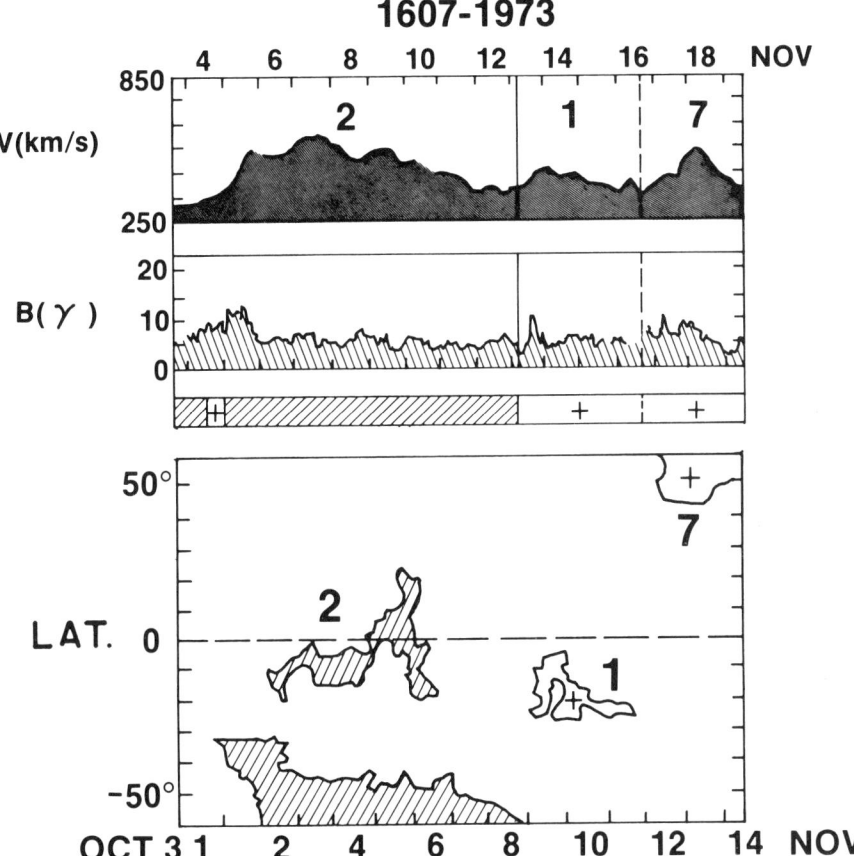

Fig. 7.3. Corotating streams, sectors, and coronal holes. (L.F. Burlaga, K.W. Behannon, S.F. Hansen, G.W. Pneuman, and W.C. Feldman, *J. Geophys. Res.,* **83**, 4177, 1978a, copyright by the American Geophysical Union.)

hole 2 was associated with an equatorial extension of a polar coronal hole in the southern hemisphere. The small, relatively slow corotating stream 1 was associated with a small coronal hole 1 just south of the equator, so that the spacecraft observed only the edge-flow from the northern part of the coronal hole 1. The small, moderate speed, corotating stream 7 was associated with the equatorward extension of a northern polar coronal hole, and the spacecraft observed only the edge-flow from the southern part of coronal hole 7.

The association between the corotating streams and coronal holes just described is supported by the solar and interplanetary magnetic field observations. Pneuman computed the coronal magnetic fields from the Kitt

Peak photospheric magnetic field observations and a potential field model with a source surface at 2.5 solar radii. His results show that the magnetic fields are open in the magnetic holes. The fields are negative (directed toward the sun) in the shaded coronal holes in Fig. 7.3 and positive (directed away from the sun) in the unshaded coronal holes. The polarity of the interplanetary magnetic field, shown in the third panel from the top of Fig. 7.3, is in agreement with the polarity of the magnetic field in the coronal hole from which the corotating stream containing the magnetic field originated. Note that the polarity changed between stream 2 and stream 1, indicating the crossing of a sector boundary. Note also that the two streams 1 and 7 were contained within a single sector, because they originated in two coronal holes with the same magnetic polarity.

The relations among the coronal holes, the footpoints of the sector boundary (the solid curve corresponding to the neutral point in the potential field calculation), and the nonradial flow velocity component on a spherical surface at ≈ 2–10 solar radii are shown in Fig. 7.4. Here the cross-hatched coronal holes correspond to a negative magnetic polarity and the solid coronal holes correspond to a positive magnetic polarity. The reason for the observed polarities of the streams in Fig. 7.4 and the sector boundary crossing between stream 2 and stream 1 is obvious from this figure. Figure 7.4 shows that two closely spaced, near-equatorial coronal holes can produce a large inclination of the footpoints of the heliospheric current sheet relative to the solar equator even if the effective tilt of the HCS is relatively small.

The inferred flow directions are shown by the arrows in Fig. 7.4. Following Pneuman (1973), Pneuman and Kopp (1970, 1971), and Parker

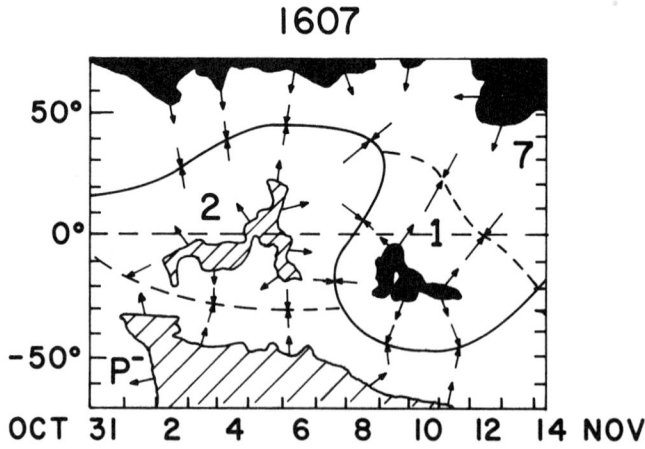

Fig. 7.4. Coronal holes, inferred flow directions, and the magnetic neutral line. (L.F. Burlaga, K.W. Behannon, S.F. Hansen, G.W. Pneuman, and W.C. Feldman, *J. Geophys. Res.*, **83**, 4177, 1978a, copyright by the American Geophysical Union.)

(1963), it is assumed that the coronal magnetic field lines diverge from the coronal holes and that the flow is along the field lines because the plasma beta is very small in the corona. A consequence of these assumptions is the existence of "convergence lines," presumably related to the footpoint of the heliospheric current sheet for the flows from two coronal holes of opposite polarity. The hypothetical convergence lines are shown by the dashed lines in Fig. 7.4 for flows from two coronal holes with the same magnetic polarity.

The convergence lines must correspond to the footpoints of a convergence surface extending through the corona into the interplanetary medium. The convergence surface corresponding to the heliospheric current sheet might be displaced somewhat from the HCS, or there might actually be two convergence lines corresponding to the two boundaries of the heliospheric plasma sheet (Burlaga et al., 1990b), but these are higher order effects that we will not pursue here. The convergence line between coronal holes 1 and 7 is noteworthy because Whang and Sheeley (1990, 1992) and Sheeley et al. (1991) predicted that the speed will be highest where flows converge. One does not see the predicted increase in speed in Fig. 7.3 corresponding to the crossing of this convergence line.

7.1.5 Boundaries of Corotating Streams

Since corotating streams originate in coronal holes, and since the boundaries of coronal holes observed in X-rays or the He 10830 Angstrom line have a width of $10°$ or less (Krieger et al., 1973; Neupert and Pizzo, 1974), one might expect the boundaries of corotating streams to be relatively thin. The latitudinal width of the boundary of a corotating stream between 0.3 AU and 1 AU was determined by comparing the IMP-7 and IMP-8 observations with the Helios observations of a stream originating at Carrington longitude $\approx 140°$ on Carrington rotation 1625 (Schwenn et al., 1978). The interplanetary and coronal observations are shown in Fig. 7.5, together with the projections of the trajectories of IMP and Helios. IMP-7 and IMP-8 passing at $-5°S$ through the equatorial extension of the south polar coronal hole at CR $140°$ observed a broad fast corotating stream, whereas Helios passing at $+5°N$ latitude near the northern limit of the coronal hole observed a fragmented stream. The difference between the IMP and Helios speed profiles was interpreted as evidence that the latitudinal width of the northern boundary of the corotating stream was $\approx 10°$ (Schwenn et al., 1978). This implies a latitudinal velocity shear of 30 km/s/deg or more at the northern boundary of the corotating stream, possibly even 100 km/s/deg, between 0.3 AU and 1 AU.

The latitudinal width of the northern boundary of a corotating stream between the corona and 0.35 AU was determined by Burlaga et al. (1978a) and Burlaga (1979) using Helios plasma data and coronal hole observations based on the He 1030 Å line measurements of J. Harvey. The interplanetary and coronal observations show that the projection of the trajectory of

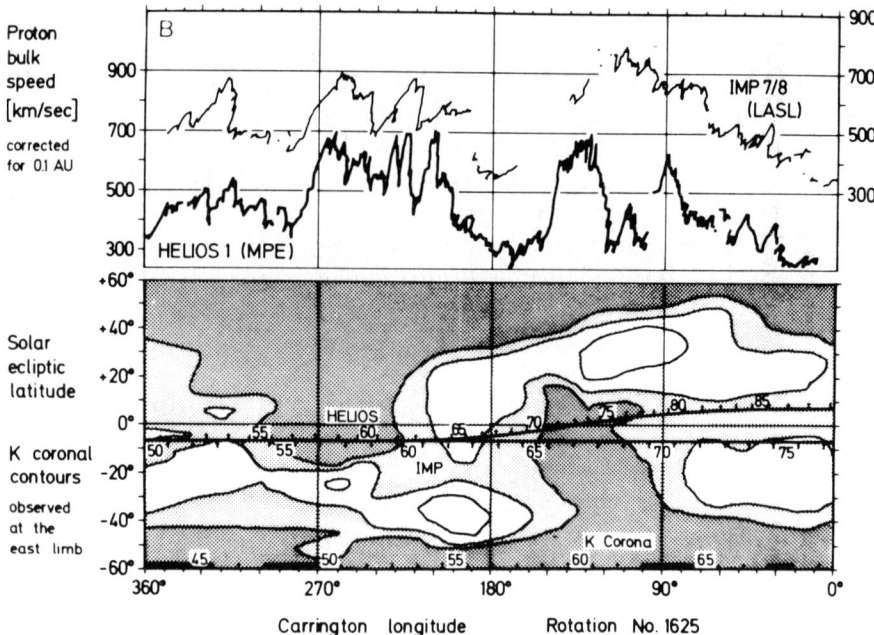

Fig. 7.5. Latitudinal dependence of corotating streams. (R.M. Schwenn, D. Montgomery, H. Rosenbauer, and H. Miggenrieder, K.H. Mulhauser, S.J. Bame, W.C. Feldman, and R.T. Hansen, *J. Geophys. Res.*, **83**, 1011, 1978, copyright by the American Geophysical Union.)

Helios 1 passed ≈ 15°N of the northern edge of the south polar coronal hole on March 15–17, but no fast stream was observed. This observation implies that the latitudinal width of the northern boundary of the corotating stream between the corona and 0.35 AU was 15° or less, which is consistent with the observations of Schwenn et al. (1978) of the width of the northern boundary of a corotating stream between 0.3 AU and 1 U.

The longitudinal width of the boundary of a corotating stream A3 at 0.3 AU was also determined by Burlaga et al. (1978b). The speed increased by 350 km/s in 2° of heliographic longitude, indicating a velocity shear of 130 km/s/deg or more at 0.3 AU. The velocity shear of the rear boundary of the stream A3 was found to be ≈ 20 km/s/deg at 0.3 AU (Burlaga et al., 1978b). The difference in the velocity shear at the front and rear boundaries of the stream was attributed to kinematic steepening of the front boundary and kinematic broadening of the rear boundary between the sun and 0.3 AU. Projecting from Helios to the sun kinematically, they estimated that both the front and rear boundary had a width of 7.4 ± 4.5° at 2.5 solar radii. Assuming a longitudinal divergence factor of 3 between the lower corona and 2.5 solar radii, they estimated that the width of the boundary of the coronal hole was ≈ 2.5 ± 1.5°.

7.1.6 Kinematic Models of Corotating Interaction Regions and Streams

Since the solar wind moves radially and since the speed within a corotating stream is higher than that ahead of the stream, the fast plasma will advance toward the slow plasma ahead, leading to a steeper leading edge profile. This process is referred to as "overtaking" or "kinematic steepening" (Parker, 1963). Similarly, the fast plasma moves away from the slower plasma behind, causing a broader "trailing edge." Kinematic steepening leads to a compression of the plasma and magnetic field in the leading edge of the stream, because the plasma at the leading edge tends to be confined to a smaller volume at later times. Similarly, the motion of the fast plasma away from the slow plasma at the rear of the stream leads to a rarefaction, because the plasma in the trailing part of the stream is distributed over a larger volume. The processes of compression and rarefaction are a result of kinematic effects, as modeled by Sarabhai (1963).

The simple kinematic picture just described has to be modified in two respects in the light of recent observations. First, the leading edge of a stream near the sun is much narrower than originally supposed, so that the interaction between a fast, low density corotating stream and the slow, high density material ahead of it resembles a collision in many respects (Burlaga et al., 1971). Thus, a compression wave would form even if the leading edge of the stream were infinitely thin. Second, it is known that the density within corotating streams is low because the streams originate in low density coronal holes. Thus, the low density in corotating streams is not entirely a consequence of a rarefaction; the effects of rarefaction are superimposed on the low density source signal. Nevertheless, the tendencies for kinematic steepening and rarefaction are always present. Since corotating streams are stationary, the kinematic steepening is represented by the nonlinear term $(\mathbf{V} \cdot \nabla)\mathbf{V}$ in the equation of motion.

Either a collision or kinematic steepening would cause an increase in the magnetic field strength, because the field is frozen in to the plasma. Similarly, a kinematic rarefaction would cause a decrease in the magnetic field strength. These effects were understood by Parker, as indicated by a sketch in his book (1963, Fig. 11.1).

The kinematic effects of the radial evolution of an initially sinusoidal corotating stream on an initially uniform density and magnetic field were calculated by Burlaga and Barouch (1976). The variation of the density and magnetic field strength are basically determined by a projective transformation of a volume element in the kinematic limit; the nonradial sides of the volume element are bounded by lines originating at a point, the sun.

The fundamental kinematic equations in the Lagrangian approach are the equation of continuity and Walen's equation (Boyd, 1969)

$$\frac{d(\mathbf{B}/\rho)}{dt} = [(\mathbf{B}/\rho) \cdot \nabla]\mathbf{V} \qquad (7.1)$$

from which one derives the solution

$$\frac{\mathbf{B}}{\rho} = \left[\left(\frac{\mathbf{B}_0}{\rho_0}\right) \cdot \nabla_0\right]\mathbf{X} \tag{7.2}$$

where $\mathbf{X}$ is the displacement vector and the subscripts indicate values at the initial time.

Using equations (7.1) and (7.2), Burlaga and Barouch (1976) derived the kinematic approximation for the compression of B in the leading part and the rarefaction of B in the trailing part of a corotating stream at 1 AU. The maximum field is significantly larger than that observed, indicating that some other process is competing with kinematic steepening. The rarefaction in B is small in any case. The change in the magnetic field direction as a result of the kinematic effects in the streams is small, only about 15°. Small changes in the initial magnetic field direction relative to that for a stationary spiral magnetic field can cause significant changes in the magnetic field profiles. This effect was also considered by Bieber et al. (1993) in relation to the large-scale magnetic field. Compression and rarefaction of the density as a result of kinematic effects were also computed by Burlaga and Barouch (1976). These effects are independent of the initial magnetic field direction. The kinematic models overestimate the enhancement of density and magnetic field strength in the interaction region. Nevertheless, they demonstrate that the compression and rarefaction of the magnetic field and density observed in corotating streams are basically kinematic effects.

Three-dimensional kinematic effects were studied by Barouch and Burlaga (1976) for an initially sinusoidal corotating stream with constant density and magnetic field strength. The most important point illustrated by their calculation is that the 3-D magnetic field strength and density profiles depend on the latitudinal variation of the bulk speed. Barouch and Burlaga considered two illustrative cases:

$$V \approx 325 + 75[1 + \cos(4\phi_0)] \cos\theta$$

$$V \approx 325 + 75[1 + \cos(4\phi_0)] \exp\left[-\left(\frac{\theta - \theta_0}{\sigma}\right)^2\right]$$

The resulting magnetic field strength contours on a sphere at 1 AU are shown at the top and bottom of Fig. 7.6, respectively. In both cases the compression is greatest near the solar equatorial plane, because the compression is ultimately caused by solar rotation, and the effect of solar rotation is greatest near the equator. There is no compression or rarefaction at the poles, where the effect of solar rotation is zero. For a corotating stream that extends from the pole to the equator, as in the case of a stream originating in an equatorial extension of a polar coronal hole, the

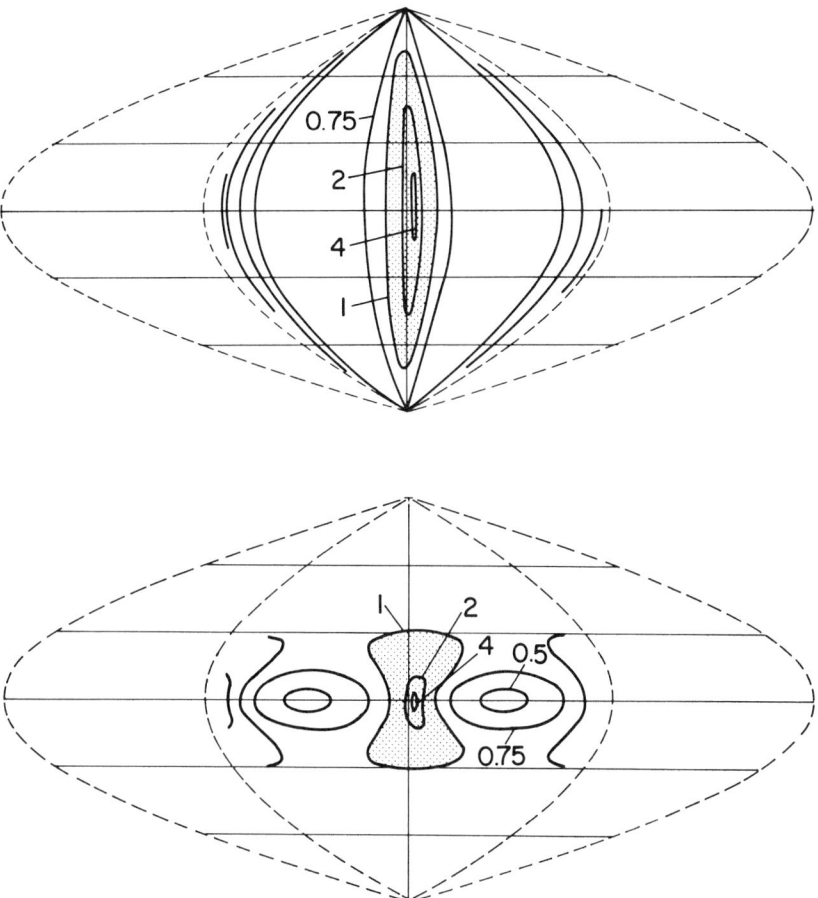

Fig. 7.6. A kinematic model of three-dimensional magnetic field strength. (E. Barouch and L.F. Burlaga, *J. Geophys. Res.*, **81**, 2103, 1976, copyright by the American Geophysical Union.)

compression extends to high latitudes, but the amplitude of the compression decreases with increasing latitude. For an equatorial coronal hole (bottom of Fig. 7.6) the compression and rarefaction regions are confined to near the equatorial region.

In view of the polytropic relation between the proton pressure and density ($p \propto \rho^\gamma$), and considering that $p = NkT$, it is clear that one should expect an increase in the proton temperature T in the interaction region where the density is high owing to compression by the increasing speed. The temperature increases with speed on a large scale (Burlaga and Ogilvie, 1970a). Subtracting the large-scale variation of the temperature determined from the speed profiles from the temperature profiles in the streams, Burlaga and Ogilvie (1973) found that the proton temperature is indeed relatively high in the interaction regions. No appreciable increase in the electron temperature in interaction regions was found by Burlaga et al. (1971), using the observations of the electron temperature from Vela 3 made by Las Alamos Scientific Laboratory (LASL). This is consistent with the weak dependence of electron pressure on density given by the polytropic relation. More recent observations, most notably those from Helios, confirm that the electron temperature is relatively constant across corotating streams (e.g., see the extensive review of corotating streams by Schwenn, 1990).

Up to this point we have been considering only kinematic effects, that is, the effects of the term $\rho(\mathbf{V} \cdot \nabla)\mathbf{V}$ in the equation of motion. From a physical point of view, this is a momentum flux that does work on the gas in the interaction region. The term $\rho(\mathbf{V} \cdot \nabla)\mathbf{V}$ is a nonlinear term, and the nonlinearity is crucial in the evolution of corotating streams. Considering this term alone allows one to compute its effects in the nonlinear limit. Since the kinematic effects produce an increase in N, B, and T at the leading edge of a corotating stream, they produce an increase in the total pressure P. In other words, corotating streams produce interaction regions as a result of the kinematic effects of overtaking and steepening.

On the sunward side of an interaction region there is a gradient in the total pressure which reacts back on the corotating stream that produced it, thereby decelerating the front part of the stream. Inside of 1 AU this deceleration appears as a rounding of the high speed part of the velocity profile, thereby creating the condition for kinematic steepening to act further. On the opposite side of an interaction region, there is a gradient of the total pressure that acts to accelerate the material ahead of the stream, thereby reducing the net speed gradient at the leading edge of the stream. The overall effect of the pressure gradients bounding the interaction region within 1 AU is to maintain the conditions for kinematic steepening, but at an increasingly reduced level with increased distance from the sun.

The kinematic models might be valid close to the sun, and they produce qualitatively correct results at 1 AU. However, their main value is in demonstrating that the formation of an interaction region is driven by kinematic (inertial) effects. At 1 AU the dynamical effects in the interaction regions are at least comparable to the kinematic effects, and beyond 1 AU the pressure gradients are dominant. Thus, a model of the structure and evolution of interaction regions must include both the kinematic effects and the dynamical effect resulting from the gradient in the total pressure, which is a consequence of the kinematic effects.

7.1.7 Dynamical Models of Corotating Interaction Regions and Streams

The evolution of a corotating stream near and beyond 1 AU is basically a competition between the kinematic steepening and the dynamical reaction produced by the interaction region (Burlaga et al., 1983). The equation of motion for a stationary corotating stream is thus

$$(\mathbf{V} \cdot \nabla)\mathbf{V} = -\nabla P \tag{7.3}$$

where P is the sum of the magnetic pressure and the plasma (proton and electron) pressure. The magnetic curvature force is negligible, but the magnetic pressure gradient is at least as important as the plasma pressure gradient near 1 AU.

The early models of corotating streams neglected the magnetic pressure, assumed that the fast streams existed near the sun, and followed the evolution of the plasma parameters with increasing distance from the sun. These early gas dynamic models are reviewed in the excellent book by Hundhausen (1972). Beginning with a low speed and a temperature perturbation at the inner boundary, one can produce a hot, fast stream with the proper relation between the temperature and density at 1 AU (Burlaga et al., 1971; Hundhausen and Burlaga, 1975). The temperature perturbation guarantees that the corotating streams will be hot as observed, and it gives the proper relation between the density and temperature in the interaction region.

The preceding results show that a model of corotating streams should include the following ingredients:

1. the gradient in magnetic pressure must be considered as well as the gradient in plasma pressure (i.e., one must have an MHD model).
2. The stream should be generated by a temperature perturbation at the inner boundary.
3. The density at the source must be relatively low.
4. The magnetic field should be constant across the region producing the stream.

The model should also be at least two-dimensional, to ensure the reproduction of the east–west flow deflections in the leading edge of the stream that are known to be produced by the interaction of corotating stream and the slow, dense plasma ahead of it (see the review of this subject in the book by Hundhausen, 1972). Ideally, the model should include a discontinuous stream interface across which there is a velocity shear, since such a shear can reduce the stress, slow the growth of interaction regions, and delay the formation of shocks, but the present models do not include this effect.

MHD models of corotating streams were developed by Goldstein and Jokipii (1977), Whang (1981), Whang and Chien (1981), and Pizzo (1982). We shall focus on the stationary, 2-D MHD model of Pizzo (1982), because it incorporates all the essential features listed above, it uses the Helios data from Rosenbauer et al. (1977) at 0.3 AU to provide a realistic input, and it computes the parameters that are observed at 1 AU (except T_e). The panels

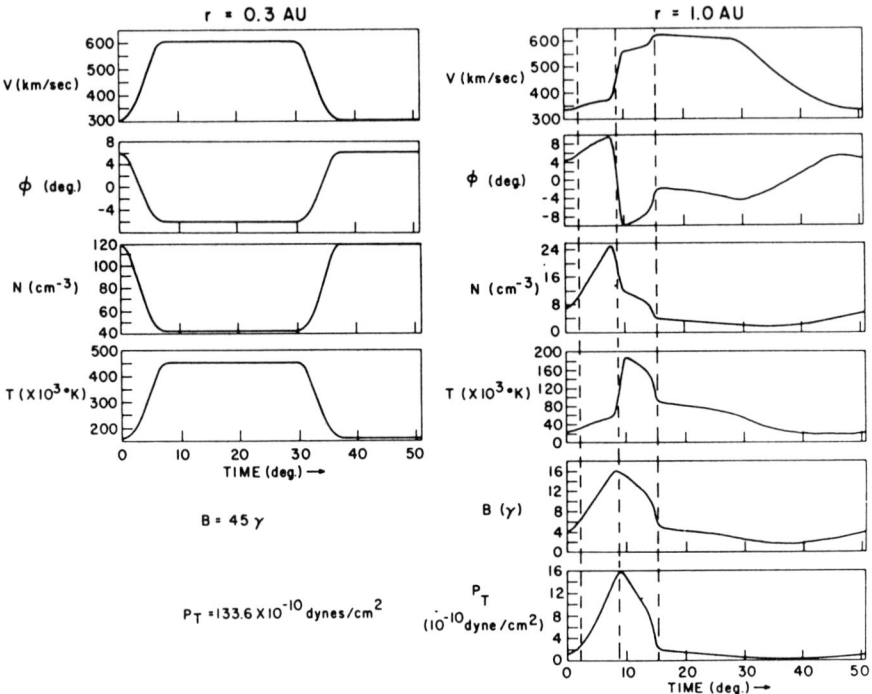

Fig. 7.7. MHD evolution of corotating streams and interaction regions. (V.J. Pizzo in *Solar Wind Five*, edited by M. Neugebauer, p. 675, NASA Conf. Publ. 2280, Washington, DC, 1983a.)

on the left of Fig. 7.7 show the input functions that Pizzo (1982) used. Note the mesa-like profiles of V, N, and T, as well as the enhanced temperature and the decreased density in the stream. The parameters were chosen such that the total pressure and magnetic field strength were constant across the stream at 0.3 AU; hence no interaction region was present at 0.3 AU.

The output of the model of Pizzo (1982) for the conditions at 1 AU is shown in the panels on the right of Fig. 7.7. The front boundary of the stream, while steep at 0.3 AU, has become steeper at 1 AU, leading to the formation of a relatively thin stream interface (see Burlaga, 1974). The thinning of the front boundary of the interaction region was modeled with a gas dynamic code by Hundhausen and Burlaga (1975). Note that the signature of the stream interface—an abrupt decrease in density and abrupt increases in temperature and speed—is basically a consequence of the initial condition. The magnetic field is maximum at the stream interface, because the gradient in speed is greatest there.

An interaction region has been created in front of the stream at 1 AU in Fig. 7.7. The interaction region is clearly the result of the kinematic effects discussed above, because initially there was no pressure gradient. The pressure in the interaction region is maximum at the stream interface. Thus, the dynamical effect of the pressure gradient is negligible where the kinematic effect is greatest, at the stream interface. Clearly, the stream dynamical effects do not alter the motion of the stream interface near 1 AU. The front and rear of the interaction region are characterized by an abrupt increase and decrease in pressure, as shown in Fig. 7.7. Since the code of Pizzo (1982) is a finite difference scheme with artificial viscosity and a finite grid, it does not model shocks very accurately. Thus, one cannot say that the shocks are actually present in Fig. 7.7, but they are clearly forming if not present at the boundaries of the interaction region at 1 AU. The acceleration of the slow flow ahead of the stream interface by the pressure gradient in front of the interaction region is evident in Fig. 7.7, which also shows a deceleration of the flow behind the interface, leading to "erosion" (Burlaga et al., 1985b) of the corotating stream.

An interaction region originates in a relatively narrow region near the sun, and it grows wider with increasing time. The front and rear of the interaction region tend to move away from the stream interface at the magnetoacoustic speed (Burlaga, 1975). Since the magnetoacoustic speed is always greater than the sound speed, the width of an interaction region is always greater in an MHD model than in a gas dynamic model. This result was demonstrated using both 2-D MHD and 2-D gas dynamic models by Pizzo (1981). Since the width of an interaction region is always underestimated in a gas dynamic model, the density is overestimated in such a model, by at least a factor of 2. The greater width of the interaction region in an MHD model as compared to a gas dynamic model implies smaller pressure gradients at the front and rear of the interaction region, thereby delaying the formation of forward and reverse shocks, consistent with the observations. Gas dynamic models tend to produce corotating shocks within 1 AU, whereas these are seldom observed at 1 AU.

7.2 Compound Streams Near 1 AU

7.2.1 Definition and Classification of Compound Streams

Most of the early studies of the solar wind focused on isolated flows: corotating streams, the slow "quiet" solar wind, and transient ejecta (Parker, 1963; Hundhausen, 1972). Noting that many high speed flows are contiguous or even superimposed on one another, Burlaga (1975) and Burlaga and Ogilvie (1973) identified three classes of fast flows: (1) "simple streams," which include both isolated corotating streams and isolated ejecta; (2) "compound streams," in which the leading or trailing edge of a simple stream is interrupted by a substantial increase in V from another simple stream; and (3) "irregular variations," in which changes in V smaller than the magnetoacoustic speed and lasting less than a day or two are superimposed on a fast flow. The study of compound streams is basically the study of the interaction between two simple streams and the interaction between a shock and a simple stream. Given an understanding of these basic processes, one can also consider more complex interactions involving additional simple streams and shocks.

7.2.2 Stream–Stream Interactions

There is some confusion in the literature concerning the meaning of "stream." In this book, the word is synonymous with simple stream, and it is a category that includes both corotating streams and transient ejecta. Thus, stream–stream interaction can be of three types: (1) the interaction between two corotating streams, (2) the interaction between two ejecta, and (3) the interaction between a corotating stream and an ejection.

The terminology for noncorotating streams is confusing and still in a state of flux. The term "ejecta" was used in Hundhausen's book for both one nonsteady, noncorotating flow (and more than one). The use of the terms "ejection" for a single noncorotating stream and "ejecta" for more than one ejection removes this ambiguity. The universally accepted term "coronal mass ejection" (CME) describes a class of transient flows observed in the solar corona. Unfortunately, many authors also use "CME" for an interplanetary mass ejection, based on the unproven assumption that all interplanetary mass ejecta are related to high density coronal mass ejections.

7.2.3 Corotating Stream–Corotating Stream Interaction

The radial evolution of a compound stream consisting of a fast corotating stream overtaking a slow corotating stream at 1 AU is discussed by Burlaga et al. (1985b). The authors identify this flow as a compound stream for four reasons. First, the speed profile observed by IMP-8 shows two maxima. Second, the magnetic polarity of the slower component of the compound stream was negative, while that of the faster component was positive. The slower negative component originated in a small equatorial coronal hole, and the faster positive component originated in an equatorial extension of the north polar coronal hole. Third, there were two interaction regions in the compound stream at 1 AU, as shown by the magnetic field strength. Fourth, there were two stream interfaces at 1 AU, one corresponding to each of the interaction regions. The radial evolution of this compound stream is discussed in Section 8.2.3.

7.2.4 Shock–Corotating Stream Interaction

A corotating stream and its corotating interaction region can interact with a transient shock, producing a variety of flow profiles depending on the relative orientation of the stream and shock (Ogilvie and Burlaga, 1974). Shock–corotating stream interactions were invoked by Heineman and

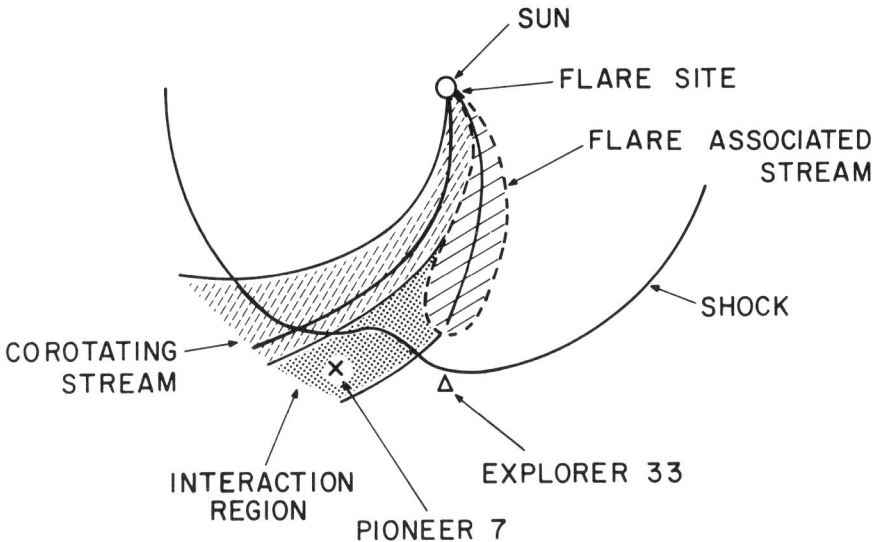

Fig. 7.8. Interaction of a shock with a corotating stream and interaction region. (L.F. Burlaga and J.D. Scudder, *J. Geophys. Res.*, **80**, 4044, 1975, published by the American Geophysical Union.)

Siscoe (1974) to explain the broad distribution of shock normals observed by Chao and Lepping (1974) and others referenced by them.

Three complementary models of shock–stream interactions have been introduced. A linear model for the perturbation of strong shocks was presented by Heineman and Siscoe (1974). They found that the shock shape is distorted kinematically as it is convected faster by the fast stream than the slower flow. Hirshberg et al. (1974) presented a numerical model in which a disturbance produced by a narrow stream moved into a broader stream. Since the narrow stream did not steepen into a shock, these authors did not, strictly speaking, simulate a shock–stream interaction, but they did include nonlinear effects.

The more realistic case of a discontinuous shock of moderate strength moving through a corotating stream was simulated by Burlaga and Scudder (1975) using Whitham's method. They found that in addition to the kinematic effect, the shock strength becomes weaker as it passes through a high pressure interaction region. A shock driven by a fast transient ejection interacting with a CIR and a corotating stream will tend have the shape illustrated in Fig. 7.8.

7.2.5 Corotating Stream–Magnetic Cloud Interaction

A relatively complicated compound stream is illustrated in Fig. 7.9. The central feature is a magnetic cloud indicated schematically by the dashed

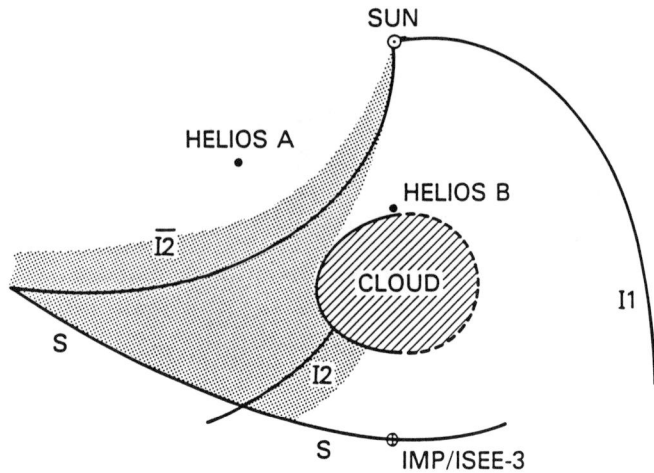

APRIL 24, HOUR 23, 1979

Fig. 7.9. Interaction of a magnetic cloud with a corotating stream and interaction region. (L.F. Burlaga, K.W. Behannon, and L.W. Klein, *J. Geophys. Res.* **92,** 5725, 1987a, copyright by the American Geophysical Union.)

region in the magnetic field profile. The magnetic cloud was moving relatively fast and drove a forward shock S ahead of it. The sheath between the magnetic cloud and the shock was characterized by exceptionally high density, temperature, and magnetic field strength, owing to compression by the shock and the magnetic cloud. The magnetic field strength increased from the shock to the magnetic cloud. The magnetic cloud and the shock were approaching or just entering the trailing edge of a corotating stream interface $I1$. The magnetic cloud was also interacting with a corotating stream interface $I2$, and it was possibly interacting with a third corotating stream with interface $\overline{I2}$.

The geometry of the compound stream at hour 23 on April 24, 1979, is shown to scale in Fig. 7.9. The magnetic cloud was observed by Helios B and IMP/ISEE-3 but not by Helios A, so that part of the boundary of the magnetic cloud was between the longitude of Helios A and Helios B, as illustrated by the solid curve. The other part of the boundary of the magnetic cloud was undetermined, and it is shown arbitrarily as a dashed curve in the figure. The position of the interface of stream 2, which was observed by Helios A, is shown by the curve marked $I2$. The magnetic cloud and its shock S were advancing into the corotating interaction region containing the interface $I2$, thereby forming a merged interaction region (MIR). This is probably the reason for the exceptionally high values of N, T, and B between the shock and the magnetic cloud.

The shock S driven by the magnetic cloud extended from the longitude of IMP/ISEE-3 to the longitude of Helios A, since it was observed by all three spacecraft. This implies the existence of a merged interaction region with high pressures resulting from the coalescence of interaction regions associated with $I2$, $\overline{I2}$, and the shock S. The MIR is indicated by the shaded region in Fig. 7.9. Additional examples of a magnetic cloud interacting with another flow are given in Behannon et al. (1991).

7.3 Structure of Interaction Regions and Streams >1 AU

7.3.1 Corotating Shock Pairs

The existence of a corotating forward shock produced by a corotating stream at 5 or 6 AU was predicted by Parker (1963). The existence of a reverse shock as well as a forward shock, a "shock pair" at 1 AU, was inferred by Sonnett and Colburn (1965) on the basis of SI^+–SI^- pairs in the geomagnetic record at 1 AU. They inferred that the positive sudden impulse is caused by a forward shock and that the negative sudden impulse is caused by a reverse shock. This is one instance in which the inference concerning the existence of a structure based on indirect evidence was incorrect. An SI^+–SI^- pair at 1 AU is probably the result of a transient forward shock followed by a tangential discontinuity across which the density drops (Burlaga and Ogilvie, 1969).

Forward–reverse shock pairs are rarely observed at 1 AU, but occasionally either a corotating forward shock or a corotating reverse shock is observed at 1 AU. For example, a corotating forward shock was identified at 1 AU by Lazarus et al. (1970) and a corotating reverse shock was identified by Burlaga (1970a). Other examples of isolated corotating forward or reverse shocks at 1 AU or less have since been identified, particularly in the Helios data (Schwenn, 1990).

Corotating forward–reverse shock pairs are commonly observed beyond 2 or 3 AU. The existence of corotating shock pairs beyond 1 AU was demonstrated by Smith and Wolfe (1976, 1977, 1979) using their Pioneer 11 magnetic field and plasma data. The rate at which reverse shocks form is significantly smaller than that for forward shocks, and the number of forward and reverse shocks declines between 5 AU and 6 AU (Smith and Wolfe, 1977). These results were confirmed by the plasma observations from Voyagers 1 and 2 (Gazis and Lazarus, 1983).

The first attempt to compare a model of the evolution of a corotating stream and the formation of shock pairs with observations beyond 1 AU was made in the important papers of Hundhausen and Gosling (1976) and Gosling et al. (1976b) using simultaneous speed observations from Pioneer 10 and IMP-7. Their 1-D gas dynamic model was successful in describing the evolution of the speed profile between 1 AU and 5 AU, but it was less successful in modeling the shock pairs. For example, in one of the cases discussed by Gosling et al. (1976b) (day 217–239, 1973) two shock pairs were predicted ahead of a corotating stream, whereas only one was observed at 4.5 AU. The second (unobserved) shock pair was produced in the model by a relatively small speed fluctuation at the leading edge of the stream. This false prediction of the 1-D gas dynamic model was probably a consequence of neglecting the effects of the magnetic field. Shocks will form only when the speed difference in the stream is greater than the magnetoacoustic speed; speed differences smaller than the magnetoacoustic speed are "smoothed out" (Burlaga, 1975). The small speed perturbation in the leading edge of the stream discussed by Gosling et al. was probably submagnetoacoustic, leading to a smoothing-out of the speed fluctuation rather than to the formation of a shock pair.

A 1-D MHD code was used to simulate the evolution of a corotating stream from Pioneer 11 at 2.8 AU to Pioneer 10 at 4.9 AU (Dryer et al., 1978b). The authors predicted the velocity, density, temperature, and azimuthal component of magnetic field. Their pioneering 1-D MHD model was fairly successful in explaining the observed profiles, but it should be noted that the calculations began at 2.8 AU, where the shock pair was already present, and extended only over 2.1 AU.

7.3.2 Stream Erosion and Pressure Waves

The amplitude of a stream decreases with increasing distance from the sun (Collard and Wolfe, 1974; Mihalov and Wolfe, 1978; Collard et al. 1982; Gazis and Lazarus, 1982). The flow ahead of a stream is accelerated during this process. At large distances the corotating stream structure is no longer present, and the distribution of speeds tends to be relatively narrow.

We noted in reference to Fig. 7.7 that as the rear of the interaction region expands into the advancing stream, the pressure gradient decelerates the front of the stream. This effect was predicted by the gas dynamic models (e.g., Hundhausen and Gosling, 1976). The deceleration of a corotating stream by the pressure gradient in the rear of an interaction region was called "stream erosion" by Burlaga et al. (1985b).

A narrow stream moving through an isolated unstructured slow flow evolves more rapidly than a broad stream moving through the same flow. This effect, called "filtering," was demonstrated by Hundhausen in unpublished work based on a 1-D gas dynamic model. His results were reported in a review by Holzer (1979). The model of an isolated stream moving many AU through an unstructured medium is valuable as an illustration of filtering in its simplest form, but most streams ultimately interact with other streams or interaction regions in the outer heliosphere.

The total destruction of a corotating stream was identified by Burlaga et al. (1980) using simultaneous observations from four spacecraft. A corotating stream was observed by 1 AU by IMP-8 and within 1 AU by Helios 1 and 2, but the corotating stream was not observed by Voyagers 1 and 2 at 1.6 AU. Only the stream interface was observed at 1.6 AU.

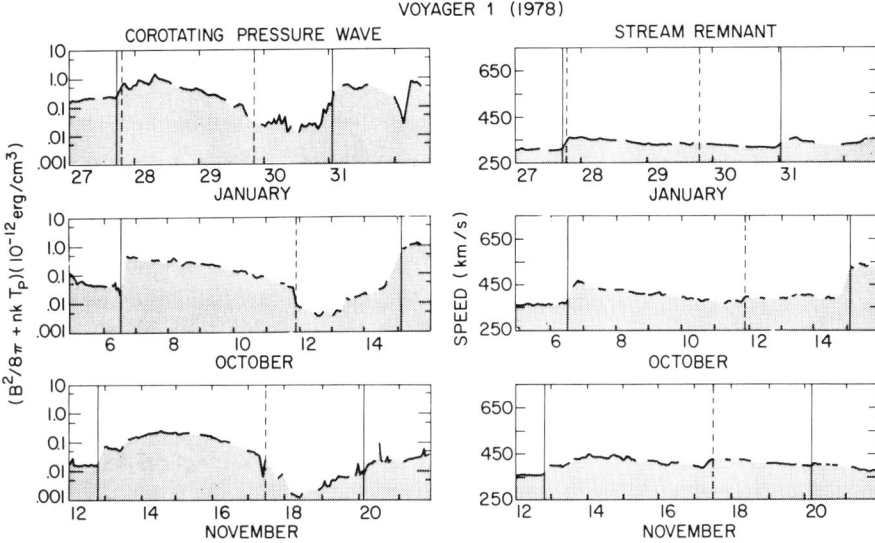

Fig. 7.10. Corotating pressure waves without streams. (L.F. Burlaga, *J. Geophys. Res.*, **88**, 6085, 1983, published by the American Geophysical Union.)

Corotating interaction regions without corotating streams were observed between 2 AU and 4 AU by Burlaga (1983), who called them "corotating pressure waves" to emphasize their physical nature and geometry. Corotating pressure waves are now commonly referred to simply as interaction regions, but it should be remembered that they are a subset of the interaction regions defined above. Three examples of corotating pressure waves and the corresponding stream remnants are shown in Fig. 7.10. The term "pressure wave" emphasizes that this type of interaction region is a physical object that has an independent existence beyond 1 AU, without the presence of a corotating stream. The term also stresses that the interaction regions without streams are nonlinear pressure waves that tend to expand inward and outward. In many cases the expansion rate is determined by the speed of the shocks that bound an interaction region. When no shock is present, shocks will tend to form, because the high pressure plasma expands faster than the low pressure plasma.

The destruction of streams and the creation of corotating pressure waves was modeled by Burlaga et al. (1985b) using the stationary 2-D MHD code of Pizzo (1982). The input to the model was a set of plasma and magnetic field parameters for coronal hole associated corotating streams observed at 1 AU by IMP-8. One stream that they modeled is the November 1977 stream discussed by Burlaga et al. (1980) (referred to above), which disappeared between 1 AU and 1.6 AU. The evolution of a second simple corotating stream was also modeled by Burlaga et al. (1985b). The interaction region grew in amplitude and width with increasing distance

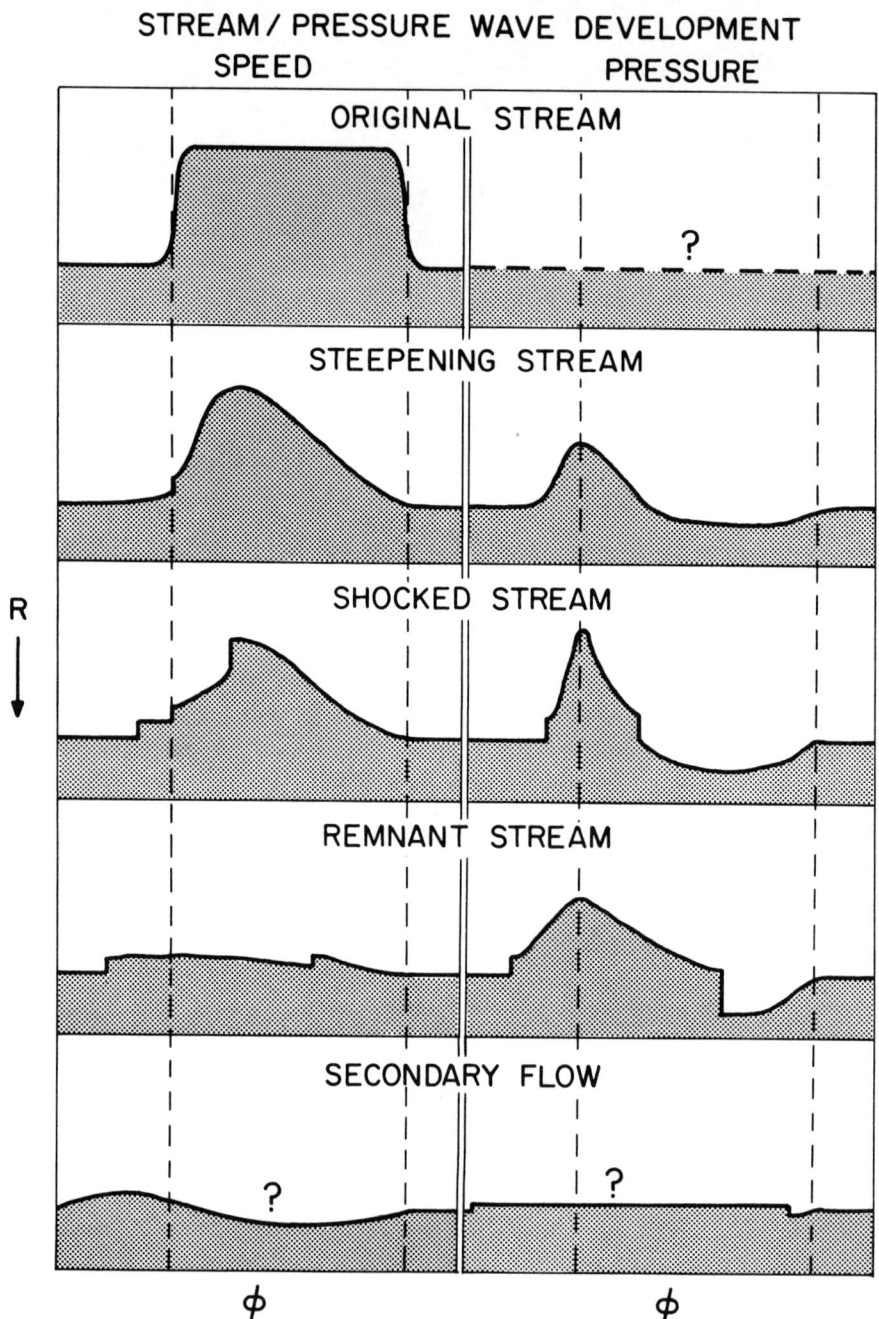

Fig. 7.11. Death of a stream and birth of a pressure wave.

from the sun, gradually eroding the corotating stream by its gradient in pressure. At 4 AU the stream was absent, and only an interaction region bounded by a forward shock and a reverse shock remained. A corotating pressure wave was created as a result of a competition between the stream (kinematic steepening) and the interaction region (pressure gradient). The stream that gave birth to the interaction region was ultimately destroyed by the interaction region. This fundamental process is illustrated in Fig. 7.11.

7.3.3 Dynamics of the Outer Heliosphere

The process by which a stream is destroyed and a corotating pressure wave bounded by a shock pair is formed is dominant between approximately 1 AU and 10 AU. The process is nonlinear and irreversible. One cannot reconstruct the streams in the inner heliosphere from observations of interaction regions (corotating pressure waves) in the outer heliosphere. Thus, one arrives at the important conclusion that as the solar wind moves to the outer heliosphere, memory of the source conditions is lost (Burlaga, 1983, 1984).

Beyond several AU, the dominant mesoscale dynamical structures are pressure waves, that is, the interaction regions having an existence independent of streams (Burlaga, 1983, 1984). Fast corotating streams are rarely observed beyond 15 AU. Thus, the dynamics of the heliosphere beyond about 15 AU is dominated by the interaction of corotating pressure waves and shocks during much of the solar cycle.

The dynamics of the outer heliosphere is fundamentally different from the dynamics of the inner heliosphere. The dynamics of the inner heliosphere is momentum driven by streams, while the dynamics of the outer heliosphere is driven by the evolution and interaction of interaction regions and shocks.

8

Merged Interaction Regions

8.1 Definition and Classification of Merged Interaction Regions

When two or more interaction regions interact and coalesce, the resulting structure is called a "merged interaction region" or MIR (Burlaga et al. 1983, 1985; Burlaga, 1987). MIRs can be produced by the coalescence of interaction regions of various types, including those associated with corotating streams, magnetic clouds, forward and reverse shocks, and interplantary ejecta. Thus, there are many kinds of MIR, depending on origin. However, it is convenient to classify MIRs in three broad categories: corotating MIRs (CMIRs), local MIRs (LMIRs), and global MIRs (GMIRs), as discussed by Burlaga et al. (1993a).

A corotating MIR is a quasi-steady pressure wave with a spiral geometry resembling that of the spiral magnetic field. A corotating MIR is ultimately formed by the interaction and coalescence of one corotating interaction region with one or more interaction regions of various types. Some processes by which a corotating MIR can be formed are discussed in Section 8.2. A corotating MIR tends to recur with a frequency equal to the solar rotation frequency. Qausi-periodic CMIRs are discussed in Section 8.3.

A local MIR is a MIR formed by the interaction of a transient ejection with other flows, which could include corotating streams or pressure waves, shocks, ejecta, and magnetic clouds. LMIRs are localized in both longitude and latitude, but there is no unique size or shape associated with a LMIR. Local MIRs are discussed in Section 8.4.

Global MIRs extend 360° in longitude and up to relatively high latitudes. The topology of a GMIRs is like that of a shell (Burlaga et al., 1984b, 1993a), although an infinite variety of specific geometrical configurations is possible. GMIRs are formed by the interaction of several ejecta (from either a single long-lasting solar active region or many active regions around he sun) with other streams and interaction regions. GMIRs are discussed in Section 8.5.

8.2 Formation of a Corotating Merged Interaction Region

Corotating MIRs can be formed in several ways, including (1) the interaction between two distinct streams, (2) the evolution of a compound stream, (3) the interaction between a corotating pressure wave (CIR without a stream) and a corotating stream that overtakes it, and (4) the interaction between two corotating pressure waves. The stream–stream interactions can be of at least three types: (a) a fast corotating stream overtaking a slower corotating or transient flow, (b) two identical corotating streams, "twin streams," and (c) a narrow, fast stream followed by a broad, slower stream. Cases a and b are discussed in this section, and case c is discussed in Section 8.3.2. Stream–stream interactions produce a compound stream, which then gives rise to a CMIR. The evolution of a compound stream is just one aspect of the interaction between two initially distinct streams, but this case is identified separately because one often has observations of a compound stream at 1 AU, say, but no observations of the flows within 1 AU that produced the compound stream. Sometimes the two streams are sufficiently far apart that each is destroyed and produces a pressure wave before a compound stream is formed. In this case the two corotating pressure waves can interact to form a CMIR, without the influence of streams on the interaction.

8.2.1 Fast Corotating Stream Overtaking a Slower Corotating Stream

The evolution of a fast corotating stream overtaking a slower corotating stream within 1 AU was discussed by Burlaga et al. (1983, 1984b). The observations of two distinct streams and their respective interaction regions at Helios 1 within 1 AU are shown on the left of Fig. 8.1. The interaction regions (which are the regions of enhanced pressure) are correlated with the regions of enhanced magnetic field strength, so that one can use the magnetic field strength as a proxy for pressure in the identification of interaction regions. The corresponding observations made by Voyager 1 at 8.5 AU, which was nearly radially aligned with Helios 1, are shown on the right of Fig. 8.1. The slower stream was destroyed; its remnant is marked by a forward shock observed on day 146. The faster stream was severely eroded. It produced both a forward shock (which was seen at the beginning of day 150) and a reverse shock (seen on day 159). The reverse shock propagated completely through the faster stream, causing most of the erosion of the stream. A reverse shock on day 152 was produced by the interaction region of the slower stream. This reverse shock passed through the forward shock from the faster stream and entered the remnant of the fast stream, causing further erosion of the stream.

The two distinct narrow interaction regions observed by Helios 1 inside of 1 AU evolved to a single merged interaction region at 8.5 AU, containing two forward shocks and two reverse shocks. Each of the interaction regions

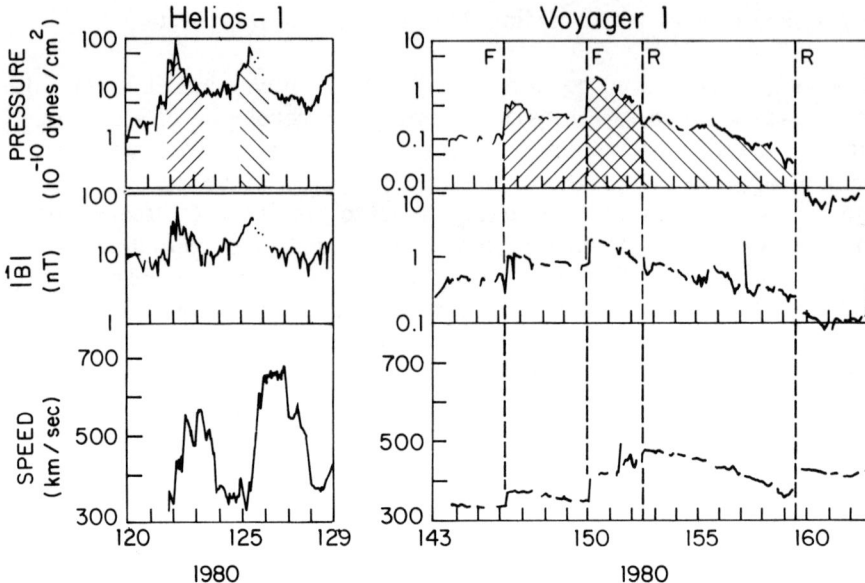

Fig. 8.1. Merged interaction regions. (L.F. Burlaga, R. Schwenn, and H. Rosenbauer, *Geophys. Res. Lett.*, **10**, 413, 1983, copyright by the American Geophysical Union.)

seen by Helios 1 was without a shock. But each interaction region produced a CIR bounded by a forward–reverse shock pair, and the two resulting CIRs overlapped as the reverse shock from the first CIR passed through the forward shock from the second CIR. Thus all of the plasma in the CMIR observed by Voyager 1 was shocked at least once, and the plasma between the forward shock and the reverse shock in the middle of the CMIR was shocked twice. The shock signature of this CMIR at 8.5 AU is FFRR. The formation of a CMIR from the interaction of the two CIRs bounded by shock pairs is fundamental in the dynamics of the outer heliosphere.

8.2.2 Twin Stream Interactions

The evolution of two identical streams between 0.3 AU and 5 AU was modeled by Dryer and Steinolfson (1976) using a MHD model based on a finite difference scheme and artificial viscosity. These authors showed the formation of two interaction regions presumably bounded by shock pairs (indicated by the enhancements and jumps in the azimuthal component of the magnetic field). They also showed the two interaction regions just beginning to interact at 7.5 AU. The authors did not present the curves for the additional fields such as density, temperature, and speed, so that one cannot verify the existence of shocks. The use of a finite difference scheme and artificial viscosity means that the shocks and shock interactions were not modeled accurately. Nevertheless, the calculation suggests that the reverse shock from one interaction region can interact with the forward shock from a following interaction region.

Twin streams are commonly used as initial conditions in simulations of the evolution of recurrent streams. However, even small departures from symmetry can have major qualitative effects on the evolution of corotating streams, as demonstrated in Section 8.3. The nonlinear evolution of twin streams shows a very sensitive dependence on small perturbations in the initial conditions. The flow pattern associated with twin streams is unstable. Twin streams are not generic, and their evolution is not likely to be observed in the outer heliosphere.

8.2.3 Compound Stream Evolution

The radial evolution of a compound stream consisting of a fast corotating stream overtaking a slow corotating stream at 1 AU is shown in Fig. 8.2. There are four reasons to identify this as a compound stream. First the speed profile observed by IMP-8 (bottom left of Fig. 8.2) shows an anomalous maximum in the leading edge of the fast corotating stream. Second, the polarity of the slower component of the compound stream was negative, while that of the faster component was positive. The slower negative component originated in a small equatorial coronal hole, and the fast positive component originated in the equatorial extension of the north polar coronal hole. Third, there were two interaction regions in the compound stream at 1 AU, as shown by the two maxima in the magnetic field strength measured across the stream at the bottom right of Fig. 8.2. Fourth, there were two stream interfaces at 1 AU, one corresponding to each of the interaction regions.

The radial evolution of the compound stream and its interaction regions at 1 AU was computed by Burlaga et al. (1985a) using the stationary 2-D MHD model of Pizzo (1982). At 2 AU the fast stream had both steepened and overtaken the slower stream (Fig. 8.2). The two corotating interaction regions that were observed at 1 AU coalesced to form a single merged interaction region at 2 AU. This, a fundamental qualitative change occurred between 1 AU and 2 AU. Such a change is irreversible, since one cannot start with the observations of a CMIR at 2 AU and reconstruct the two CIRs at 1 AU. Memory is lost as the result of the nonlinear processes (Burlaga, 1983; Burlaga et al., 1983).

8.2.4 A Corotating Stream Overtaking a Transient Shock

The formation of a merged interaction region by a corotating stream overtaking a transient shock was modeled by Burlaga et al. (1985b) using the MHD code of Pizzo (1982) with input data from IMP-8. The input data obtained at 1 AU are shown in the left-hand panels of Fig. 8.3. The

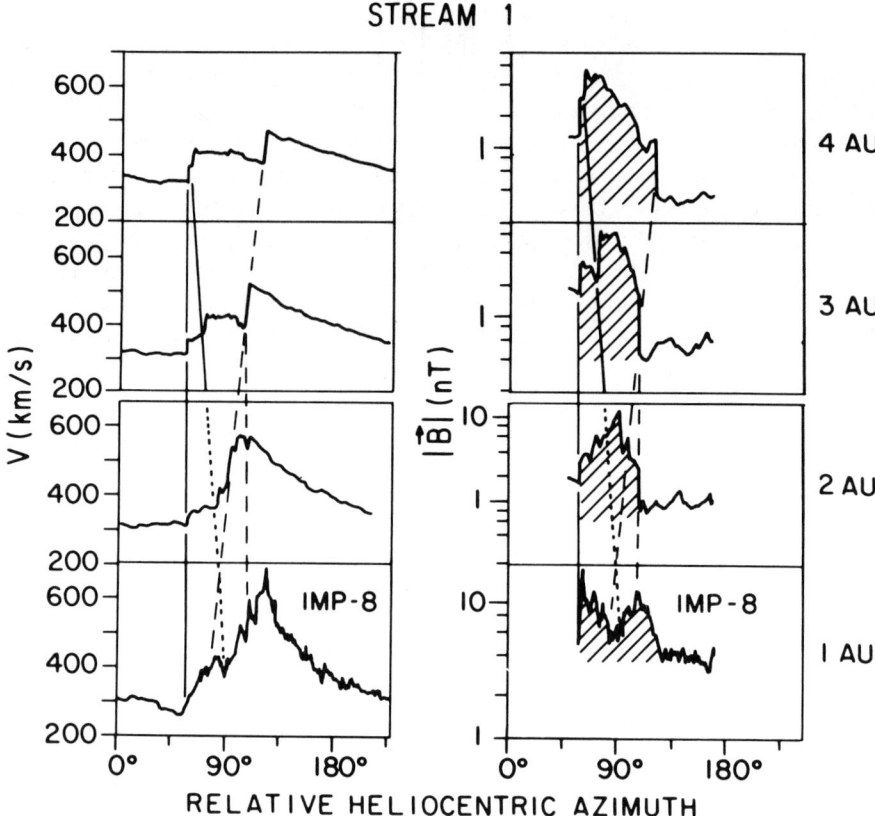

Fig. 8.2. Merging of two corotating interaction regions: fast stream overtaking slow stream. (L.F. Burlaga, V. Pizzo, A. Lazarus, and P. Gazis, *J. Geophys. Res.*, **90,** 7377, 1985b, copyright by the American Geophysical Union.)

development of this flow was studied by Burlaga et al. (1980) using simultaneous data from Helios 1, Helios 2, IMP-8, and Voyager 2. A shock $F1$, formed by the coalescence of two shocks, was observed by IMP-8 on November 25, 1977, and it was associated with a relatively large pressure wave shown in the bottom-left panel of Fig. 8.3. The shock $F1$ was associated with a flare at W66°, and it was probably driven by an ejection associated with the flare, although the ejection was not actually observed (see Fig. 8.4). A stream interface was observed by IMP-8 on November 26, 1977, after it had convected past Helios 1 and Helios 2. The interface was followed by a corotating stream and was associated with an interaction region of the same magnitude as that produced by the shock $F1$. The two interaction regions were distinct at 1 AU.

The radial evolution of the flow shown on the left of Fig. 8.3 between 1 AU and 2 AU was modeled using a stationary 2-D MHD code. One cannot model transient flows with stationary models in general. However, a

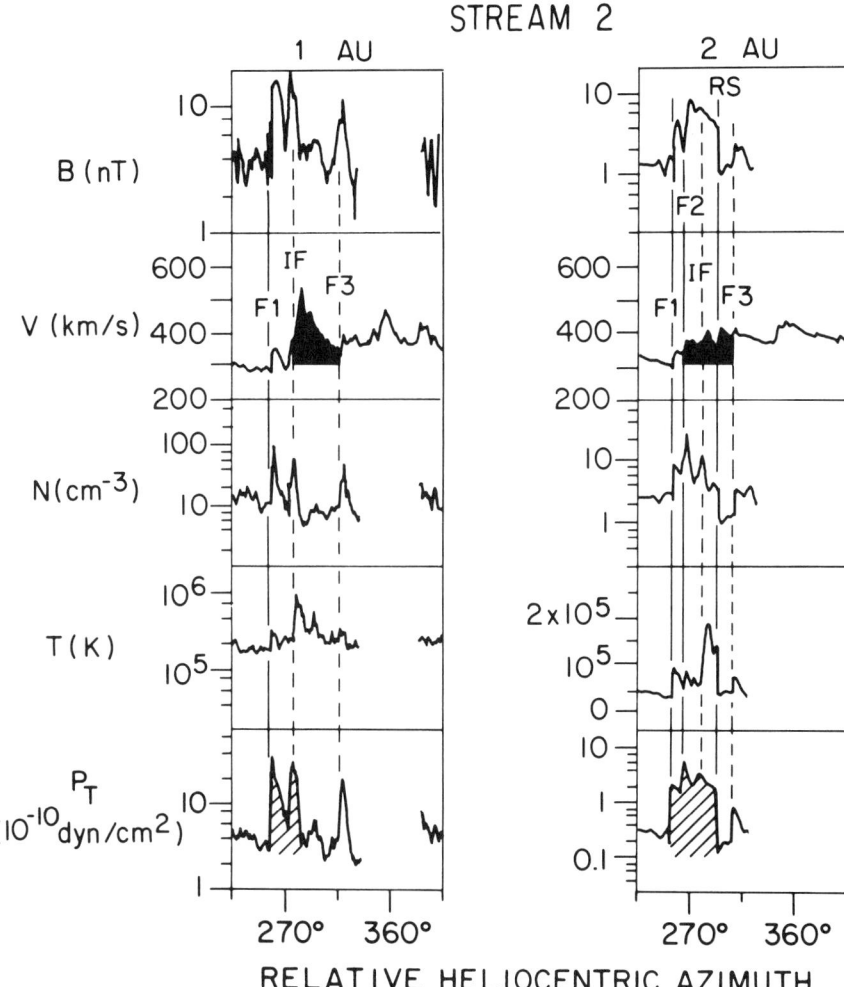

Fig. 8.3. Merged interaction region forming a CIR and a transient interaction region. (L.F. Burlaga, V. Pizzo, A. Lazarus, and P. Gazis, *J. Geophys. Res.*, **90**, 7377, 1985a, copyright by the American Geophysical Union.)

stationary model can be used to follow the evolution along a radial line extending through the position of IMP-8 to good approximation, because the flow is predominantly radial. This is confirmed by comparison of the results of a model with the observations from Voyager 2 at 1.6 AU close to a radial line through IMP-8 (Burlaga et al., 1985b, Fig. 14). The theoretical profile at 2 AU is shown on the right of Fig. 8.3. A corotating forward shock $F2$ formed ahead of the interface IF, and a reverse shock RS formed behind the interface. (A third shock $F3$ from a subsequent flow is approaching the reverse shock in the figure and will eventually interact with

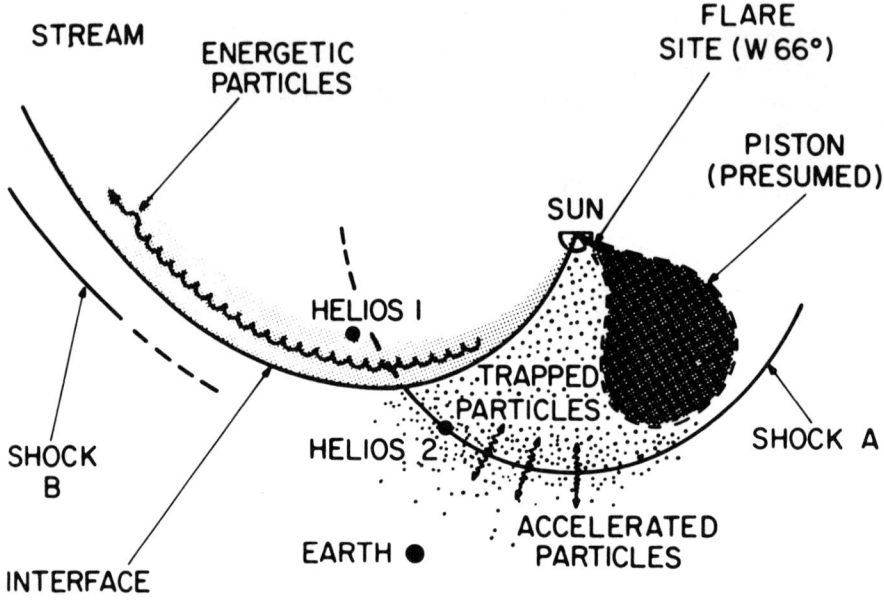

Fig. 8.4. Sketch of shocks and ejection. (L.F. Burlaga, R. Lepping, R. Weber, T. Armstrong, C. Goodrich, J. Sullivan, D. Gurnett, P. Kellogg, E. Keppler, F. Mariani, F. Neubauer, H. Rosenbauer, and R. Schwenn, *J. Geophys. Res.*, **85**, 2227, 1980, copyright by the American Geophysical Union.)

it, but at 2 AU *F*3 remains a part of a separate flow.) The most important conclusion demonstrated by Fig. 8.3 is that the two distinct types of interaction region at 1 AU coalesced to form a single merged interaction region at 2 AU. This CMIR was produced by a corotating interaction region overtaking an interaction region associated with a transient shock.

8.2.5 *A Corotating Stream and Interaction Region Overtaking a Corotating Pressure Wave*

A corotating pressure wave with no associated stream followed by a corotating interaction region driven by a fast corotating stream was

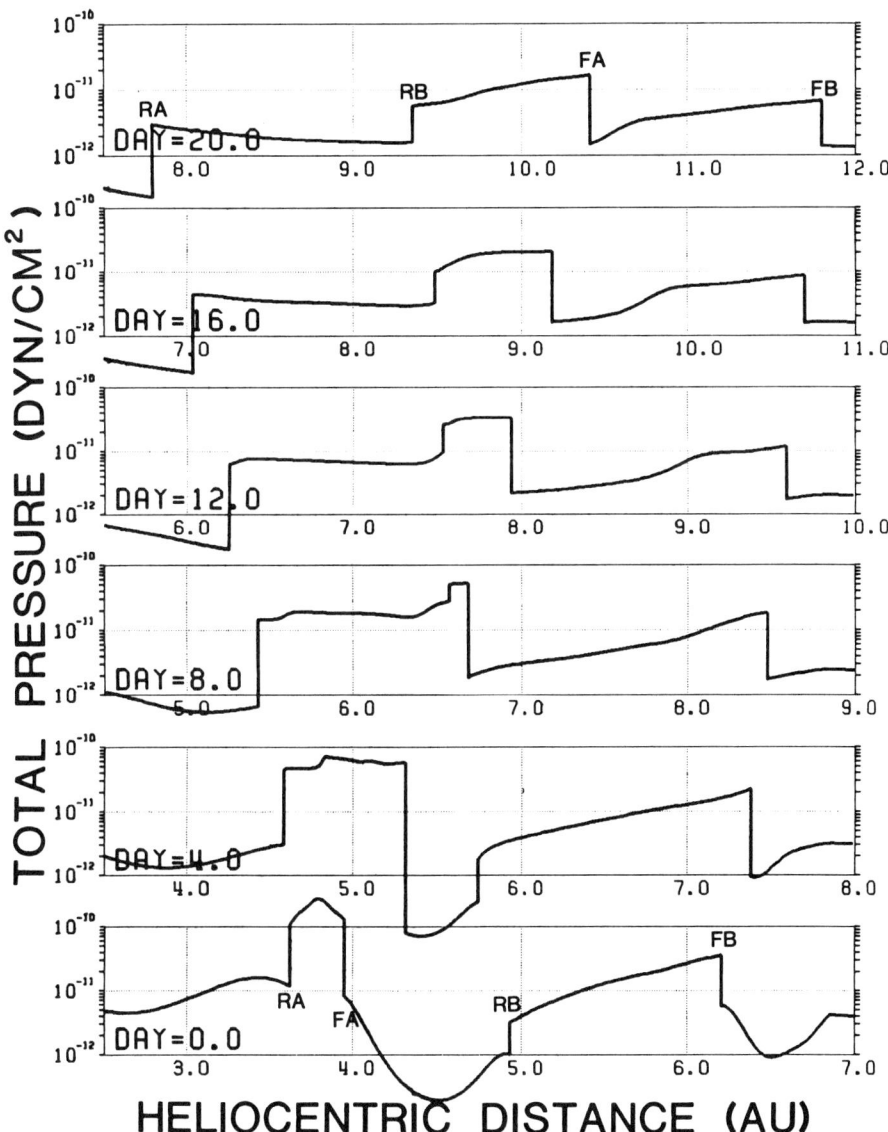

Fig. 8.5. Merged MIR forming from the coalescence of two merged interaction regions. (Y.C. Whang and L.F. Burlaga, *J. Geophys. Res.*, **90**, 221, 1985a, copyright by the American Geophysical Union.)

observed by Voyagers 1 and 2 at ≈4.0 AU in October 1978. A fit to the pressure observations is shown in the bottom panel Fig. 8.5. The corotating pressure wave is between the forward and reverse shocks *FB* and *RB*, respectively. The density, proton temperature, and magnetic field strength were high in the corotating pressure wave and low behind it. The bulk speed U declined monotonically between the forward shock and the reverse shock, and it remained less than 400 km/s for 4 days between the reverse shock RB and the fast shock FA of the corotating pressure wave driven by a fast stream. The shock pair $FB-RB$ was well separated from the shock pair $FA-RA$ at ≈ 4.0 AU.

The radial evolution of the flow just described was modeled by Whang and Burlaga (1985b) using the 1-D time-dependent MHD code of Whang (1984), which is based on the method of characteristics and treats shocks as discontinuities that satisfy the Rankine–Hugoniot equations. A 1-D MHD model is a good approximation beyond 4 AU, because the spiral magnetic field is nearly normal to the radial direction, so that the magnetic field acts simply as an additional pressure. The model can accurately describe the evolution of the flow along a radial line through the initial observation point.

The computed radial evolution of the two corotating interaction regions described above is also shown in Fig. 8.5. The absicissa shows distance rather than time, so that the corotating interaction region now appears to the left of the corotating pressure wave. As time goes on, the forward shock FA driven by the corotating stream and the reverse shock RB from the corotating pressure wave move toward each other. They collide at about 6.2 AU. In this process each shock is weakened, hence propagates more slowly, and a contact surface is created between RB and FA. This process was discussed in the context of gas dynamics by Parker (1963, p. 110). The shocks FA and RB move through each other, so that the two interaction regions overlap after day 12, forming a corotating merged interaction region. After the collision, FA and RB move away from each other. Meanwhile the forward shock FB of the corotating pressure and the reverse shock RA of the stream-associated interaction region move away from each other, so that the width of the CMIR increases with time.

8.2.6 A Magnetic Cloud Overtaking a CIR

An unusual merged interaction region formed by the overtaking of a CIR by a magnetic cloud was observed by Voyager 2 at 11 AU (Burlaga et al., 1985a). A stream interface was observed at the end on August 2, 1982, indicating the presence of a corotating structure, but the corotating stream had evidently been destroyed, because no fast stream was present. A forward shock F was observed on July 31, and a reverse shock R was observed on August 9. However, the CIR bounded by this shock pair does not show the expected mesa-like magnetic field profile because it was interacting with a magnetic cloud that had apparently advanced through the corotating stream. The magnetic cloud has the usual signature: enhanced magnetic field strength, smooth rotation of the magnetic field direction, and low proton temperature (Burlaga et al., 1981b). The magnetic cloud was the largest ever observed, with a radial cross section of ≈ 1 AU, since it passed the spacecraft in 4 days at a speed of 500 km/s. This is approximately the size expected if the magnetic cloud expanded at the usual rate of approximately half the Alfvén speed during most of its time in transit to 11 AU (Klein and Burlaga, 1982). This magnetic cloud is also the most distant ever observed, and it proves that magnetic clouds can be stable for at least 40 days, assuming that the mean speed of the magnetic cloud between the sun and 11 AU was 500 km/s.

The magnetic field strength in the magnetic cloud is larger than expected at this distance (Osherovich et al., 1993c), but that is probably because the reverse shock from the CIR passed entirely through the magnetic cloud, compressing the magnetic field of the magnetic cloud. No decreasing velocity was observed during the passage of the magnetic cloud, indicating that it was no longer expanding at 11 AU. The reverse shock might also have caused the reduction of the speed of the magnetic cloud to the ambient value.

The magnetic field strength and density at the stream interface are also relatively large compared to the nominal values B_P and N_P, respectively, at 11 AU. These enhancements could have been produced in part by shock compression, if the shock on August 1 were driven by the magnetic cloud and passed through the stream interface. Additional compression of the magnetic field and plasma between the stream interface and the magnetic cloud could have been produced by the collision of the magnetic cloud with the CIR.

Summarizing, the region of enhanced pressure observed by Voyager near 11 AU from July 31 to August 8, 1982, was probably a merged interaction region resulting from the overtaking of a CIR (bounded by a forward–reverse shock pair) by a magnetic cloud driving a shock. The shock signature of this MIR is FFR.

8.2.7 Catastrophe Theory Classification of Shocks in MIRs

The preceding examples show that merged interaction regions can be formed in several ways and that a variety of shock signatures is possible in the MIRs, depending on the types of interaction involved. MIRs can interact with other MIRs or shocks to produce more complex MIRs (merged MIRs!). The MIRs bounded by shocks can be classified according to their shock signatures. The structure of MIRs and the qualitative evolution of their shock signatures can be described and classified using catastrophe theory (Burlaga, 1990b). For reviews of catastrophe theory, see Gilmore (1981) and Arnold (1972, 1986).

The basic idea of Burlaga (1990b) is to associate shocks with the extrema of polynomials. A fast shock is associated with the maximum of a quadratic polynomial $r = -x^2$. A reverse shock is associated with the minimum of a quadratic polynomial $r = x^2$. A forward–reverse shock pair is associated with the minimum and maximum of a cubic $r(x; a) = -x^3/3 + ax$, where $a < 0$, etc. The formation, evolution, and destruction of shocks is controlled by a path determined by the variation of the parameters such as a in a space called "control space."

A cubic polynomial describes the formation of a shock pair as follows. When $a > 0$ there is no maximum or minimum in the cubic, hence no shock. When $a < 0$ the cubic has a maximum corresponding to the existence of a forward shock and a minimum corresponding to the existence of a reverse shock; that is, $a < 0$ corresponds to the existence of a forward–reverse shock pair. The path in the 1-D control space from $a > 0$ to $a < 0$ describes the formation of a forward–reverse shock pair at $a = 0$. The point $a = 0$ is the separatrix of the catastrophe A_2. The subscript indicates the multiplicity of the singularity, which is the number of critical points and the number of shocks in the model under consideration.

In the general case, one considers a polynomial whose degree is one more than the maximum number of shocks in the MIR, because the number of maxima and minima is determined by the number of zeros of the derivative. The number of maxima and minima is determined by the degree of the polynomial and its parameters. The evolution of the maxima and minima is determined by a path in the control space. The classification of merged interaction regions by their shocks corresponds to polynomials of various types. The formation, evolution, and destruction of the shocks in a MIR is described by the qualitative changes in a polynomial as its parameters change along a path in the control space. The change in the character of a polynomial as its parameters change is the subject of catastrophe theory.

Consider the case of a transient forward shock and a corotating shock pair as described in Section 8.2.4. The appropriate polynomial is fourth order with two parameters $r(x; a, b) = -x^4/4 - ax^2/2 - bx$. The control space is two-dimensional (a, b) as shown in Fig. 8.6. There is a set of values of (a, b) at which $r(x; a, b)$ has a degenerate critical point called "the separatrix of the dual cusp catastrophe," A_3, which is shown by the dashed cusp in Fig. 8.6. The qualitative form of the polynomial depends on the position in parameter space. For example, at point 1 in Fig. 8.6 the polynomial consists of a single maximum, the single forward transient shock $F1$. At point 3 a degenerate critical point forms in the polynomial, corresponding to the formation of an additional maximum and minimum in the polynomial (i.e., the formation of a forward–reverse shock pair). Inside the cusp, one finds the MIR consisting of the transient forward shock $F1$ and the forward–reverse shock pair $F2$–$R2$. The forward shock advances toward the reverse shock and the two eventually coalesce. The coalescence occurs on the Maxwell set of the dual cusp catastrophe, which is the vertical dashed line in the cusp region. On the Maxwell set the two maxima are equal, hence $F1$ and $F2$ are the same, representing a shock $F12$ formed by the coalescence of shocks $F1$ and $F2$.

FORWARD SHOCK AND SHOCK PAIR

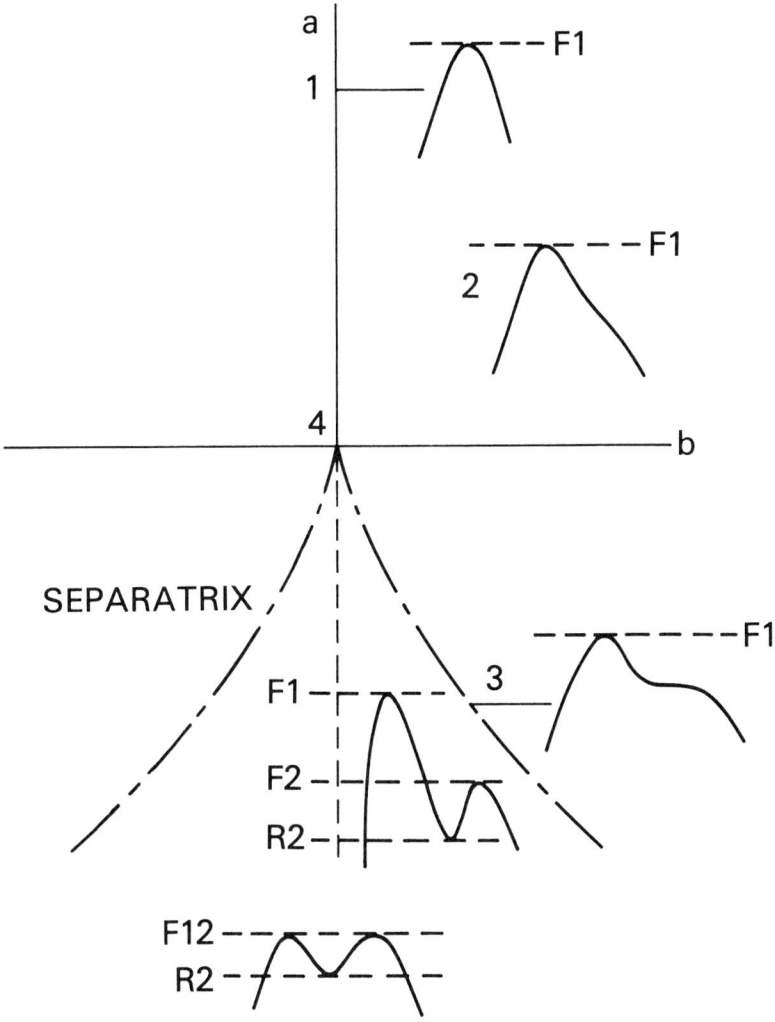

Fig. 8.6. Cusp catastrophe illustrating the formation and coalescence of a shock pair. (L.F. Burlaga, *Geophys. Res. Letters.*, **17**, 1633, 1990b, published by the American Geophysical Union.)

As a final example, consider the swallowtail catastrophe, A_{+4}, represented by the polynomial $r(x; a', b', c') = x^5/5 + a'x^3/3 + b'x^2 + c'x$. The control space is now three-dimensional, and the separatrix of the swallowtail is a self-intersecting surface whose two-dimensional section is the dot–dashed curve in Fig. 8.7. Again, the number of maxima, hence the shock

TWO SHOCK PAIRS

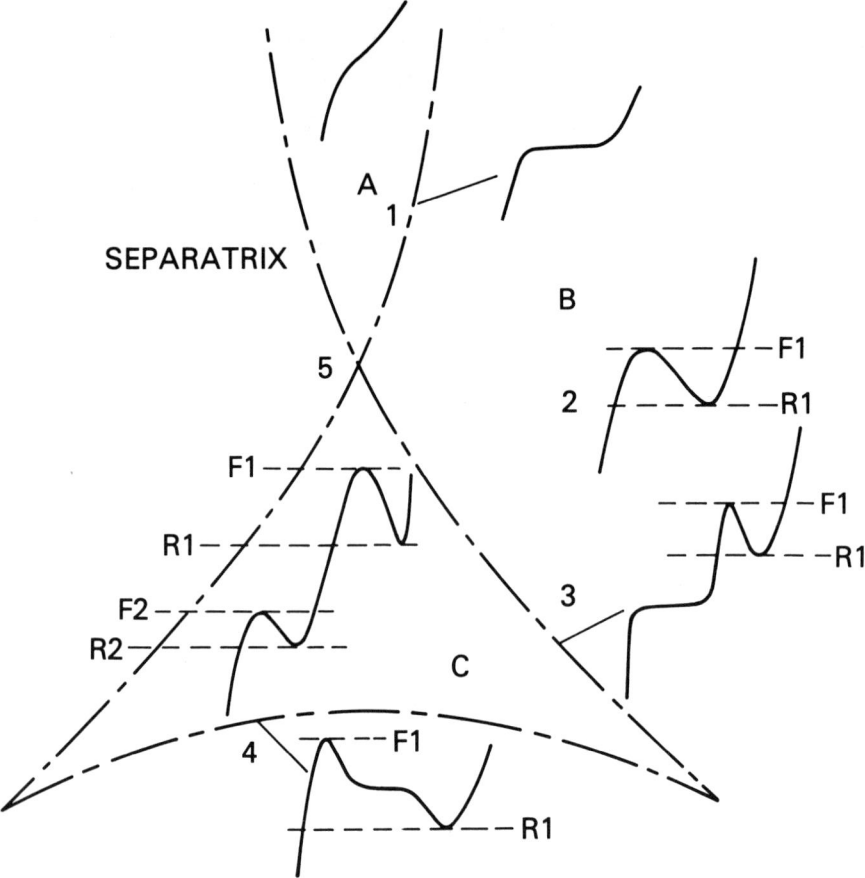

Fig. 8.7. Swallowtail catastrophe illustrating the formation and coalescence of shock pairs. (L.F. Burlaga, *Geophys. Res. Lett.,* **17,** 1633, 1990b, published by the American Geophysical Union.)

configuration, depends on the position of a point in control space. The evolution of the shocks corresponds to a path in control space, and qualitative changes in the shock configuration occur as the path crosses a separatrix. Many cases are described by the swallowtail, but we mention just a few. The region *A* in the control space corresponds to a polynomial with no maxima (i.e., no shocks). Crossing the separatrix at point 1 corresponds to a degenerate critical point, representing the formation of a forward–reverse shock pair. Crossing the separatrix at point 3 corresponds to the formation of a second shock pair, which is seen in the tail *C*. The shocks in region *C* can interact and coalesce in various ways that are described by the Maxwell set of the swallowtail catastrophe. For example, the minimum of *R*2 may coincide with the maximum of *F*2, representing the interaction of shock *R*2 with shock *F*1. Another possibility is that the two maxima coincide, which represents the coalescence of *F*1 and *F*2. Similarly, the two minima can coincide, representing the coalescence of *R*1 and *R*2. Finally, the two maxima can coincide and the two minima can coincide, representing the coalescence of the two forward shocks and the coalescence of the two reverse shocks. The result is that a single shock pair is formed from two separate shock pairs.

One could continue in this way to describe the configurations and interactions involving more shocks. For example, the merging of two merged interaction regions each consisting of two forward–reverse shock pairs is described by the catastrophe A_8, which has a control space of seven dimensions describing the formation and coalescence of eight shocks. One can no longer picture the situation very simply, but the various interactions are described by the Maxwell set of A_8. Catastrophe theory provides a powerful and beautiful means of organizing the shock configurations and shock interactions that can occur in the outer heliosphere. Our discussion assumed planar shocks, but the formation and coalescence of curved shocks could be analyzed using the results of catastrophe theory that describe the metamorphoses of moving caustics (e.g., see Arnold, 1986, p. 34).

8.3 Quasi-Periodic Corotating MIRs

8.3.1 Existence

Corotating interaction regions have frequently been observed at 1 AU. Corotating interaction regions (CIRs) bounded by shock pairs were observed beyond 1 AU in the 1973–74 Pioneer 10 data (Smith and Wolfe, 1979). Quasi-periodic corotating merged interaction regions (CMIRs) were first identified in the Voyager observations from 6.9 AU to 8.2 AU (Burlaga et al., 1984a) and at 11 AU during 1982 and 1983 (Burlaga et al., 1983, 1985a). The period of the CMIRs was one solar rotation period in the latter case and half the solar rotation period in the former case. The presence of such quasi-periodic MIRs in the outer hemisphere was confirmed by Gazis (1987) using data from Pioneers 10 and 11 obtained from 1975 to 1983 in the region between 8.7 AU and 30.4 AU.

Quasi-periodic CMIRs are illustrated in Fig. 8.8 by the data obtained from Voyagers 1 and 2 between 12 AU and 22 AU during 1984 and 1985. The top left panel of Fig. 8.8 shows the magnitude of the magnetic field (which again serves as a proxy for the total pressure in identifying MIRs) observed in 1984 by Voyager 2 from 13.2 AU to 15.9 AU near the ecliptic.

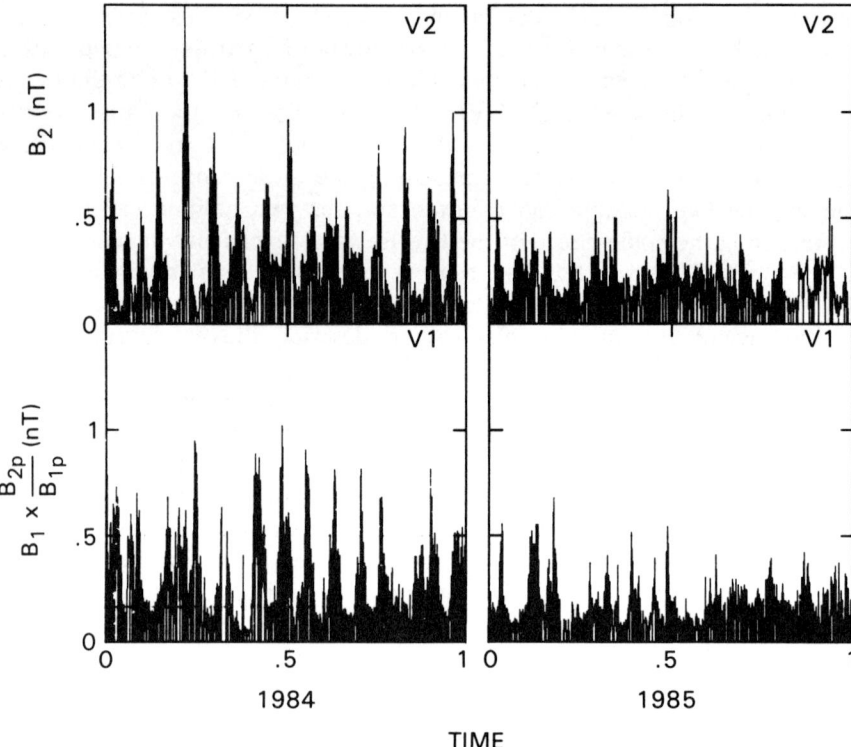

Fig. 8.8. Quasi-periodic MIRs. (L.F. Burlaga, N.F. Ness, and F.B. McDonald, *J. Geophys. Res.,* **92,** 13647, 1987b, copyright by the American Geophysical Union.)

The lower left panel of Fig. 8.8 shows the magnitude of the magnetic field observed in 1984 by Voyager 1 between 18.4 AU and 21.9 AU from 21.8°N to 24.8°, multiplied by a factor based on the Parker spiral field model in order to compare it with the Voyager 2 data. The existence of quasi-periodic CMIRs is evident in both the Voyager 1 and the Voyager 2 data. A period close to the solar rotation period is apparent to the eye, and spectral analysis shows a peak at 26 days (see below: Fig. 9.6). However, the amplitude of the fluctuations varies appreciably from one rotation to the next, and the CMIRs are not exactly in phase.

The right-hand side of Fig. 8.8 shows the corresponding observations during the following year, 1985. Quasi-periodic CMIRs were not very apparent during this interval. In fact, no large-amplitude MIRs were present in the Voyager data during 1985, near solar minimum. Thus, the quasi-periodic CMIRs represent just one state of the outer heliosphere, which is present during the declining phase of solar activity but not near solar minimum.

8.3.2 Period Doubling

Corotating interaction regions recurring with a period of approximately 13.4 days were observed by IMP-8 at 1 AU from July 28 to November 26, 1984 (Fig. 8.9, bottom). Corotating merged interaction regions recurring with a period of approximately 25 days were observed between 15.2 AU and 16.2 AU by Voyager 2 during the corresponding interval from September 1984 to January 27, 1985 (Fig. 8.9, top). Thus, the period of the corotating interaction regions doubled between 1 AU and 15 AU. This effect was called "period doubling" by Burlaga (1988b), who analyzed the observations just described. Period doubling is another example of the formation of large-scale structures from multiple small-scale structures. In this case, both the large-scale structure and the small-scale structure are quasi-periodic, but the periods are different.

The data at 1 AU for the period discussed above were not sufficiently complete to provide the input conditions for a model of period doubling. However, the evolution of quasi-periodic corotating streams and interaction regions between 1 AU (IMP data), 4.58 AU (Pioneer 11 data), and 5.83 AU (Pioneer 10 data) provide a set of input data that can be used to model period doubling (Burlaga et al., 1990b). Two streams per solar rotation at 1 AU coalesced to form a single compound stream at 5.83 AU (Fig. 8.10). However, two CIRs per solar rotation were observed at both 1 and 5.83 AU. Period doubling did not occur within 6 AU in this instance.

A description and qualitative analysis of the radial evolution of the flow in Fig. 8.10 was presented by Burlaga et al. (1990b). The speed profile at 1 AU (Fig. 8.10, top) does show a tendency for the streams to be alternately fast and slow. However, the speed observation from Pioneer 11 at approximately 4.6 AU and by Pioneer 10 at approximately 5.8 AU (suitably shifted to allow for the propagation of the solar wind from one spacecraft to another) show that the fast streams did not overtake the slower streams ahead of them. Rather, the slow streams appeared to overtake and merge with the faster streams ahead! The formation of compound streams in this way is caused by an additional asymmetry of the streams, their unequal widths. The fast streams were steeper and narrower than the slower streams at 1 AU. Consequently, the fast streams produced stronger CIRs than the slow streams, and they were eroded more rapidly than the slow streams. This is an extension to the case of interacting streams of the concept of "filtering" introduced in the unpublished work of Hundhausen (Holzer, 1979) for isolated streams. The fast streams decelerated more rapidly than the slower streams between 1 AU and approximately 5 AU, allowing the slow streams to eventually overtake the streams that were originally faster. Burlaga et al. (1990b) suggested that the asymmetry in the widths of the streams caused period doubling of the interaction regions beyond 6 AU.

The radial evolution of the interaction regions associated with the streams in Fig. 8.10 was modeled by Whang and Burlaga (1990a) using

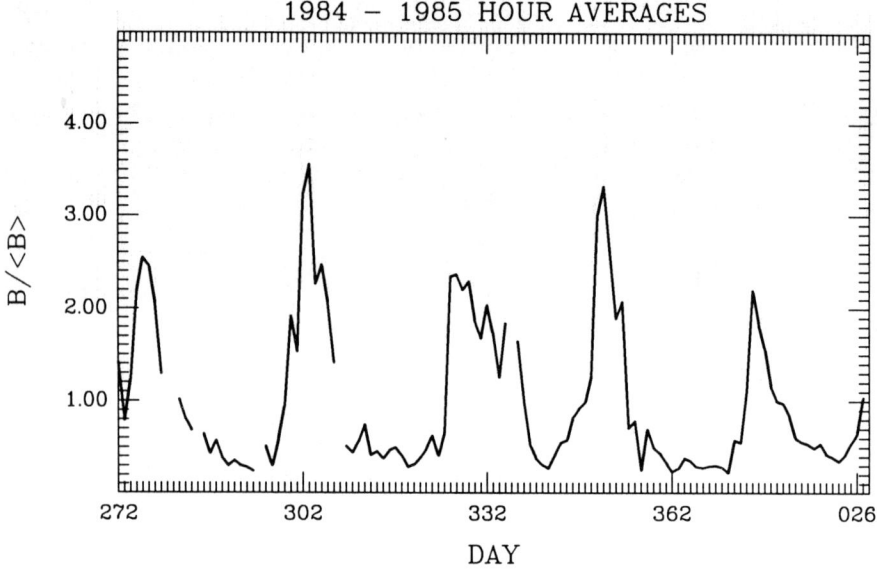
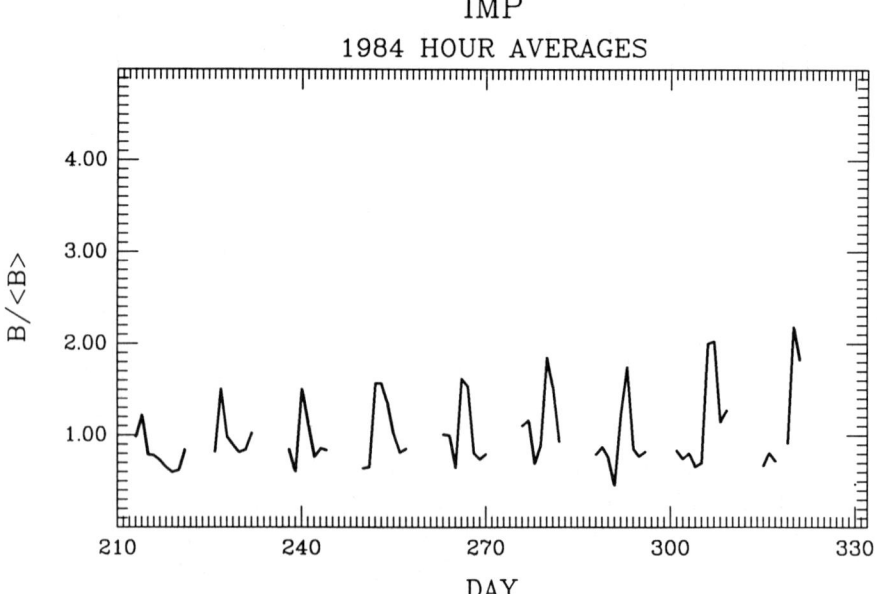

Fig. 8.9. Period doubling. (L.F. Burlaga, *J. Geophys. Res.,* **93**, 4103, 1988b, published by the American Geophysical Union.)

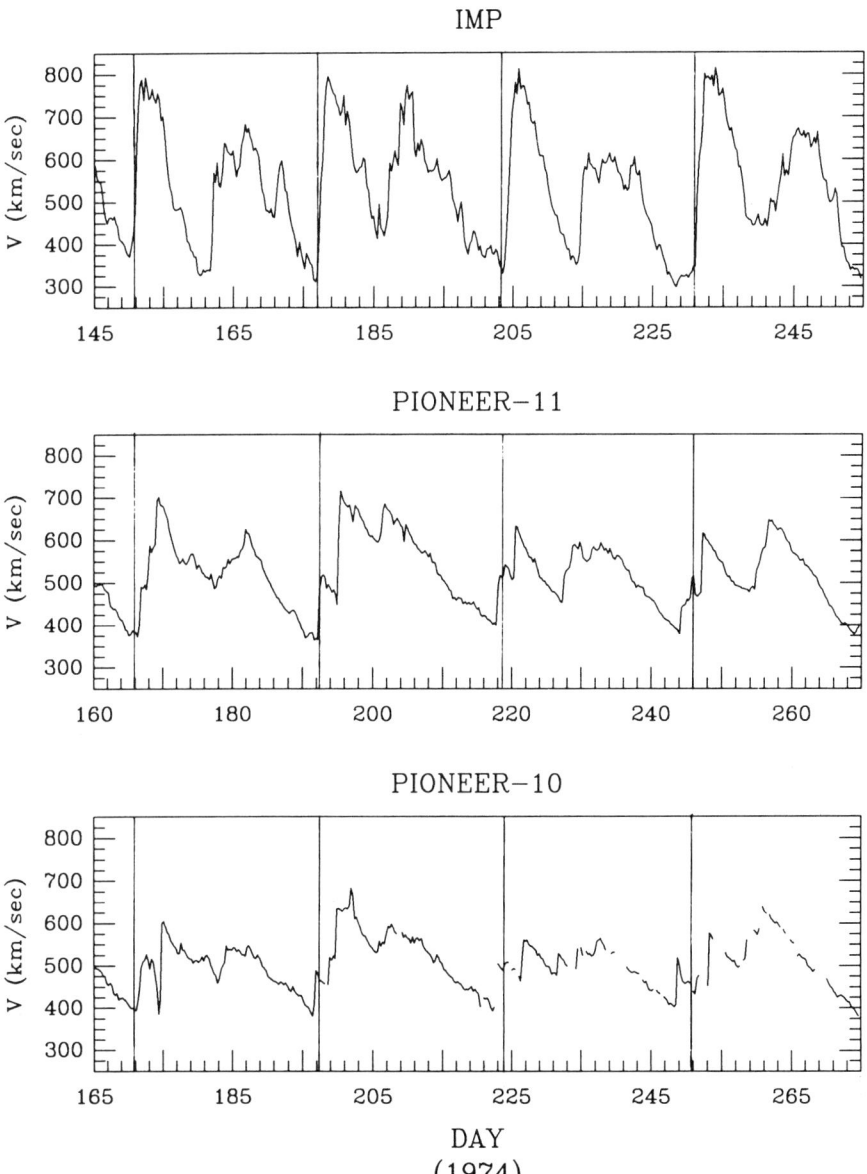

Fig. 8.10. Asymmetric recurrent streams. (L.F. Burlaga, W.H. Mish, and Y.C. Whang, *J. Geophys. Res.*, **95,** 4247, 1990b, copyright by the American Geophysical Union.)

Whang's 1-D MHD code with the Pioneer 11 data at 4.58 AU as the initial condition. Pioneer 11 observed the interaction regions recurring with a period of approximately 13.5 days, like IMP-8 in the period July 28 to November 26, 1974, as discussed at the beginning of this section. However, the corotating interaction regions at 4.58 AU during 1974 were bounded by shock pairs. The calculations confirmed that the steep, narrow, fast corotating streams were eroded more rapidly by their interaction regions than the slow broad corotating streams. The coalescence of the respective interaction regions between 1 AU and 15 AU leading to period doubling was demonstrated by this model (see Fig. 8.11). Figure 8.11 is a space–time diagram showing the corotating interaction regions as shaded areas and the forward and reverse shocks as the heavy curves. At 4.58 AU there were two CIRs bounded by shock pairs on each solar rotation. Each successive pair of CIRs merged to form a single CMIR per solar rotation somewhere between ≈ 7 AU and ≈ 10 AU. In each case, the reverse shock bounding the interaction region of a fast stream passed through the forward shock bounding the interaction region of the following slower stream.

The asymmetries in the widths and heights of the recurrent streams are crucial to the formation of corotating merged interaction regions and to period doubling. Period doubling cannot be predicted by the conventional

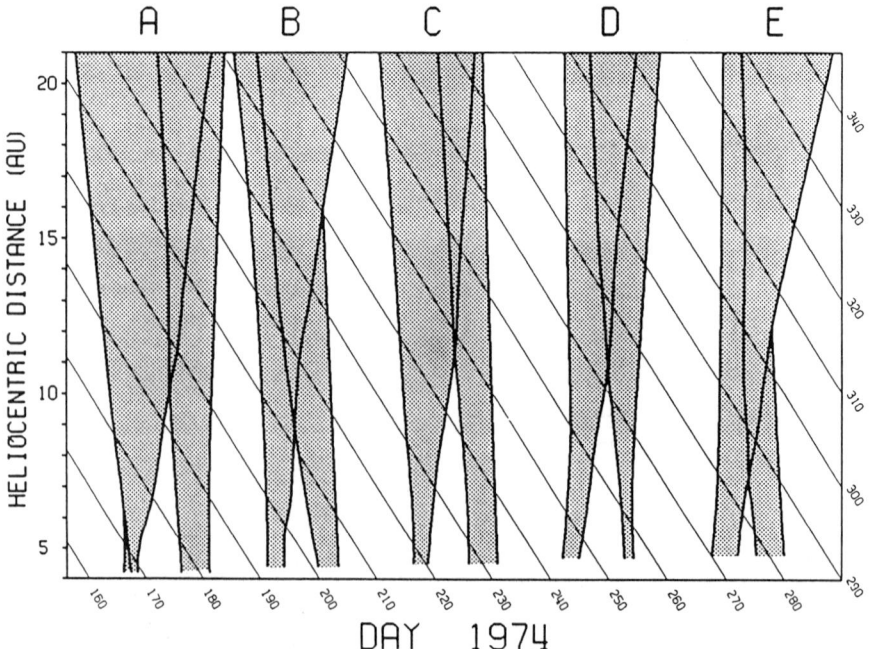

Fig. 8.11. Formation of corotating MIRs from the streams in Fig. 8.10. (Y.C. Whang, and L.F. Burlaga, *J. Geophys. Res.*, **95**, 20663, 1990a, copyright by the American Geophysical Union.)

"twin stream" models. In such models, there will always be two interaction regions and two shock pairs per solar rotation at all distances from the sun if there are two interaction regions per solar rotation at 1 AU. The corresponding highly symmetric shock configuration in the interplanetary medium as viewed from above the ecliptic is illustrated in Burlaga and Klein (1986a). The shocks and interaction regions from twin streams at 1 AU do interact, but the resulting pattern is unstable. Small perturbations in the widths and heights of the streams and other asymmetries such as asymmetries of the densities in the heliospheric plasma sheet ahead of the streams can produce qualitative changes in the structure of the outer heliosphere (Burlaga et al., 1990b).

8.3.3 Growth of Quasi-Periodic CMIRs from Aperiodic Interaction Regions

There is a strong tendency for quasi-periodic CMIRs to form from a variety of possible flow conditions at 1 AU, not necessarily periodic, throughout most of the solar cycle. Quasi-periodic CMIRs represent a stable state, an attractor, toward which the system evolves from many different initial conditions. This type of evolution is illustrated by the Voyager 1 magnetic field observations between 1 AU and 8.2 AU shown in Fig. 8.12. Here the magnetic field strength identifies the interaction region. The magnetic field strength is normalized by the nominal Parker spiral value B_p so that the amplitude of the interaction regions can be compared at various distances from the sun. Each panel shows 10-hour averages of the normalized magnetic field strength over an interval of 170 days. Thus, we are discussing the radial evolution of "large-scale fluctuations" (see Chapter 9). Between 1 AU and 2.6 AU the fluctuations are aperiodic, there are many interaction regions, the width of each interaction region is narrow, and the amplitude of each interaction region is relatively small. Between 4 AU and 5.2 AU there are fewer interaction regions, the interaction regions are broader, and they have larger amplitudes than those near 1 AU. Between 6.9 AU and 8.2 AU there are six quasi-periodic CMIRs, with a recurrence period of approximtely 26 days, the solar rotation period. The widths and amplitudes of the MIRs between 6.9 AU and 8.2 AU are larger than those of the interaction regions between 1.0 AU and 2.6 AU. The number of MIRs at the largest distances is smaller than the number of interaction regions at the smaller distances. Thus the smaller aperiodic interaction regions near 1 AU coalesced to form larger quasi-periodic MIRs near 8 AU. The formation of ordered large structures from irregular smaller structures is apparent in Fig. 8.12. This was the first example of "order out of chaos" in the outer heliosphere.

The formation of large ordered structures from smaller irregular structures with increasing distance from the sun is illustrated in a different way in Fig. 8.13. Simultaneous data from the experiment of Smith et al. on

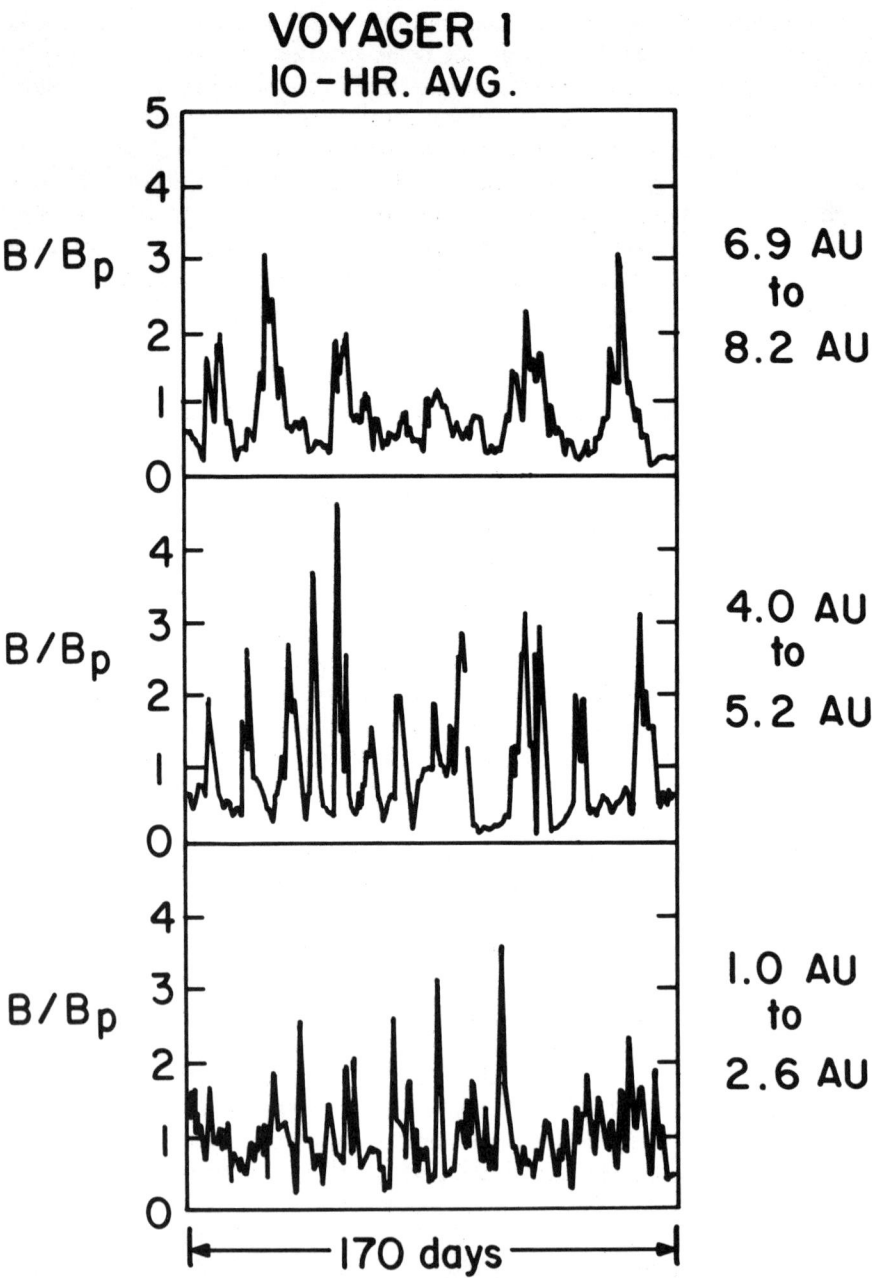

Fig. 8.12. Formation of merged interaction regions with increasing distance from the sun. (L.F. Burlaga, L. Klein, R.P. Lepping, and K.W. Behannon, *J. Geophys. Res.,* **89,** 10659, 1984a, copyright by the American Geophysical Union.)

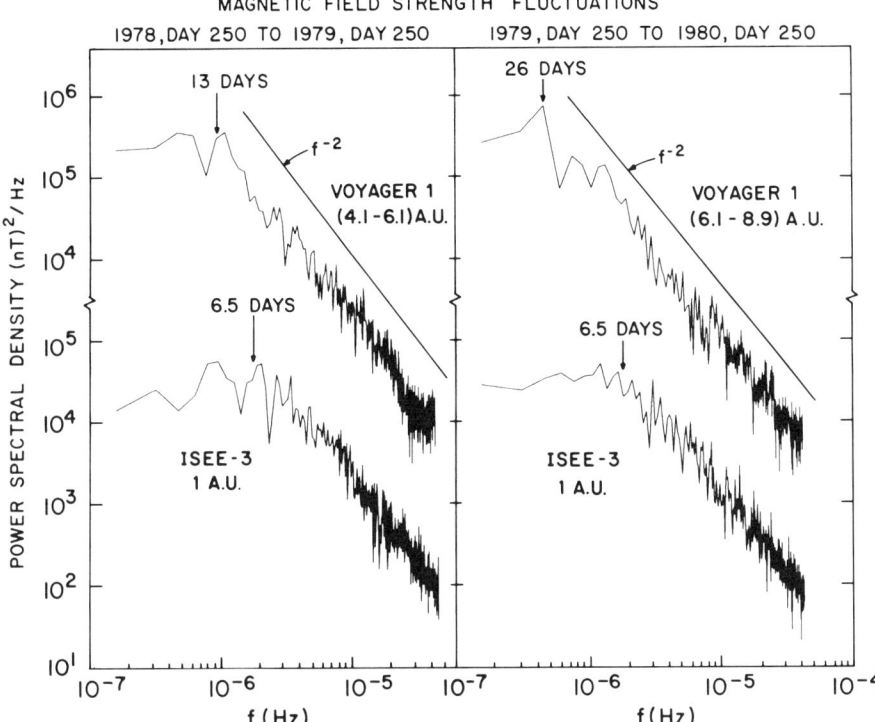

Fig. 8.13. Spectra illustrating period doubling. (L.F. Burlaga and W.H. Mish, *J. Geophys. Res.*, **92,** 1261, 1987, published by the American Geophysical Union.)

ISEE-3 at 1 AU and from the experiment of Ness et al. on Voyager 1 were used by Burlaga and Mish (1987) to eliminate the possibility that the evolution shown in Fig. 8.12 is a temporal effect, rather than an effect of the radial evolution. Power spectra are shown rather than time series, so that one can clearly see the emergence of quasi-periodic structures with lower periods, hence larger sizes.

Consider the evolution of the power spectral density of the magnetic field strength fluctuations. The left panel of Fig. 8.13 shows that at 1 AU there were structures with periods of 13 days (two interaction regions per solar rotation) and 6.5 days (four structures per solar rotation). Between 4.1 AU and 6.1 AU, the dominant peaks are at 26 days and 13 days, indicating the formation of subharmonic MIRs with a period of 26 days and the coexistence of interaction regions with a period of 13 days.

The panel on the right of Fig. 8.13 shows a maximum power spectral density in the range 13 days to a few days at 1 AU; the peaks are less well defined than in the period described above, indicating perhaps a greater contribution from noncorotating streams. The observations at 6.1–8.9 AU show the emergence of a single dominant peak at 26 days, representing the formation of quasi-periodic recurrent MIRs with a period equal to the solar rotation period.

The observations discussed above show the formation of a few large-scale ordered structures from many smaller irregular structures with increasing distance from the sun. The formation of ordered, large-scale structures from less ordered, smaller scale structures has been observed in many other driven, nonlinear, dissipative systems—particularly those involving a periodic driving force. In the heliosphere the fundamental driving force is provided by the rotating sun. The evolution of the quasi-stationary flows is such that subharmonics appear, and the solar rotation period emerges as the dominant period for a wide variety of initial conditions.

8.3.4 A Model of the Formation of CMIRs from Complex Initial Conditions

The results of the preceding section show a general tendency for quasi-periodic MIRs to evolve from complex initial conditions, but they do not show the specific mechanisms involved. Although the formation of ordered, large-scale structures from the competition among smaller, less ordered structures is a general characteristic of many periodically driven, highly nonlinear, dissipative systems, the heliosphere cannot be described directly by the usual results of dynamical systems theory and nonlinear hydrodynamics because the flow is supersonic. This section summarizes the results of a calculation by Whang and Burlaga (1985b), shown in Fig. 8.14, which demonstrate how nonlinear, dissipative processes involving shocks can produce recurrent MIRs with a period approximately equal to the solar rotation period from a mixture of corotating streams, compound streams, and transient ejecta near 1 AU.

The input to the model is a set of plasma and magnetic field data from IMP at 1 AU. Stream B is the compound corotating stream discussed in Section 8.2.3, and stream C is the corotating stream overtaking a shock that was discussed in Section 8.2.4.

The evolution of the various shocks and interaction regions was followed out to 16 AU using a model that treats shocks as discontinuities satisfying the Rankine–Hugoniot conditions without the use of artificial viscosity. The model is the 1-D MHD model of Whang (1984) discussed above. Shock–shock interactions are treated exactly in this model, but the restriction to a 1-D model implies that the evolution near 1 AU is treated only approximately, so that the results are qualitative. In particular, it is assumed that the shocks observed by Voyager 1 at 1.4 AU were present even at 1 AU, which was not the case for most of the shocks.

The results of the calculation are conveniently summarized by the space–time diagram in Fig. 8.14. Stream C evolved by first forming a forward–reverse shock pair; the forward shock of this pair overtook and coalesced with the forward shock from a transient ejection that was ahead

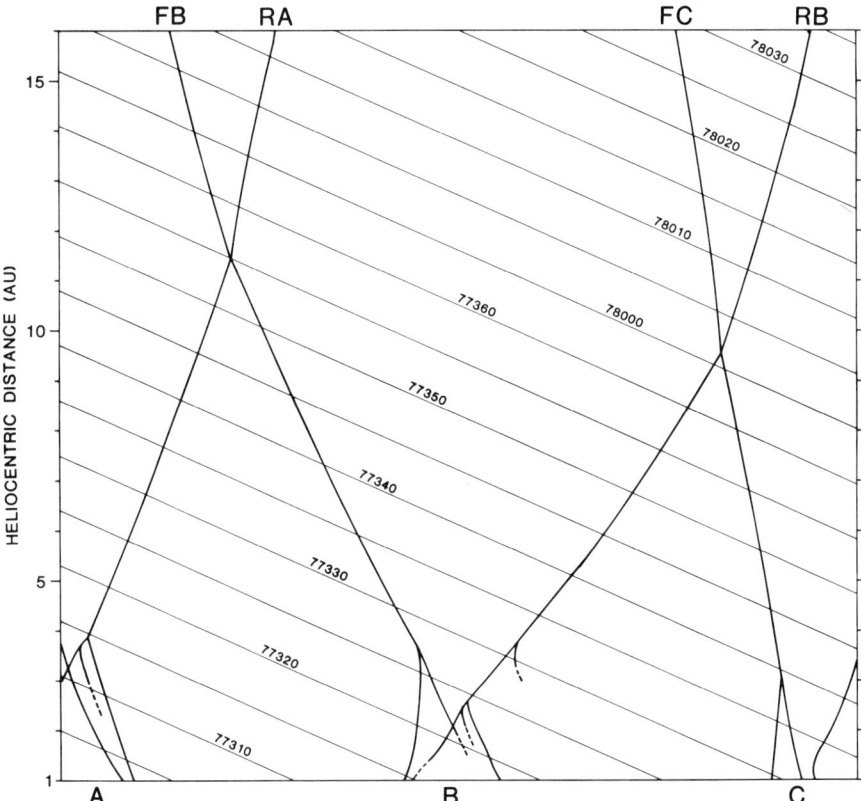

Fig. 8.14. Merging of several shocks. (Y.C. Whang and L.F. Burlaga, *J. Geophys. Res.*, **90**, 10765, 1985b, copyright by the American Geophysical Union.)

of it to form a single fast forward shock *FC* at 3 AU. The compound corotating stream *B* evolved by first forming two corotating shock pairs. The reverse shock of the first shock pair passed through the forward shock of the second shock pair, as discussed in Section 8.2.4. The two forward shocks coalesced near 4 AU to form a single forward shock *FB*, and the two reverse shocks coalesced near 3 AU to form a single reverse shock *RB*. Finally, the compound stream *A* evolved by forming two shock pairs. The two reverse shocks coalesced to form a single reverse shock *RA* near 4 AU. Thus, by 4 AU, nonlinear interactions among the shocks from the streams *A*, *B*, and *C* had already considerably simplified the flow. Three merged interaction regions formed from at least five interaction regions at 1 AU. The coalescence of shocks is a key factor in the simplification of the flow with increasing distance from the sun. The process is irreversible. Memory of the source conditions is lost when shocks merge and interaction regions coalesce (Burlaga, 1983).

The subsequent evolution of flow shown in Fig. 8.14 is relatively simple, but it is important and representative of the interactions that take place between 4 AU and 15 AU. The merged interaction region produced by the compound stream B at 1 AU broadened with increasing distance from the sun as the forward shock FB and the reverse shock RB bounding it moved apart. The forward shock FC from the MIR produced by stream C interacted with the reverse shock RB near 10 AU, so that the two MIRs coalesced to form a merged MIR. Similarly, the reverse shock RA from the MIR produced by stream A interacted with the forward shock FB from the MIR produced by stream B near 11.5 AU to form a second merged MIR. The two merged MIRs bounded by $FB-RA$ and $FC-RB$, respectively, were separated by approximately 20 days and increased in size as they moved away from the sun.

8.4 Local MIRs

A local merged interaction region is defined as a localized MIR formed from the interaction of one or more transient ejecta and interaction regions with other streams and interaction regions. A near-radial alignment of Helios 2 at ≈ 0.85 AU and Voyager 1 at ≈ 6.2 AU showed the formation of a compound stream and two MIRs from the interaction of five streams observed by Helios 2. The speed observations from the two spacecraft are shown in Fig. 8.15. The principal cause of the interaction and the formation of the LMIR was an exceptionally fast transient stream E with a speed of at least 1270 km/s at Helios 2 and another fast transient stream D with a speed of at least 1030 km/s. Figure 8.15 shows that three transient streams A, D, and E as well as two corotating streams B and C observed by Helios 2 coalesced to form a single, large compound stream at Voyager 1 that passed the spacecraft over an interval of more than 20 days. One can assume that an interaction region was associated with each of the streams observed by Helios. Thus, five interaction regions were present at 0.85 AU but only two MIRs were observed by Voyager 1 (Burlaga et al., 1986).

A kinematic analysis (Burlaga et al., 1986) suggests that the first MIR was produced by the merging of the interaction regions from streams A and B and that the second MIR was produced by the merging of the interaction regions from streams C, D, and E. Kinematics can be a useful guide to the qualitative evolution of streams and flows in the outer heliosphere, given observations at two points, but it is physically inadequate and it fails to provide a quantitative description, because the pressure gradient forces play a major role in the dynamics of the outer heliosphere. Kinematics is an essential part of dynamics, but dynamical effects cannot be derived from kinematics alone.

A kinematic model of the evolution of a series of flare-associated streams and shocks in the outer heliosphere, based on the method of Hakamada and Akasofu (1982), was presented by Akasofu and Hakamada (1983a,b) and Akasofu et al. (1985a,b) and Olmstead and Akasofu (1985).

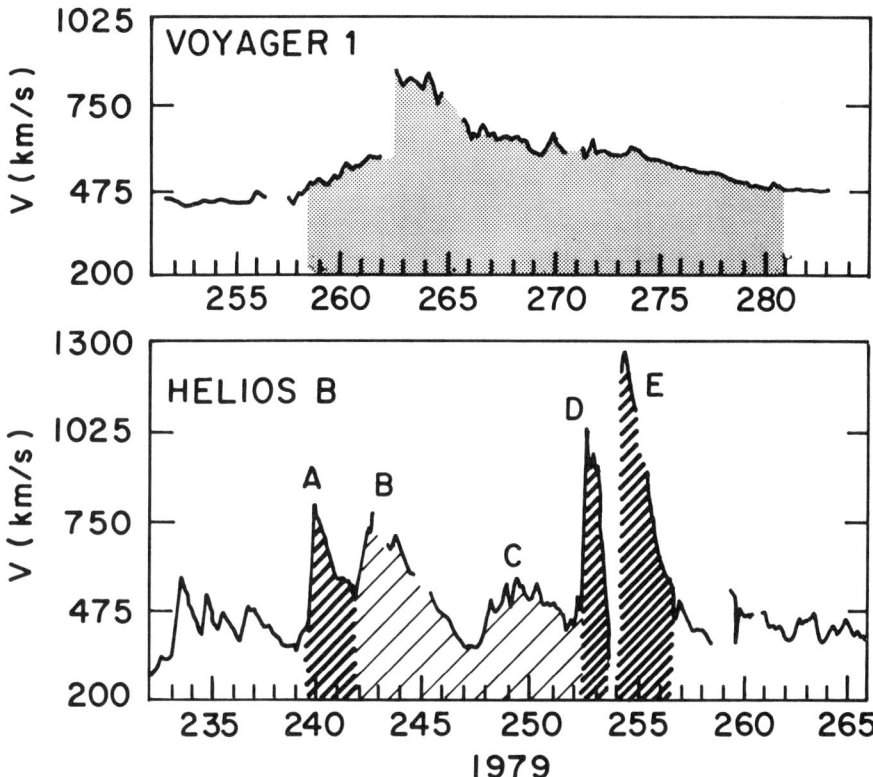

Fig. 8.15. Five streams merge to form one compound stream. (L.F. Burlaga, F.B. McDonald, and R. Schwenn, *J. Geophys. Res.*, **91**, 13331, 1986, copyright by the American Geophysical Union.)

Their kinematic model "matches the solar wind data, by design," since there are many free parameters that can be chosen to reproduce the measured results. This approach was criticized by Pizzo (1983b), referring to the point made by Burlaga (1983) that dynamical effects involving shocks and pressure waves are dominant in the outer heliosphere. The importance of dynamical effects in the outer heliosphere should be clear to the reader from the results presented earlier in this chapter and in Chapter 7. Pizzo also noted tht the corotating background flow assumed in the kinematic work of Akasofu and colleagues is "unrealistic inside about 10 AU and not even qualitatively appropriate beyond that point." A reply to Pizzo's criticism was published by Akasofu (1983).

The kinematic approach of Akasofu and his colleagues has the advantage that it can readily produce appealing images of the magnetic field configuration, but these images can be very misleading. The smooth Archimedean magnetic field lines in the transient flows are incorrect, and the merging of corotating interaction regions in the outer heliosphere is not

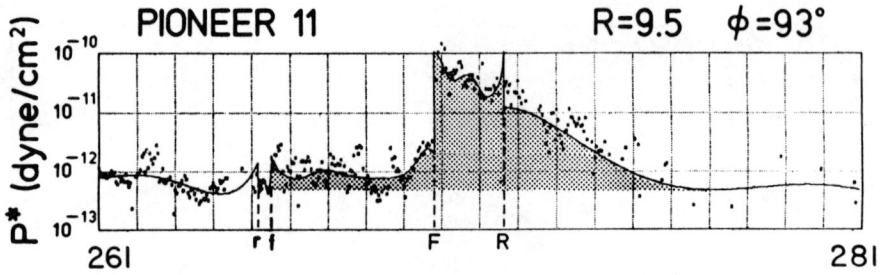
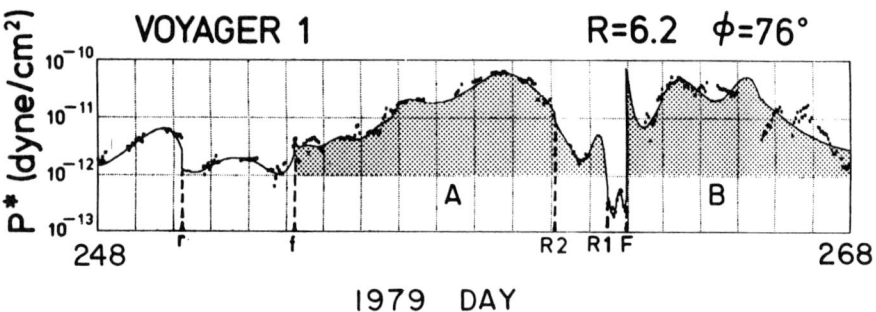

Fig. 8.16. Two MIRs coalesce to form one merged MIR. (Y.C. Whang, and L.F. Burlaga, *J. Geophys. Res.*, **91**, 13341, 1986, copyright by the Americal Geophysical Union.)

represented. The kinematic approach also has the advantage of simplicity. However, it provides no physical insight concerning the evolution of flows in the outer heliosphere, and it has no predictive value because the parameters are chosen to give the desired results. The results of kinematic models in the outer heliosphere should be interpreted very cautiously.

The two MIRs observed by Voyager 1 at 6.2 AU coalesced to form a single merged MIR: a local merged interaction region (LMIR) observed by Pioneer 11 at 9.2 AU (Whang and Burlaga, 1986). The LMIR formed in this way is illustrated in Fig. 8.16. The coalescence of the two MIRs observed at 6.2 AU to form a single LMIR at 9.5 AU was modeled by Whang (1986) using the code of Whang (1984) with the Voyager 1 data as input. The computed profiles at 9.5 AU were qualitatively in agreement with the observed profiles, but the shocks arrived somewhat earlier than predicted.

The key to the formation of an LMIR is a very fast transient stream that overtakes other flows and interaction regions. The example of a magnetic cloud interacting with a CIR discussed in Section 8.2.6 was probably of this nature, although the speed observations were not available near 1 AU to verify this conjecture. Fast transient streams preceded by strong shocks are unusual events beyond ≈ 15 AU, but they are observed and they are very important. Such a fast stream was found in the Pioneer 10 data at 30 AU by Kayser (1985), for example.

8.5 Global MIRs

8.5.1 Definition and Existence

A global merged interaction region (GMIR) is defined as a large, shell-like MIR with intense magnetic fields that encircles the sun and extends to high latitudes (Burlaga et al., 1984b, 1993a). An observation of a GMIR in the outer heliosphere is shown on the right of Fig. 8.17. The Voyager 2 observations were made near the ecliptic at approximately 30 AU, and the Voyager 1 observations (Fig. 8.17, left) were made near 30°N heliographic latitude at approximately 40 AU. The GMIR is the region of higher than average magnetic field strength passing Voyager 1 over an interval of approximately 100 days and Voyager 2 over an interval of approximately 150 days. Thus, the radial cross section of the GMIR is of the order of 20 AU beyond 30 AU.

The observations of cosmic rays >70 MeV/nuc, from the experiment of Stone et al. (1977), plotted in the middle panels of Fig. 8.17, showed a

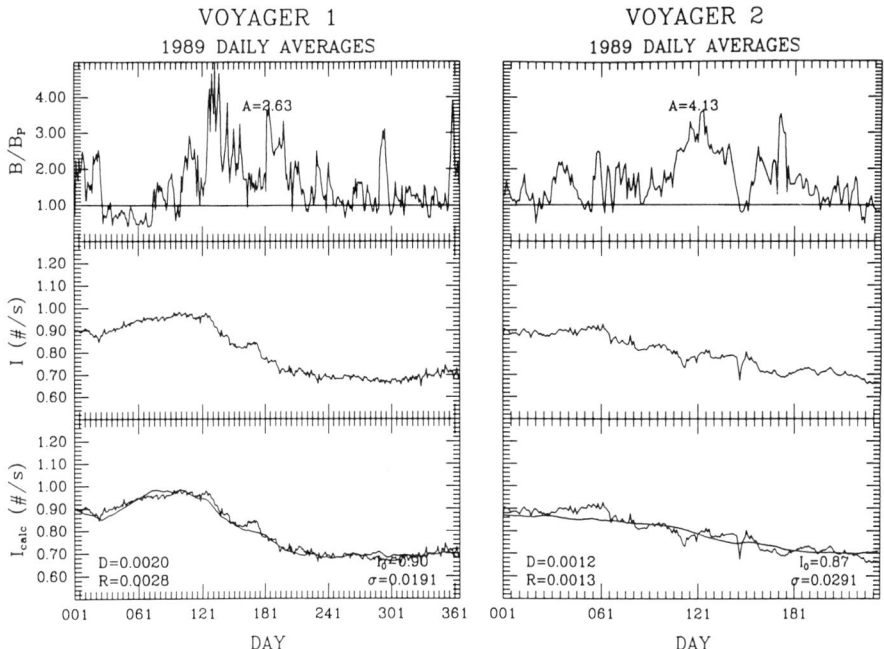

Fig. 8.17. Global merged interaction region. (L.F. Burlaga, F.B. McDonald, and N.F. Ness, *J. Geophys. Res.*, **98**, 1, 1993a, copyright by the American Geophysical Union.)

correlation with the MIR. As indicated by the model curve and the observations in the lower panels of Fig. 8.17, there is a close relation between the decreases in the cosmic ray intensity and the enhancements of the magnetic field strength in the GMIRs (Burlaga et al., 1993a). A decrease in the cosmic ray intensity was observed by Pioneer 10 (McDonald et al., 1991) near the ecliptic and opposite to Voyager 2 in longitude. The observations of decreases in the cosmic ray intensity on opposite sides of the sun, both near the ecliptic and near 30°N latitude, suggests that the GMIR that caused the cosmic ray intensity decreases was indeed a global structure with a shell-like geometry (Fig. 8.18).

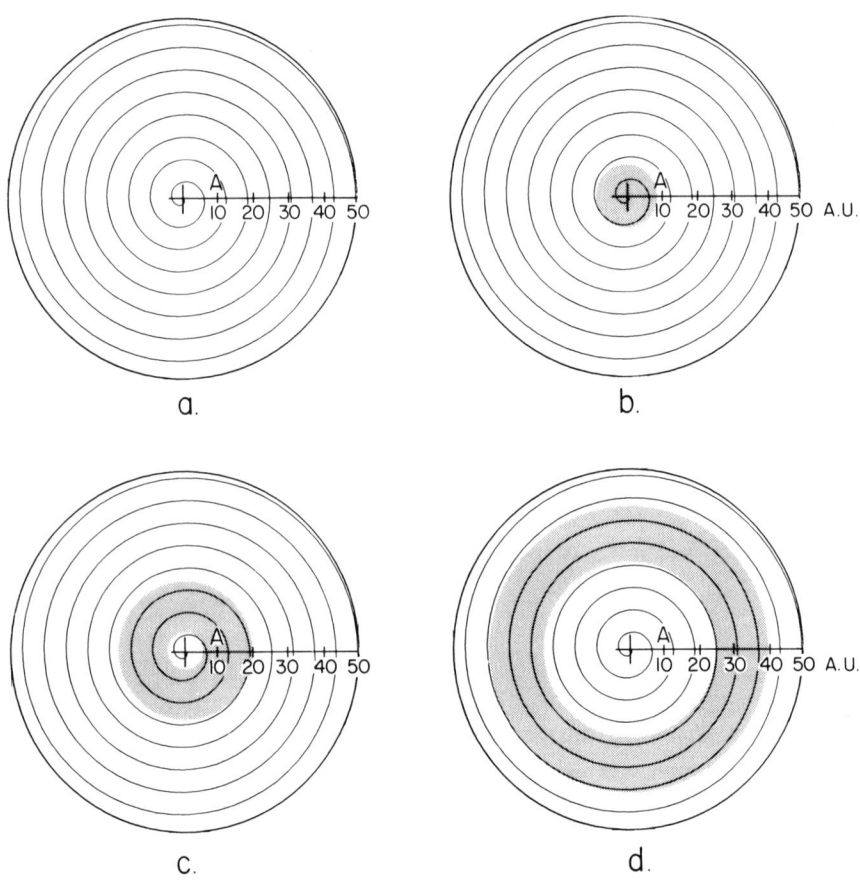

▭ SYSTEM OF TRANSIENT FLOWS

Fig. 8.18. Shell corresponding to global merged interaction regions; for discussion of (a)–(d), see text. (L.F. Burlaga, F.B. McDonald, N.F. Ness, R. Schwenn, A.J. Lazarus, and F. Mariani, *J. Geophys. Res.*, **89,** 6579, 1984b, copyright by the American Geophysical Union.)

8.5.2 Earlier Indirect Evidence for Shell-like Modulating Regions

The decreases in the cosmic ray intensity shown in Fig. 8.17 appear as large steplike decreases in the cosmic ray intensity when viewed on a scale of several years. Morrison (1956) noted that the decline in cosmic ray intensity from sunspot minimum in 1945 to sunspot maximum in 1947 took place in two sharp falls, each followed by a slow partial return to the mean. Morrison (1956) suggested that the modulating region responsible for such steplike decreases in the cosmic ray intensity is a large diffusive shell with a radial extent of tens of AU formed from magnetized plasma (transient ejecta) ejected from the sun when it is active. In the following solar cycle, during 1954–58, Lockwood (1960) observed a series of large and sudden drops from which only partial recovery occurred. Lockwood (1971) stressed the importance of "long duration Forbush decreases" (lasting weeks or months) for the 11-year modulation. He suggested that the modulation region is possibly a thick, hollow, turbulent shell. Barouch and Burlaga (1975) showed that these long-lasting Forbush decreases are associated with a series of enhancements in the magnetic field strength at 1 AU.

Steplike decreases in the cosmic ray intensity were observed again from 1977 to 1981 (at lower energies) by McDonald et al. (1981). These authors used simultaneous observations at 1 AU and in the outer heliosphere to show that the steplike decreases propagate away from the sun at the solar wind speed.

8.5.3 Cosmic Ray Steps, Systems of Transients Flows, and GMIRs

The cause of the three major steplike decreases in the galactic cosmic ray intensity from 1977 to 1981 was investigated using the plasma and magnetic field data from Helios 1 and Voyagers 1 and 2. The Voyager spacecraft were within 12 AU during this time, so that MIRs and GMIRs independent of streams were not observed. These regions were not fully developed at the distance in question. However, Burlaga et al. (1984b) found that the steps in cosmic ray intensity were related to systems of transient flows and enhanced magnetic fields. They suggested that the each step was caused by a shell consisting of a system of transient flows and strong magnetic fields.

The formation of a shell (GMIR) related to the step observed by Voyager 2 from January 1, 1978, to July 1, 1978, is illustrated very schematically, but approximately to scale, in Fig. 8.18. Assume that initially (Fig. 8.18(a)) there are no transient flows in the heliosphere on a scale of 50 AU, implying no major solar activity for about 200 days. Next, assume that the sun becomes very active, ejecting shocks, magnetic clouds, and other transient magnetic field configurations that fill the shaded region in Fig. 8.18(b). For an example of such a situation, see Cliver et al. (1987).

This active state continues for about two solar rotations, and then the sun returns to a quiet, stationary state (Fig. 8.18(c)). The system of transient flows and magnetic fields forms a large shell that moves out through the heliosphere as shown in Fig. 8.18(d). When this shell passes a spacecraft, it causes a decrease in the cosmic ray intensity that persists as long as the shell remains in the heliosphere. As the shell moves away from the sun, the streams within it are eroded, the shocks and interaction regions form complex MIRs, ultimately (at 30 AU) forming a GMIR like that in Fig. 8.17.

9

Large-Scale Fluctuations

9.1 Introduction

9.1.1 Definition of Large-Scale Fluctuations

The "large-scale fluctuations" in the heliosphere are defined as fluctuations in the plasma and magnetic field having periods from several hours to the solar rotation period (Burlaga and Goldstein, 1984). A distinction between low frequency and intermediate frequency large-scale fluctuations was made by Burlaga et al. (1987b, 1989). "Low frequency fluctuations" are the fluctuations for which $3 \times 10^{-7}\,\text{Hz} < f < 3 \times 10^{-6}\,\text{Hz}$, and "intermediate frequency fluctuations" are those for which $3 \times 10^{-6}\,\text{Hz} < f < 3 \times 10^{-5}\,\text{Hz}$. The study of large-scale fluctuations requires at least several months of observations for valid statistical analysis, and it can be carried out using hour averages.

At periods shorter than ≈ 10 hours (frequencies greater than about $3 \times 10^{-5}\,\text{Hz}$), there is a change in the character of the fluctuations (see, e.g., Barassano et al., 1982; Denskat and Neubauer, 1983; Roberts et al., 1987). There is often a change in the slope of the spectrum near this frequency at ≤ 1 AU. The coherence and phase between **B** and **V** show an almost perfect anticorrelation at frequencies above $2.4 \times 10^{-5}\,\text{Hz}$ but the coherence is low below $2.4 \times 10^{-5}\,\text{Hz}$ (Denskat and Neubauer, 1982). Denskat and Neubauer suggested that the lower frequency fluctuations are generated by larger scale dynamical processes. A transition from an anticorrelation between the magnetic and thermal pressure on scales of the order of an hour or less to a correlation between the magnetic pressure on scales of 2 days or more was noted by Burlaga and Ogilvie (1970b).

The presence of microscale turbulence in the solar wind was suggested by Coleman (1968), and the existence of Alfvénic fluctuations was demonstrated by Belcher and Davis (1971). Since the extensive literature on the high frequency fluctuations, including turbulence and Alfvénic fluctuations, is discussed in several recent reviews (e.g., Behannon and Burlaga, 1981; Marsch, 1991; Roberts and Goldstein, 1991), it is not included as a topic of this book. This chapter discusses primarily the large-scale fluctuations in the interplanetary magnetic field and plasma. There are several reasons to identify large-scale fluctuations as a special object of study, which are discussed briefly in the following paragraphs.

Since the interplanetary medium extends to 40 AU, one must consider at least ≈ 160 days of data in order to see the plasma and magnetic fields that occupy this region. The distance to the termination shock is probably at least 100 AU, so that one should consider the data for a period of at least one year in order to model the structure of the heliosphere at any instant (Burlaga and Goldstein, 1984). Earlier chapters show that the character of the solar wind changes with the solar cycle. Corotating streams and corotating interaction regions can be dominant for a year or more during the declining phase of the solar cycle. Transient ejecta and local interaction regions are important during the years near the maximum of solar activity. Streams and interaction regions can be absent or weak during the years around solar minimum. The large-scale fluctuations (in speed, pressure, etc.) during these various periods are different, and the fluctuations within a given epoch always have a statistical component. A description of the fluctuations associated with many streams and/or interaction regions requires data for at least several months.

The sun is variable, and it emits a mixture of corotating streams and transient ejecta that changes with solar activity. In other words, the source function on a scale of many solar rotations can be an irregular function reflecting the complexity of the corona and the solar activity. It is impossible to determine the initial conditions for such a system globally. Consequently, it is impossible to develop a deterministic model for interplanetary and heliospheric dynamics in general. Statistical models of heliospheric dynamics are needed to obtain a description of the heliosphere for such complex source functions (Burlaga, 1975). To date, however, there is no statistical model of heliospheric structure and dynamics. Until recently, even the description of large-scale fluctuations in the solar wind was largely neglected, and little was known about the statistical character of large-scale fluctuations.

9.1.2 Early Studies of Large-Scale Fluctuations

Most of the early studies (say up until 1982) of the variations in the interplanetary plasma and magnetic field were concerned with the distribution functions of N, T, V, and B, and in particular with the mean and variance of the distribution functions; for a survey of these results, see, for example, Chapter II in Hundhausen (1972). It was known that the distribution functions for the plasma and magnetic field strength are not symmetric and have "tails", but no attempt was made to describe them quantitatively. There were few studies of the properties of time series of measurements extending for a year or more. Gosling and Bame (1972) studied the fluctuations in 3-hour averages of the bulk speed from 1964 to 1967. They found that the autocorrelation coefficient becomes very small at a lag of the order of 100 hours, indicating that the solar wind speed at 1 AU is not steady on the scale of the solar wind expansion (Gosling and Bame, 1972).

Using a 621-day magnetic field data set from IMP, Matthaeus and Goldstein (1982, 1983) showed that the average magnetic field and the second-order moments of the magnetic field are "stationary," which means that the ensemble average does not depend on time. "Strict stationarity" requires that all moments be independent of time. It is more correct to say that the interplanetary magnetic field at 1 AU is a weakly stationary function.

The distribution of the hour averages of magnetic field strength at 1 AU from 1963 to 1975 is approximately log normal, as shown by the dashed histogram in Fig. 9.1. Note that the tail of the distribution is approximately exponential from the peak of the distribution to two decades below the peak (solid curve in Fig. 9.1). It is generally difficult to distinguish between an exponential distribution and the tail of a log-normal distribution when the data set is small.

The distribution of the differences between successive 3-hour averages of the bulk speed at 1 AU approaching solar maximum during 1967 was studied by Burlaga and Ogilvie (1970a). They showed that both the distribution of positive speed increments and the distribution of negative speed changes are exponential (Fig. 9.2). The distribution of positive speed increments is flatter than the distribution of negative speed increments because there are more small speed increases than large speed decreases, owing to the steepening of streams.

9.2 Spectral Signatures of Large-Scale Fluctuations

9.2.1 Large-Scale Magnetic Field Fluctuations at 1 AU and Near 5 AU

Two types of large-scale magnetic field fluctuation and their radial evolution were analyzed by Burlaga and Goldstein (1984) using the spectral analysis techniques developed by Matthaeus and Goldstein (1982). The observations at 1 AU are shown in the top panels of Fig. 9.3. The first interval (interval A) is from August 14, 1978, to February 5, 1979, and the second interval (interval B) is from March 29, 1979, to June 30, 1979. The spectra of the magnetic field strength fluctuations for these two intervals are shown in the lower panels of Fig. 9.3.

The radial evolution of the fluctuations in interval A is seen by comparing the Voyager 1 observations with the 1 AU observations on the left of Fig. 9.3. Since both the 1 AU magnetic field strengths and the Voyager 1 magnetic field strengths are plotted on a log scale, one can see that the relative amplitudes of the fluctuations increased with distance from

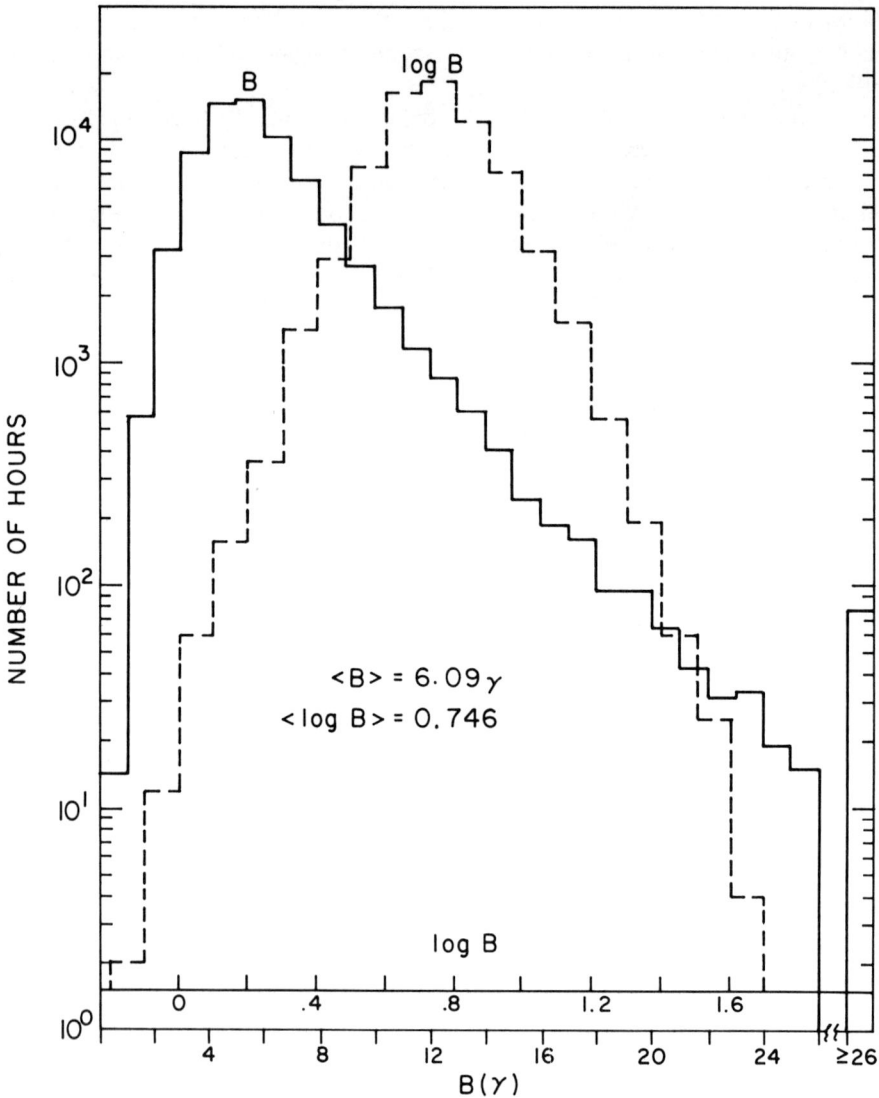

Fig. 9.1. Log-normal distribution of the magnetic field strength at 1 AU. (L.F. Burlaga and J.H. King, *J. Geophys. Res.,* **84,** 6633, 979, published by the American Geophysical Union.)

the sun. The spectra of the Voyager 1 observations at the bottom of Fig. 9.3, indicate that for interval A the turbulent $f^{-5/3}$ law extends to periods longer than 10 or 11 days. The magnetic field observed during interval A became increasing turbulent with increasing distance. The turbulent spectrum extended to lower frequencies with increasing distance from the sun.

The radial evolution of the fluctuations in interval B is rather different.

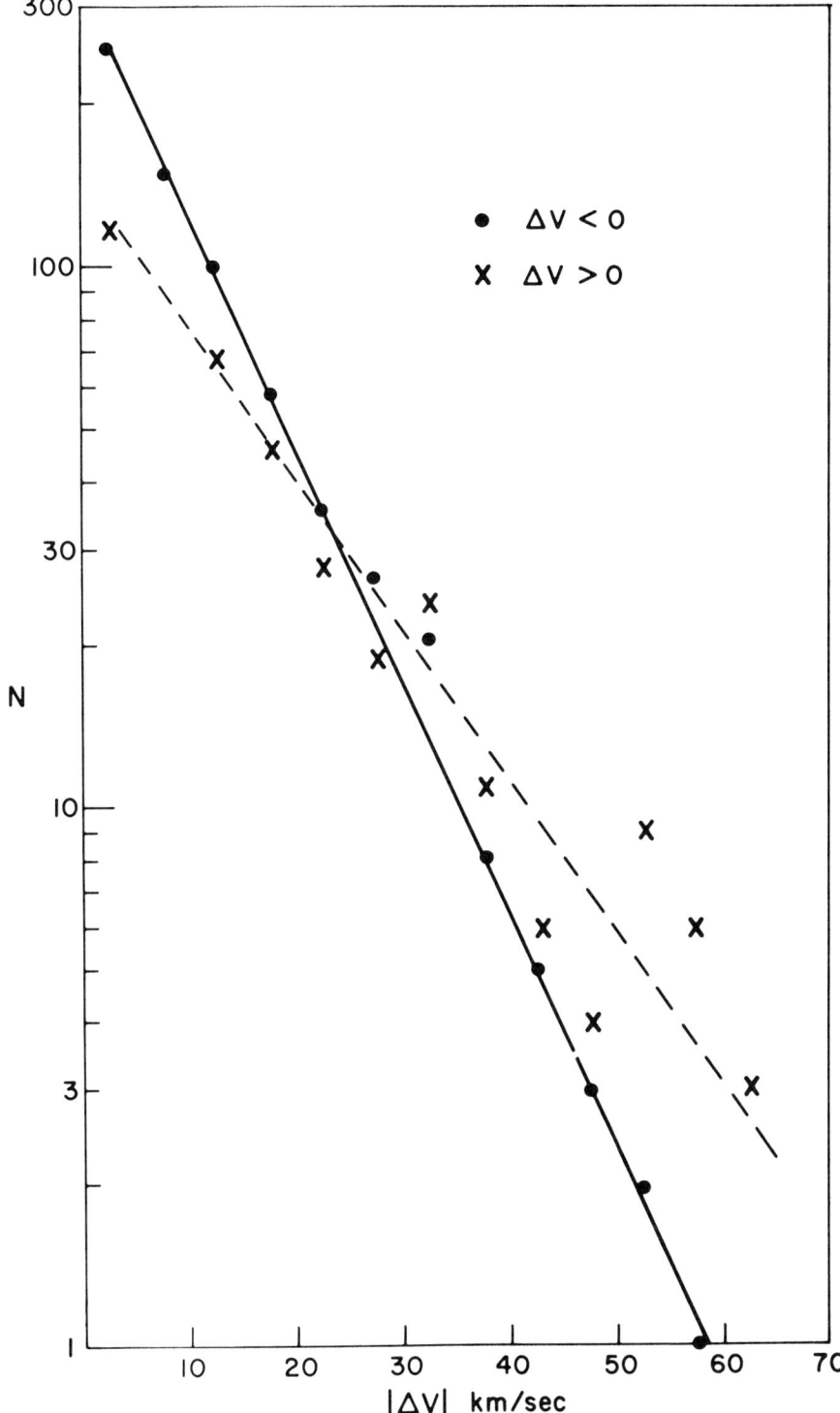

Fig. 9.2. Distribution of speed differences at 1 AU. (L.F. Burlaga and K.W. Ogilvie, *Astrophys. J.*, **159,** 659, 1970a.)

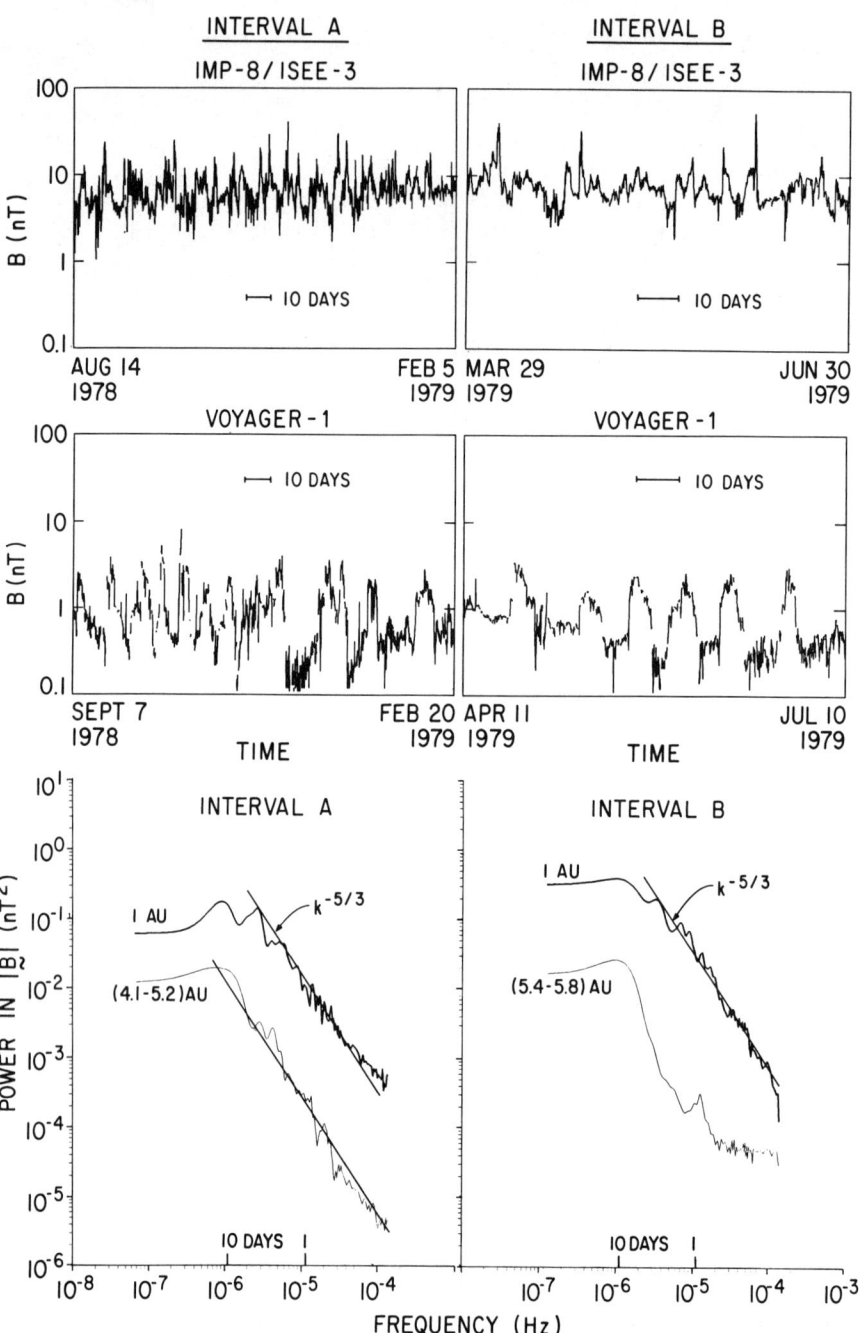

Fig. 9.3. Time series and spectra of transient and corotating streams. (L.F. Burlaga and M.L. Goldstein, *J. Geophys. Res.*, **89**, 6813, 1984, published by the American Geophysical Union.)

The fluctuations at Voyager 1 at 5.4–5.8 AU in interval B have a larger scale size and a more regular appearance than the fluctuations during interval A. The fluctuations during interval B appear to be predominantly CIRs. The spectra of the fluctuations during interval B are shown at the bottom right of Fig. 9.3. One does not see a turbulent $f^{-5/3}$ law at 5–6 AU, even at high frequencies. Instead, there is a large peak near 10 days, corresponding to the formation of CIRs and corotating rarefaction regions.

One concludes that while the large-scale magnetic field strength fluctuations in adjoining periods at 1 AU might have similar spectra, they can evolve very differently with increasing distance from the sun. In one case the fluctuations became increasingly turbulent, the turbulence extending to larger scales with increasing distance from the sun. In the other case the turbulence was not seen at larger distances, because the dominant power was produced by recurrent interaction regions. In both cases, however, larger scales became more prominent with increasing distance from the sun.

The formation of the recurrent interaction regions in interval B can be traced to the presence of corotating streams at 1 AU, although the streams were not periodic and they were mixed with transient flows (Burlaga and Goldstein, 1984). Thus, the recurrent interaction regions at Voyager 1 might have been predominantly CMIRs.

The extension of the turbulence in interval A to larger scales with increasing distance from the sun is possibly the result of two effects: an inverse cascade and the production of a larger "stirring scale." Matthaeus and Goldstein (1982) demonstrated that the magnetic helicity is conserved in the interplanetary turbulence. Thus, the magnetic energy might be transferred from intermediate scales to larger scales (Matthaeus and Goldstein, 1982, 1983; Montgomery, 1983). This transfer of energy to larger scales is called an "inverse cascade" because in ordinary turbulence, kinetic energy always cascades to smaller scales.

Evidence for an inverse cascade of helicity in interval A was found by Burlaga and Goldstein (1984), who calculated the helicity using the method of Matthaeus and Goldstein (1982). During interval A the peak in the magnetic helicity spectrum was near the correlation length, 1.6×10^{-5} Hz, at 1 AU, and it moved to lower frequencies at 5 AU. The power spectra of the components of the magnetic field in Fig. 9.3 show that at frequencies greater than the correlation length the spectra were close to the Kolmogorov law, but at lower frequencies the spectra had the form f^{-1} (Burlaga and Goldstein, 1984). The $1/f$ spectrum was identified as a source signal, attributed to the superposition of uncorrelated samples of solar turbulence that have log-normal distributions of correlation lengths corresponding to scale-invariant distribution of correlation times (Matthaeus and Goldstein, 1986). In interval A the break in the spectrum moved to lower frequencies with increasing distance from the sun, consistent with the growth of turbulence at larger scales with increasing distance from the sun.

The other possible explanation for the extension of the turbulent spectrum to larger scales is that the source of the turbulence has a larger scale at larger distances from the sun. This implies that turbulence is produced locally. The stirring scale is possibly the size of the interaction regions. Chapters 7 and 8 show that the size of the interaction regions increases with distance from the sun. Given the assumptions above, this implies that the turbulence extends to larger scales as the interaction regions grow in size. These ideas are speculative, but they suggest the value of developing models of turbulence that consider the role of MIRs in its production.

9.2.2 Differences Between Systems of Transient Flows and Corotating Flows

Using the methods described above, Goldstein et al. (1984) searched for signatures in the power spectra of large-scale magnetic field fluctuations that might allow one to distinguish between systems of transient flows and systems of corotating flows. Their work was motivated by the observation that the long-term decreases in the cosmic ray intensity tend to occur in conjunction with systems of transient flows, whereas plateaus in the cosmic ray intensity are associated with systems of corotating flows (Burlaga et al., 1984b).

At 1 AU, the correlation lengths in systems of transient flows tend to be smaller than those in corotating flows. The smaller correlation lengths for transient streams indicate that they are smaller, more complex, and less coherent than the corotating streams, consistent with the visual impression that one obtains from the time series. On the other hand, the magnetic helicity length scales in transient flow systems at 1 AU tend to be large (comparable to the scale length of the flow), whereas the magnetic helicity length scales in corotating flow systems tend to be small (much smaller than the correlation length of the flow). Goldstein et al. (1984) conclude that in transient flow systems the predominant scale of the helicity structure is part of the stream structure (e.g., owing to helical magnetic field lines as in magnetic clouds), whereas in corotating streams the magnetic helicity is associated with small-scale Alfvénic fluctuations.

At larger distances from the sun the correlation lengths for the transient flow systems and the corotating flow systems are nearly equal, owing to an extension of the turbulence in transient flows to larger scales as discussed above. At large distances the magnetic helicity scale length for transient flow systems is comparable to that in corotating flow systems. This equality is the result of an increase in the magnetic helicity scale length in corotating flow systems, possibly caused by stirring in the interaction regions as discussed above. The primary difference between the transient and corotating flow systems at large distances from the sun is in the spectrum of the magnetic field strength fluctuations (e.g., see Fig. 9.3). This effect is a manifestation of the formation of merged interaction regions, which dominate the spectra at large distances from the sun.

9.2.3 Spectra in the Outer Heliosphere

The spectra of large-scale magnetic field strength fluctuations with periods from a several hours to at least 6.5 days was studied by Burlaga and Mish (1987) as a function of distance from the sun for three 1-year intervals as Voyager moved from 1 AU to 6 AU. They found power law spectra $f^{-\alpha}$ with an exponent $\alpha = 2.0 \pm 0.05$. They suggested that the f^{-2} spectra were produced by steplike changes in the magnetic field corresponding to shocks and the boundaries of interaction regions. By identifying the jumps in the time series, Roberts and Goldstein (1987) confirmed that the f^{-2} spectra observed by Burlaga and Mish were produced by jumps in the magnetic field, but they did not determine the physical nature of the jumps.

The spectra of the large-scale bulk speed fluctuations were also studied by Burlaga and Mish (1987) as a function of distance from the sun (Fig. 9.4). They again found power law spectra, but the exponents were somewhat larger than 2, ranging from 2.1 to 2.3, indicating some tendency for persistence.

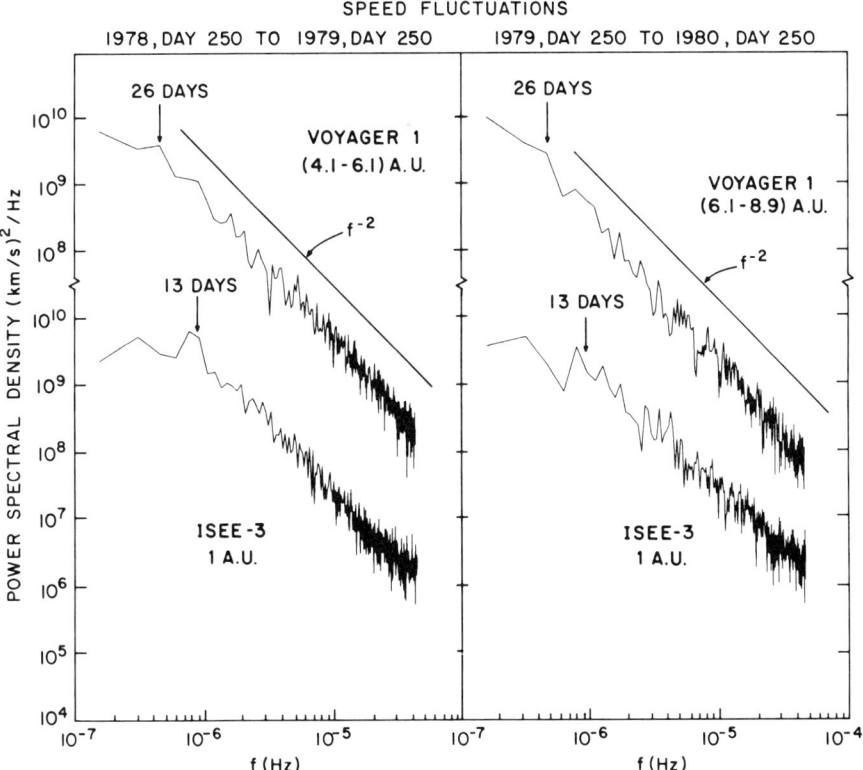

Fig. 9.4. Spectra of large-scale speed fluctuations. (L.F. Burlaga and W.H. Mish, *J. Geophys. Res.*, **92,** 1261, 1987, published by the American Geophysical Union.)

An f^{-2} spectrum can be produced by either a single large step function or many smaller steps. One cannot distinguish between these two possibilities on the basis of the spectra alone. A different approach, based on fractals (Mandelbrot, 1982), allows one to make this distinction. The method is described in detail by Burlaga and Klein (1986b), who used it to study magnetic turbulence. Briefly, the idea is to think of the magnetic field profile $B(t)$ as a curve and compute the length of the curve for various averages B_T for averaging intervals T. Strictly speaking, one is dealing with a self-affine fractal rather than a fractal curve in two dimensions, but the approach is essentially the same in the two cases. The "length" of the curve $B_T(t)$ is computed from the equation

$$L(T) = \sum \frac{|B_T(t_k + T) - B_T(t_k)|}{B_{pT}(t_k)} \qquad (9.1)$$

where the sum extends from $k = 1$ to N_T. If the fluctuations are self-affine, then $L(T) = L_0 T^{-S}$ and a plot of $\log L$ versus $\log T$ should be a straight line with slope $-S$. The slope $-S$ is related to the power spectral exponent α by $\alpha = 3 - 2S$. A plot of $\log L$ versus $\log T$ for the magnetic field strength is shown on the left of Fig. 9.5; the corresponding results for the speed data are shown on the right. In each plot there is an interval over which the points fall on a straight line, indicating self-affine behavior, that is, fluctuations with similar characteristics on many different scales. The spectral exponents deduced from the slopes of these lines are in good agreement with the spectral exponents derived from the time series by Burlaga and Mish (1987).

During 1984, recurrent CMIRs were observed by Voyager 2 between 13.2 AU and 15.9 AU near the heliographic equator and by Voyager 1 between 18.4 AU and 21.9 AU from 21.8°N to 24.8°N. During 1985, MIRs were weak or absent at Voyager 2 between 15.9 AU and 18.8 AU and at Voyager 1 between 21.9 AU and 25.4 AU from 24.8°N to 26.8°N. These observations were discussed in Section 8.3. The spectra of the temporal fluctuations of the time series in Fig. 8.8 are shown in Fig. 9.6. In the spectrum for the quasi-periodic CMIRs observed by Voyager 2 during 1984 (top left panel), a sharp peak at 26 days reveals that the CMIRs tended to recur at the solar rotation period near the ecliptic beyond ≈ 13 AU. The spectra obtained at a latitude of 25° by Voyager 1 during 1984 also show a clear peak at 26 days, but its amplitude is smaller than that of the corresponding peak observed by Voyager 2. During 1985, only a very weak peak or cutoff in the power spectral density near 26 days was observed by Voyager 1 and 2, consistent with the absence of strong CMIRs during that year.

During 1984, a broad f^{-2} spectrum was present in the low frequency range, from a period of 26 days to approximately 5 days, in both the

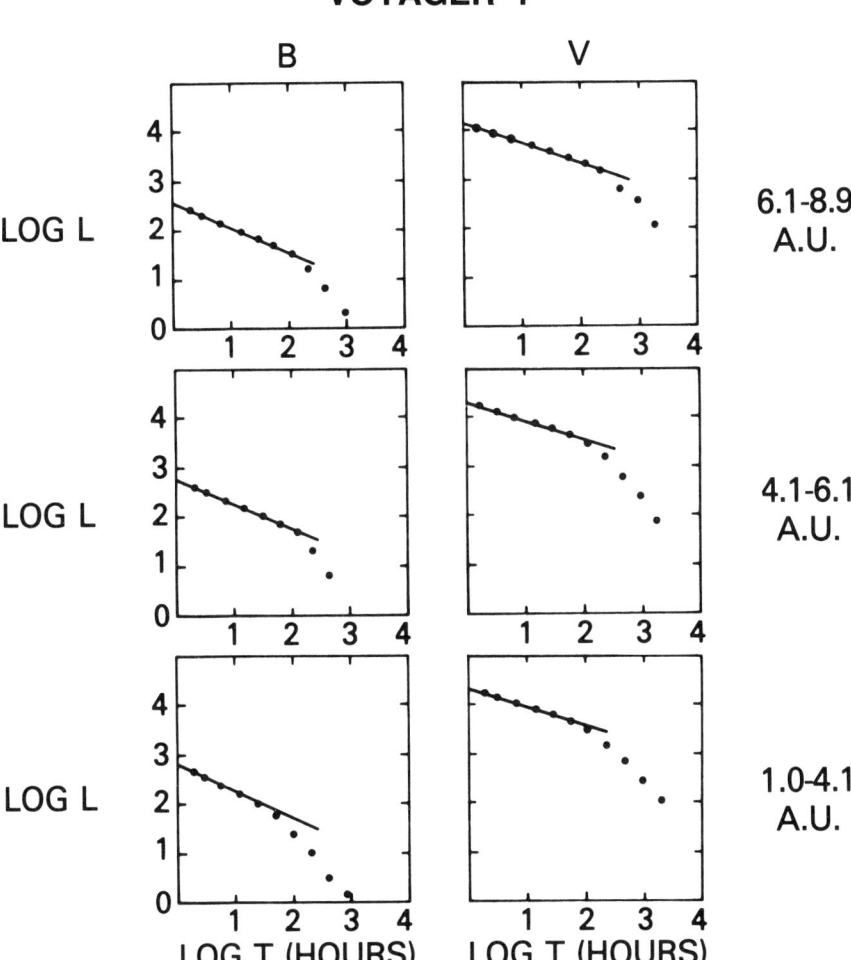

Fig. 9.5. Fractal structure of large-scale speed and field fluctuations at 1 AU. (L.F. Burlaga and W.H. Mish, *J. Geophys. Res.*, **92**, 1261, 1987, published by the American Geophysical Union.)

Voyager 1 and Voyager 2 data. This spectrum suggests the dominance of shocks and jumps in the magnetic field strength related to the CMIRs in the low frequency large-scale fluctuations. At Voyager 2 the f^{-2} spectrum also extended through the intermediate frequency fluctuations, from 5 days to several hours. At Voyager 1, however, the intermediate scale fluctuations had an $f^{-5/3}$ spectrum, indicating that the turbulence extended down to the intermediate frequency range at 22–25 AU, but the f^{-2} spectrum did not extend up to these frequencies. The power level was higher at Voyager 2 than at Voyager 1, probably because CMIRs are generally stronger near

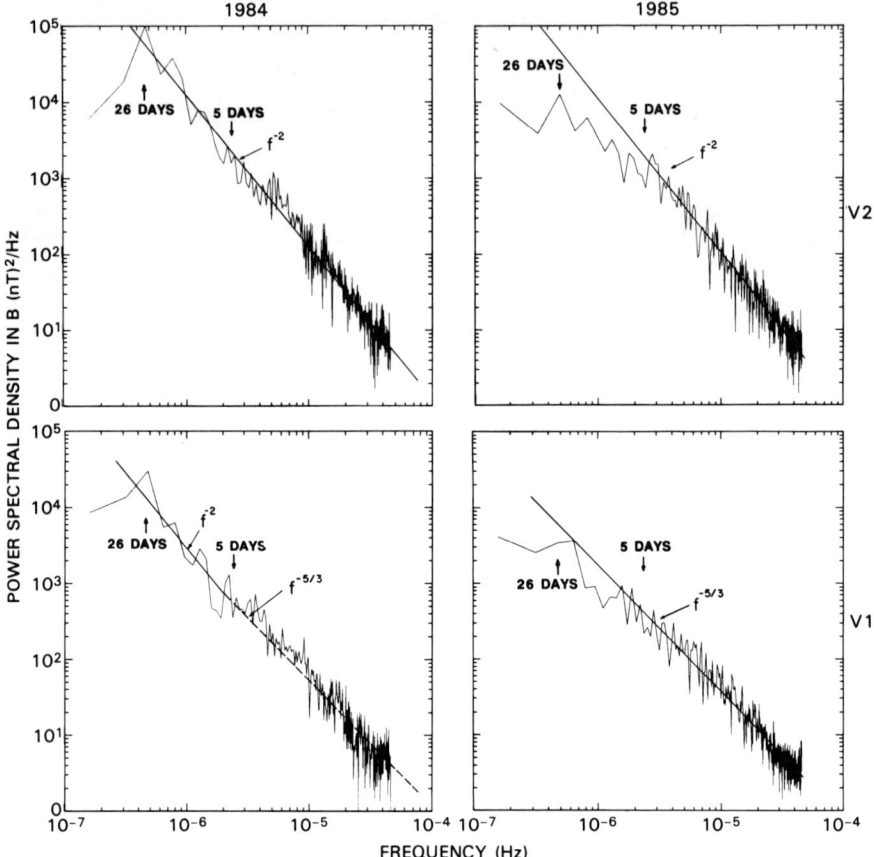

Fig. 9.6. Spectra of large-scale magnetic field strength fluctuations observed by Voyagers 1 and 2 during 1984 and 1985. (L.F. Burlaga, N.F. Ness, and F.B. McDonald, *J. Geophys. Res.*, **92**, 13647, 1987b, copyright by the American Geophysical Union.)

the equator (where the effects of solar rotation are greatest) than at higher latitudes. It is possible that the turbulence did extend into the intermediate frequency range at Voyager 2 during 1984 but was hidden by the greater power produced by the shocks and jumps in B associated with the CMIRs.

During 1985, the f^{-2} spectrum was not observed in the low frequency fluctuations by either Voyager 1 or Voyager 2 (see right-hand panels of Fig. 9.6). This indicates an absence of shocks and jumps in B, which can be attributed to the absence of large MIRs during 1985. On the other hand, the $f^{-5/3}$ spectrum extended throughout the intermediate frequency range of the large-scale fluctuations, down to periods of 5 days for both Voyager 1 and Voyager 2. This shows once again that the turbulence extends to lower frequencies and larger scales at larger distances from the sun. In this case, large MIRs were not present, suggesting that an inverse cascade rather than stirring by MIRs was responsible for the turbulence at low frequencies observed by Voyagers 1 and 2 during 1985.

9.2.4 Solar Cycle Variations of Large-Scale Fluctuations at 1 AU

The properties of the large-scale fluctuations at 1 AU from 1978 to 1982 were investigated by Burlaga et al. (1989). In the intermediate frequency range (3×10^{-6} Hz $< f < 3 \times 10^{-5}$ Hz) the spectra of both the magnetic field strength fluctuations and the bulk speed fluctuations had the form f^{-b}. For the magnetic field strength fluctuations, the average of the exponents b for each of the years is $\langle b_B \rangle = 1.92 \pm 0.06$, and for the bulk speed fluctuations $\langle b_V \rangle = 1.92 \pm 0.10$. There was essentially no temporal variation in the spectral exponent of either the magnetic field strength fluctuations or the bulk speed fluctuations in the intermediate frequency range from 1978 to 1982.

The spectral exponents just described are close to 2. The absence of a turbulent $f^{-5/3}$ spectrum in the intermediate frequency range at 1 AU is to be expected, but it is surprising that the spectrum is so close to f^{-2}. A detailed analysis of the spectra and time series for V and B show that the $f^{-1.92}$ spectra were the result of the superposition of two components (Burlaga et al., 1989). One component, due to a jump-ramp approximation to the time series, falls off faster than f^{-2} and is dominant at low frequencies. The other component, possibly related to turbulence and Alfvénic fluctuations, falls off more slowly than f^{-2} and is dominant at high frequencies.

9.3 Multifractal Fluctuations

9.3.1 Introduction to Multifractals

Spectra and self-affine fractals provide a fairly complete description of fluctuations that are homogeneous, symmetric about a mean, and nearly Gaussian on various scales. Such fluctuations relative to a mean are described by variances that are scale-dependent. However, spectra are not sufficient to describe fluctuations that are non-Gaussian. For example, fluctuations that are intermittent, bursty, spiky, nonhomogeneous, and asymmetric about the mean cannot be fully characterized by spectra and second moments. Multifractal time series have these properties, yet they possess a symmetry that gives a description of the time series in terms of a simple nonlinear (polynomial) curve. The set of coefficients describing this curve, "the multifractal spectrum" is a generalization of the single number (power law exponent) that describes either a power law spectrum or a fractal.

The literature on multifractals is very extensive. The basic theoretical results are summarized in several key papers and excellent reviews (Mandelbrot, 1972, 1989; Paladin and Vulpiani, 1987; Stanley and Meakin, 1988; Tel, 1988; Schertzer and Lovejoy, 1990). The essential concepts, which are relatively simple, are described briefly in paragraphs that follow. However, a number of technical issues should be considered, and many theoretical models for generating and explaining multifractals are available. These are discussed in the reviews but are not considered in depth here. The analysis for the velocity fluctuations differs from that for the fluctuations of a field involving some conserved quantity (mass, magnetic flux, energy, etc). These two cases are discussed separately in Section 9.4 and 9.5, respectively.

9.3.2 Multifractal Fluctuations in the Interplanetary Medium

The existence of the multifractal structure of large-scale magnetic field strength fluctuations in the outer heliosphere was established by Burlaga (1991a) and Burlaga et al. (1991a, 1993b). Subsequent studies, reviewed below, have shown that multifractal fluctuations are observed in large-scale fluctuations of other parameters and at other distances. In fact, multifractal structure appears to be a common feature of large-scale fluctuations throughout the interplanetary medium.

The multifractal character of small-scale (0.85–13.6 hours) velocity fluctuations was identified in the Voyager data by Burlaga (1991b), who showed that it was related to the presence of intermittent turbulence. Multifractal magnetic field and velocity fluctuations were also identified in the high frequency fluctuations using higher resolution Helios data by Marsch and Liu (1993), although not always of the form expected for intermittent turbulence. High frequency fluctuations are not a topic of this book. However, intermittent turbulence is observed in large-scale fluctuations associated with turbulent flow systems at 1 AU, and these results are discussed in the following section. The material in the following sections is not in the order that it appeared in the literature, but the organization chosen should provide a clearer understanding of multifractals in the heliosphere than a chronological approach.

9.4 Intermittent Turbulence and Multifractal Velocity Fluctuations

9.4.1 Scaling Laws for Multifractal Velocity Fluctuations

Multifractals arise naturally in the subject of intermittent turbulence (Mandelbrot, 1974; Paladin and Vulpiani, 1987; Meneveau and Sreenivasan, 1991; Sreenivasan, 1991). To motivate the basic equation for multifractal velocity fluctuations, consider a simple conceptual model of homogeneous, nonintermittent, nonmultifractal (Kolmogorov) turbulence following the approach of Frisch et al. (1978). The basic component of turbulence is an eddy, and an eddy has the following basic properties: size ℓ_n, speed v_n (equal to the absolute value of the difference in speed across the eddy $|\Delta V(\ell_n)|$), lifetime $t_n = \ell_n/v_n$, mass m_n, and energy $e_n \sim m_n v_n^2$.

Following Richardson (1922), assume that a large eddy gives birth to several smaller eddies and then dies ("big whorls have little whorls..."), with a resulting energy transfer from the scale of the eddy to eddies with a smaller scale at a rate e_n/t_n. The process can also be viewed as a fragmentation process. Assume that an eddy of size ℓ_n breaks into N eddies of size $\ell_n/2$. If the eddies fill space, the number of offspring of a given eddy is $N = 2^3$. Imagine a cube breaking into eight identical smaller cubes. If the largest eddy has size ℓ_0, then $\ell_n = \ell_0/2^n$, and the number eddies of size l_n is $N_n = N^n = 2^{3n} = (\ell_0/\ell_n)^3$. The result is a multiplicative process like a family tree.

Assume that the energy of an eddy is distributed equally among each of its offspring and that the energy transfer rate is the same on every scale. The total energy in the eddies of size ℓ_n is $E_n \approx N_n m_n v_n^2 \approx (\ell_0/\ell_n)^3 m_n v_n^2$. The energy transfer rate is $\epsilon_n = e_n/t_n \sim m_n v_n^2 (v_n/\ell_n) = m_n v_n^3/\ell_n$, which is assumed to be constant in the (nonmultifractal) Kolmogorov model. This equation implies that $v_n \sim \ell_n^{1/3}$, so that one should observe $|\Delta V(\ell_n)| \sim \ell_n^{1/3}$ for Kolmogorov turbulence.

In a turbulent medium, v_n is a fluctuating, non-Gaussian quantity, so that one is led to consider the average of each of its moments $\langle v_n^q \rangle = \langle |\Delta V(\ell_n)|^q \rangle$, where q is any positive real number. The function $\langle |\Delta V(\ell_n)|^q \rangle$ is called "the qth order velocity structure function" (Monin and Yaglom, 1975). For Kolmogorov turbulence $\langle |\Delta V(\ell_n)|^q \rangle \sim \ell_n^{s(q)}$, where $s(q) = q/3$, a linear relation between s and q.

The generalization to multifractals is simple: one allows $s(q)$ to be a nonlinear function of q. Thus, we arrive at the important result that for multifractals

$$\langle |\Delta V(\ell_n)|^q \rangle \sim \ell_n^{s(q)} \tag{9.2}$$

In other words, for multifractal velocity fluctuations the qth-order velocity structure function varies with scale as a power law, but the exponent is a nonlinear function of the moment q. For a fractal there is one scaling law, described by a single straight line that is characterized by its slope corresponding to a power law spectral exponent. For a multifractal, one has a family of scaling laws characterized by an infinite number of straight lines whose slopes give a simple nonlinear curve $s(q)$.

In the solar wind the spacecraft is essentially fixed and the plasma is convected past it at the solar wind speed U, so that $\ell_n = U\tau_n$. Then $v_n = |\Delta V(\ell_n)| = |V(x + \ell_n) - V(x)| = |V(t + \tau_n) - V(t)| = |\Delta V(\tau_n)|$ and

$$\langle |\Delta V(\tau_n)|^q \rangle \sim \tau_n^{s(q)} \tag{9.3}$$

Plots of $\log(\langle|\Delta V(\tau_n)|^q\rangle)$ versus $\log(\tau_n)$ for various values of q give a family of straight lines with slopes $s(q)$ over some range of τ_n if the velocity fluctuations are multifractal.

The exponent $s(2)$ is related to the exponent α of the power law spectrum $f^{-\alpha}$ by the simple equation

$$\alpha \approx 1 + s(2) \tag{9.4}$$

For homogeneous, isotropic, space-filling turbulence with a constant energy transfer rate ϵ_d, the result given above for Kolmogorov turbulence ($s(q) = q/3$) implies $s(2) = 2/3$ and equation (9.4) gives $\alpha = 5/3$, the famous "5/3 law" of Kolmogorov (1941), usually referred to as K41.

Kolmogorov (1962) generalized his K41 results by considering a lognormal distribution of ϵ_n, obtaining the quadratic relation

$$s(q) = \frac{q}{3} + \frac{q(3-q)\mu}{18} \tag{9.5}$$

where μ is a constant called the intermittency exponent. Since $s(q)$ is a nonlinear function of q, the log-normal model gives a multifractal model of intermittent turbulence in accordance with equation (9.3). Mandelbrot (1972, 1974) points out that the log-normal model has certain technical difficulties. Nevertheless, equation (9.5) provides a good approximation to intermittent turbulence in the laboratory and in the atmosphere for $q < 4$, with values of μ in the range from 0.2 (Anselmet et al., 1984) to 0.6 (Gibson et al., 1970). The best estimate of μ is 0.025 ± 0.05 (Sreenivasan and Kailasnath, 1993). In the multifractal description of observations, μ is obtained from the equation

$$\mu = 2 - s(6) \tag{9.6}$$

9.4.2 Intermittent Turbulence in Large-Scale Velocity Fluctuations: Observations

Multifractal large-scale fluctuations in the solar wind speed at 1 AU were observed during 1979, approaching solar maximum (Fig. 9.7). The solar wind speed fluctuations were irregular, aperiodic, of varying widths, and of varying amplitudes. To isolate eddies on a convection time scale of 8 hours corresponding to a scale of ≈ 0.08 AU, 8-hour averages of the solar wind speed were computed. The differences of successive 8-hour averages ΔV_8 are plotted in the second panel of Fig. 9.7, representing the fluctuations in speed across the eddies on that scale. Similarly, the velocity fluctuations across

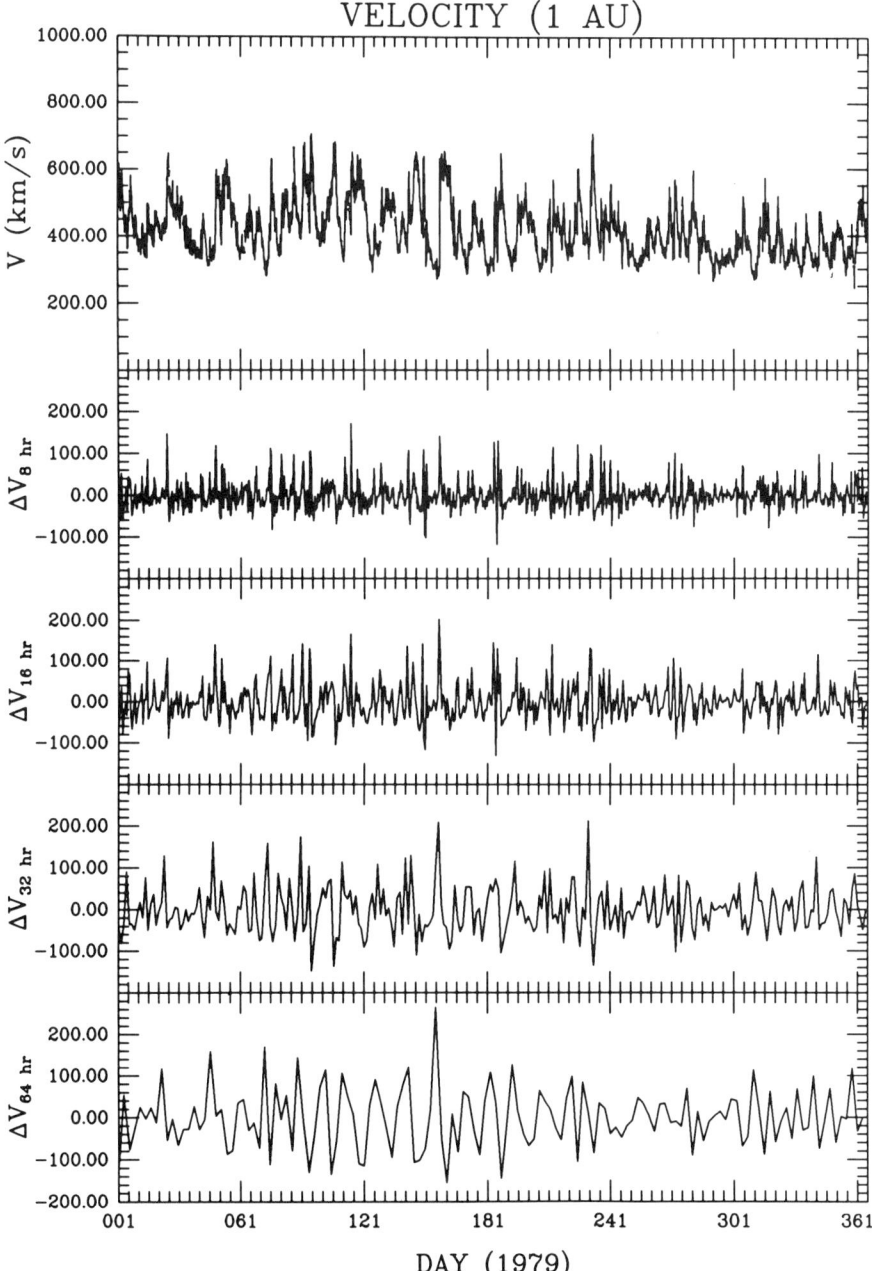

Fig. 9.7. Multifractal speed fluctuations in intermittent turbulence. (L.F. Burlaga, *J. Geophys. Res.*, **98,** 17467, 1993, published by the American Geophysical Union.)

larger eddies were computed from 16-, 32-, and 64-hour averages of the speed. These are also plotted in Fig. 9.7.

Several important qualitative features of the fluctuations in Fig. 9.7 are apparent. The fluctuations are bursty, spiky, and inhomogeneous. They are also asymmetric, the magnitude of the positive increments being greater on average than the magnitude of the negative increments in speed. Finally, the fluctuations are similar (self-affine) on different scales; the time series for the larger scales are magnified images of a part of the time series for the velocity increments on smaller scales. All these features are qualitative properties of multifractal fluctuations.

To demonstrate that the fluctuations are multifractal, Burlaga (1993) plotted $\log(\langle|\Delta V(\tau_n)|^q\rangle)$ versus $\log(\tau_n)$ for various moments q and for various averaging intervals (scales) τ_n (Fig. 9.8). The points for any given value of $q (1 < q \leq 14)$ fall on a straight line for 8 hours $< \tau_n < 64$ hours. The slope of each line and the error in that slope was computed by a least squares fit. The resulting values of $s(q)$ can be described by the following nonlinear (quadratic) equation:

$$s(q) = -(0.0024 \pm 0.0031)q^2 + (0.2676 \pm 0.0054)q + (0.195 \pm 0.078) \quad (9.7)$$

Thus, the complex time series for $V(t)$ at the top of Fig. 9.7 has been reduced to a simple polynomial. Substituting $q = 2$ in equation (9.7) gives $s(2) = 0.73 \pm 0.08$ for the 1979 large-scale velocity fluctuations. Equation (9.4) gives $\alpha = 1.73 \pm 0.08$, consistent with $\alpha = 5/3 = 1.67$, the spectral exponent of Kolmogorov turbulence. Substituting $q = 6$ into equation (9.7) gives $s(6) = 1.71 \pm 0.15$, and equation (9.6) gives $\mu = 0.29 \pm 0.02$. This result, derived from the large-scale interplanetary velocity fluctuations during 1979, is in excellent agreement with the best estimate of μ derived from Earth-based observations of intermittent turbulence, $\mu = 0.25 \pm 0.05$ (Sreenivasan and Kailasnath, 1993).

9.4.3 Intermittent Turbulence: Theory

In view of the excellent agreement between the foregoing results for the large-scale interplanetary velocity fluctuations and the laboratory measurements of α and μ, it is natural to ask whether the models of intermittent turbulence that describe Earth-based observations might also describe the interplanetary velocity fluctuations. Borgas (1992) concluded that the most satisfactory model of intermittent turbulence in Earth-based flows is the multifractal model of Meneveau and Sreenivasan (1987a,b; 1991). In that model

$$s(q) = 1 + \left(\frac{q}{3} - 1\right)D_{q/3} \quad (9.8)$$

(Sreenivasan, 1991, equation 6.1), and $D_{q/3}$ is derived from a binomial

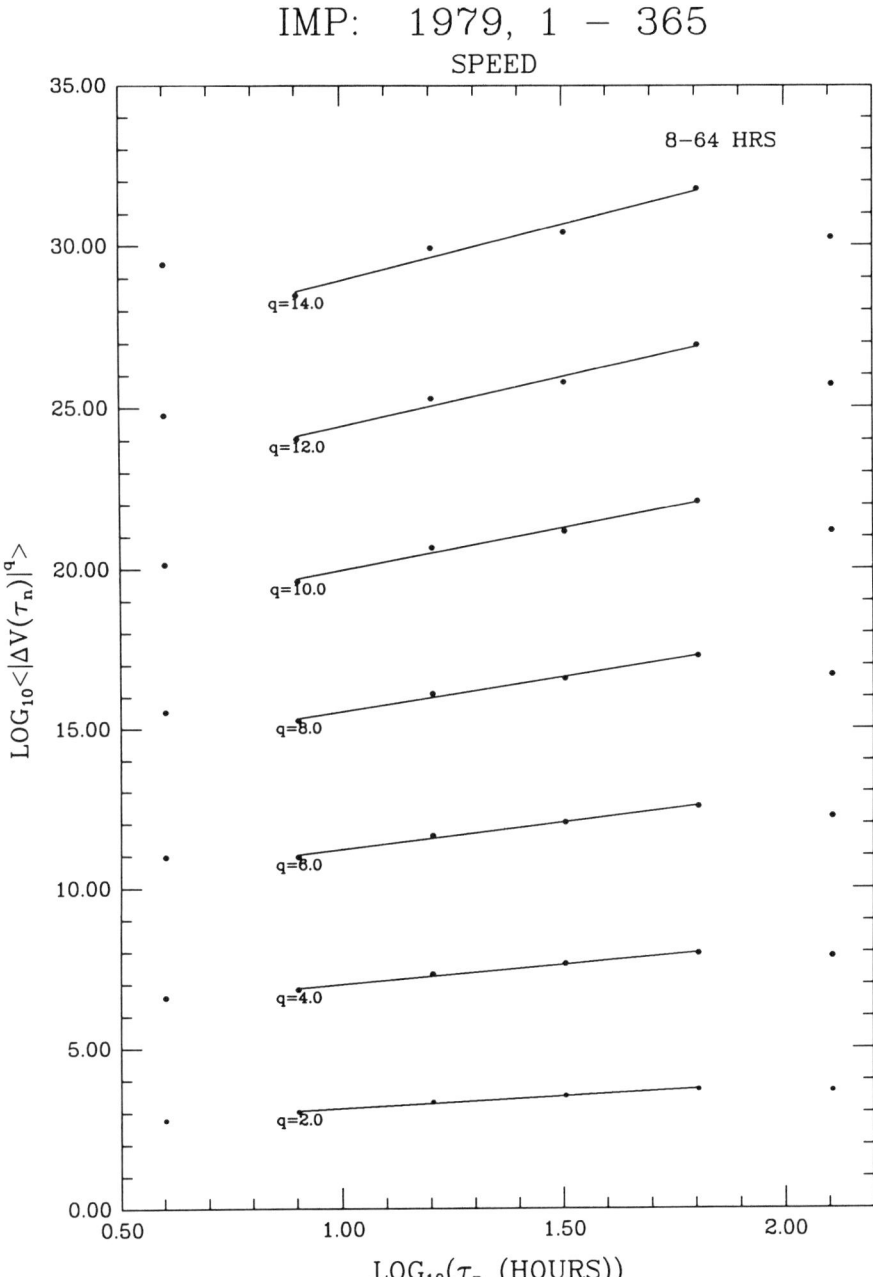

Fig. 9.8. A series of lines demonstrating multifractal scaling of large-scale speed fluctuations in turbulence at 1 AU. (L.F. Burlaga, *J. Geophys. Res.*, **98,** 17467, 1993, published by the American Geophysical Union.)

cascade model as described by Meneveau and Sreenivasan (1987a, equation 3)

$$D_{q'} = \log_2[(p^{q'} + (1-p)^{q'})]^{1/(1-q')} \quad (9.9)$$

From their laboratory data Meneveau and Sreenivasan found $p = 0.7$ and $\mu = 0.25$.

The curves $s(q)$ from equations (9.8) and (9.9) with $p = 0.7$ is plotted as the solid curve in Fig. 9.9. The binomial cascade model of intermittent turbulence with the value $p = 0.7$ derived from laboratory experiments is in excellent agreement with the results derived from the large-scale interplanetary velocity fluctuations with periods from 8 hours to 2.7 days during

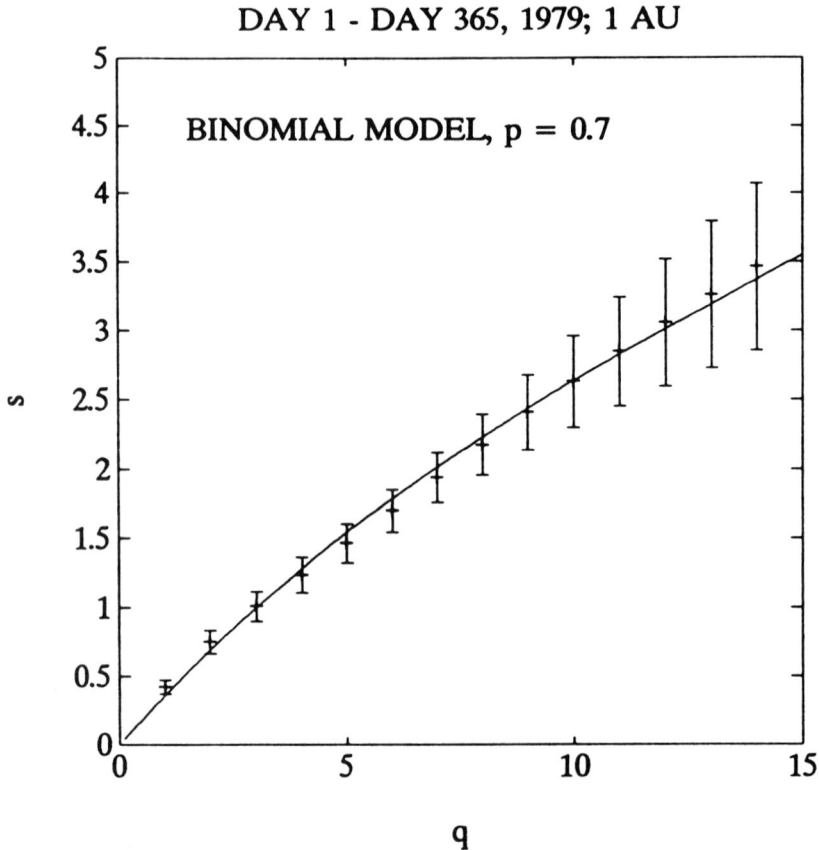

Fig. 9.9. The slope as a function of moment for the multifractal large-scale speed fluctuations in turbulent flows at 1 AU. (L.F. Burlaga, *J. Geophys. Res.*, **98**, 17467, 1993, published by the American Geophysical Union.)

1979. It is remarkable that a model derived to describe intermittent gas dynamic turbulence in a laboratory also describes MHD fluctuations on an astronomical scale in the interplanetary medium. Clearly some universal scale-independent processes are involved. The processes must be basically geometrical and kinematic, since the magnetic field and specific forces do not enter the binomial cascade model. These processes must be included in any dynamical MHD model of intermittent turbulence.

9.4.4 Multifractal Fluctuations in Corotating Streams

The results above show that the large-scale fluctuations in the bulk speed observed at 1 AU near solar maximum have the characteristics of multifractal fluctuations. The speed profile near solar maximum is generally very complex. This section examines the large-scale fluctuations in velocity during the declining phase of the solar cycle, when the quasi-stationary corotating streams are observed. It will be shown that even these simple quasi-periodic flows are associated with multifractal speed fluctuations.

The solar wind streams measured at 1 AU by the IMP spacecraft during 1974 (top panel of Fig. 8.10) evolved to compound streams near 6 AU (see Pioneer 10 data in Fig. 8.10), which led to period doubling of the interaction regions at larger distances from the sun, as discussed in Section 8.3.2. Profiles of the speed differences for various averages of the time series $V(t)$ have the same characteristics as the multifractal large-scale speed fluctuations observed during 1979 and discussed in Section 9.4.2. The fluctuations are bursty (intermittent), spiky, inhomogeneous, self-affine on different scales, and asymmetric. The bursts are associated with the regions of increasing speed, where interaction regions are observed (Burlaga, 1992a).

If the fluctuations in speed are multifractal over some range, then $\langle |\Delta V(\tau_n)|^q \rangle \sim \tau_n^{s(q)}$ [see equation (9.3) above] and plots of $\log \langle |\Delta V(\tau_n)|^q \rangle$ versus $\log(\tau_n)$ for various values of q should be linear over some range of τ_n. Such linear relations were found in the IMP data at 1 AU in the range of τ_n from 4 to 32 hours for q from 1 AU in the range of τ_n from 4 to 32 hours for q from 1 to at least 14. Linear relations were also found for the corresponding Pioneer 10 data near 6 AU in the range of τ_n from 2 to 16 hours for q from 1 to at least 14 (Burlaga, 1992b). The multifractal behavior is observed from the intermediate frequency fluctuations to the high frequency fluctuations. The corresponding slopes $s(q)$ and error bars are shown in Fig. 9.10 for both the IMP data (circles) and the Pioneer 10 data (squares). The relation $s(q)$ at 6 AU is very different from that at 1 AU, indicating significant evolution of the fluctuations from 1 AU to 6 AU, while maintaining the multifractal symmetry.

Consider the points $s(q)$ for the IMP data in Fig. 9.10. The points fall on a straight line parallel to that for the nonmultifractal model of K41 for $q > 6$ or so. However, the straight line does not extend to the origin for $q = 0$ as required by K41. Instead the curve $s(q)$ passes through $s(2) = 1.18$, corresponding to a power law spectrum $f^{-2.18}$ according to equation (9.4).

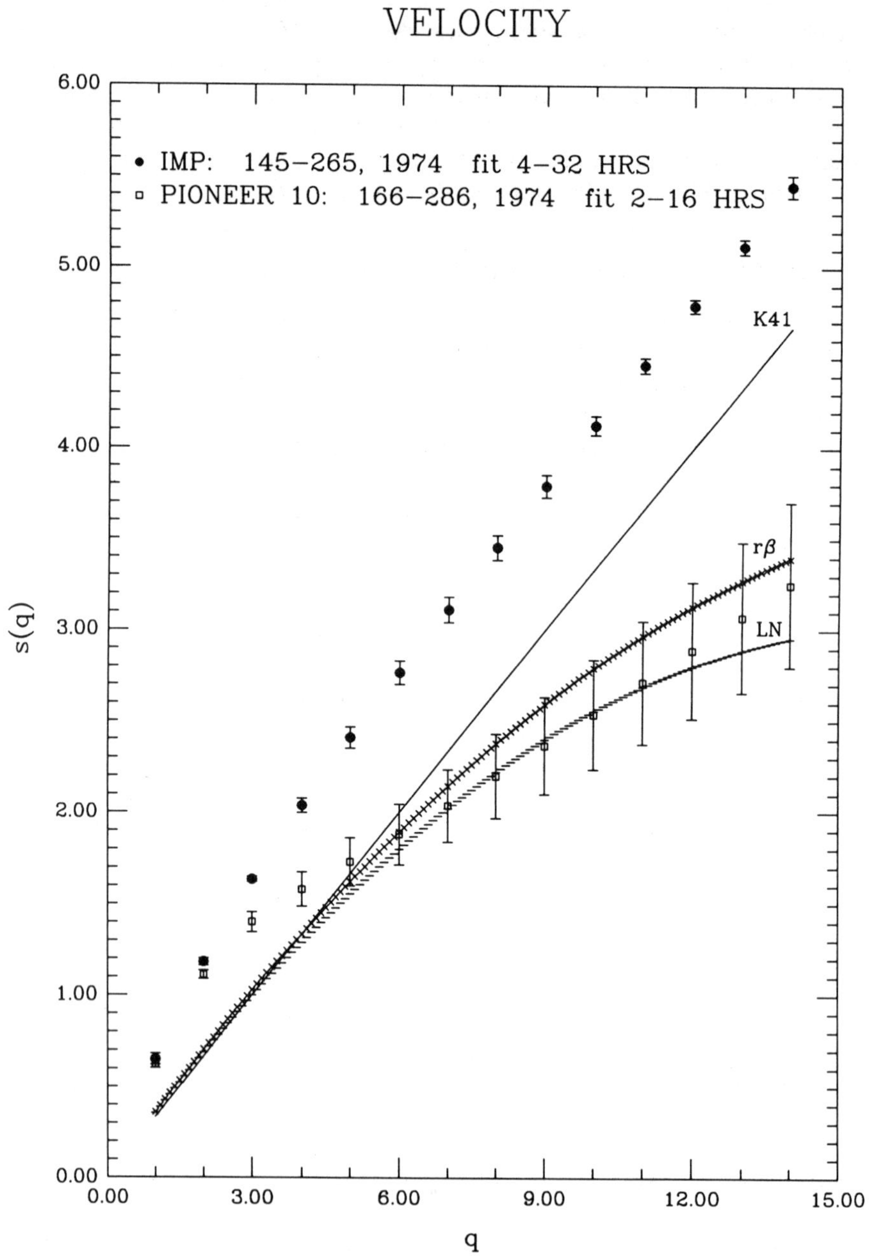

VELOCITY

Fig. 9.10. The slope as a function of moment number for the multifractal large-scale speed fluctuations in corotating flows at 1 AU. (L.F. Burlaga, *J. Geophys. Res.* **18**, 1651, 1992a, published by the American Geophysical Union.)

This exponent is consistent with the results of Burlaga et al. (1989) discussed in Section 9.2.4. Burlaga and coworkers derived this spectrum directly both by spectral analysis and by fractal analysis of the data in question. They showed that the $f^{-2.18}$ spectrum is the result of a superposition of a dominant contribution from turbulence at the shorter periods and a dominant contribution from jumps in the speed profile at larger periods. Thus, the jumps in speed on a wide range of scales are largely responsible for the deviations from a K41 law at $q \leq 4$.

The observations of $s(q)$ for the compound corotating streams observed by Pioneer 10 at 6 AU (Fig. 9.10) are very different from the results observed by IMP at 1 AU. In this case $s(q=2) = 1.1$, corresponding to a spectrum $f^{-2.1}$, consistent with the result of Burlaga and Mish (1987) for fluctuations at 6.1–8.9 AU, which they attributed to shocks and jumps in speed. However, for larger values of q the values of $s(q)$ approach the theoretical curves for intermittent turbulence derived from the log-normal model of Kolmogorov (1962) (curve LN) and from the random beta model of Paladin and Vulpiani (1987) (curve $r\beta$). Thus, the larger moments of the speed fluctuations observed by Pioneer 10 at 6 AU in the range of relatively short-period fluctuations from 2 to 16 hours behave like intermittent turbulence.

In summary, the fluctuations in the speed profiles in corotating streams differ at 1 AU and 6 AU. At both distances there is a strong contribution from jumps in the speed with a broad range of sizes, tending to make an f^{-2} spectrum or a steeper spectrum. There is also a contribution from turbulence at both distances, but the turbulence is more intermittent and extends to lower frequencies at the larger distance from the sun.

9.4.5 Speed Distribution Functions

The distribution of differences in the hour averages of the solar wind speed is of basic importance. Although there are numerous reports on the distributions of the speeds themselves, only a few observations of the distributions of the differences of speeds have been published. The exponential distribution of the speed differences for $\Delta V > 0$ and for $\Delta V < 0$ found by Burlaga and Ogilvie (1970a) were discussed in Section 9.1.2 for a period near solar maximum. The e-folding parameter was smaller for $\Delta V > 0$ than for $\Delta V < 0$, indicating an asymmetry favoring large positive increments in the speed, corresponding to the steepening of streams.

Near the following solar maximum, during 1979, there were again transient flows and intermittent turbulence, as discussed in Section 9.4.2. Again, the distribution of speed differences consists of two exponential distributions. Linear least squares fits give $N(V) \sim \exp[0.042 \, \Delta V]$ for $\Delta V < 0$, and $N(V) \sim \exp[-0.050 \, \Delta V]$ for $\Delta V > 0$. The distributions are non-Gaussian and asymmetric, with increases in V tending to be larger than decreases in V from one hour to the next. During the declining phase of solar activity in 1974, corotating flows were dominant, but the distribution of speed differences again consisted of two exponential distributions. Linear least squares fits to the points gives the following distributions:

$$N(V) \sim \exp[0.037\,\Delta V] \quad \text{for } \Delta V < 0$$
$$N(V) \sim \exp[-0.043\,\Delta V] \quad \text{for } \Delta V > 0$$

These distributions for the multifractal fluctuations in corotating streams have the same form as those for the multifractal fluctuations in transient flows, even though the latter can be described as intermittent turbulence whereas the former cannot be so described.

With the limited data set available, it is difficult to distinguish among exponential distributions, "stretched-exponential distributions," and other distributions with an exponential-like tail. The important point is that the distributions are not Gaussian; they have large exponential-like tails. This is a characteristic of both intermittent turbulence and other multifractal fluctuations.

9.5 Multifractal Magnetic Field Strength Fluctuations

9.5.1 Basic Scaling Laws for Multifractals

Assume that a stationary magnetic field is in the equatorial plane and normal to the radial direction, and assume that it is convected past a spacecraft with the solar wind speed V. Let $\tau_n = 2^n$ hours, where $n = 0, 1, 2, \ldots$ The magnetic flux that moves past the spacecraft during a τ_n-hour interval at time t_i is proportional to $(V\tau_n h)B_n(t_i)$, where $B_n(t_i)$ is the average of the magnetic field during the interval τ_n and h is a length in the direction normal to the equatorial plane. Normalizing the data so that $\langle B_n(t_i) \rangle = 1$, one sees that the total flux moving past the spacecraft during some period of observations is proportional to (NVh), in which N is the number of hour averages during the period. To each interval $\tau_n(t_i)$ one can associate a probability

$$p_n(t_i) = B_n(t_i)\frac{\tau_n}{N} \qquad (9.10)$$

Given a normalized time series $B(t_i)$, one can compute a set of probabilities $p_n(t_i)$ for the τ_n-hour averages of $B(t_i)$. By choosing different values of n ($n = 0, 1, 2, \ldots$), one obtains several sets of probabilities, one set for each value of n. The time series of the magnetic field is described by the moments of these sets of probabilities as follows.

Consider the qth moment of p_n, $M_n(q)$, defined by

$$M_n(q) := \sum \{n(p_n)p_n^q\} \qquad (9.11)$$

where $n(p_n)$ is the number of intervals with probability p_n. The moments $M_n(q)$ can be computed from the averages $\langle p_n^q \rangle$:

$$M_n(q) = \sum \{n(p_n)p_n^q\} = N_n \sum \left\{\left[\frac{n(p_n)}{N_n}\right] p_n^q \right\} = N_n \langle p_n^q \rangle \qquad (9.12)$$

where $N_n = N/\tau_n$.

A basic property of a multifractal is that the moments $M_n(q)$ have the scaling symmetry:

$$M_n(q) = \left(\frac{\tau_n}{N}\right)^{\gamma(q)} \qquad (9.13)$$

(see, e.g, Paladin and Vulpiani, 1987; Stanley and Meakin, 1988; Tel, 1988). Since the first moment $q = 1$ is just the average of p_n, it is customary to write

$$\gamma(q) = (q-1)D_q \qquad (9.14)$$

(see Hentschel and Procaccia, 1983; Coniglio, 1986; Sreenivasan et al., 1989a,b).

Substituting equation (9.10) for p_n into equation (9.12) and equating (9.12) to (9.13) gives

$$\langle B_n^q \rangle \sim \left(\frac{\tau_n}{N}\right)^{\tau(q)} \qquad (9.15)$$

where

$$\tau(q) = \gamma(q) - q + 1 \qquad (9.16)$$

and q can be positive or negative. Positive values of q describe the scaling properties of the fluctuations with magnetic field strengths greater than average. Negative values of q describe the scaling properties of the fluctuations with field strengths less than average. If the large-scale fluctuations of the magnetic field strength have multifractal structure, then equation (9.15) should be satisfied. Using the various 2^n-hour averages of the magnetic field strength and a specific value of q, a plot of $\log \langle B_n^q \rangle$ versus $\log(\tau_n)$ should give a straight line with a slope $\tau(q)$. In principle this is true for any value of q, but in practice one is limited to values of $-20 \leq q \leq 20$. Thus, for multifractal fluctuations, the different values of q give a family of straight lines with slopes $\tau(q)$. The multifractal time series is then described by a simple curve $\tau(q)$, which is typically a low order polynomial determined by a few numbers, the coordinates of a point in a low-dimensional space.

Given $s(q)$ obtained from the data as discussed above, one can describe the multifractal time series by the function

$$D_q(q) = \frac{\tau(q)}{q-1} + 1 \qquad (9.17)$$

obtained from equations (9.14) and (9.16). From $D_q(q)$ one can compute the following two functions:

$$\alpha(q) = \frac{d}{dq}[(q-1)D_q(q)] \qquad (9.18)$$

and

$$f(q) = q\alpha(q) - (q-1)D_q(q) \qquad (9.19)$$

The geometrical significance of these quantities is discussed in many papers (e.g., Halsey et al., 1986; Pietronero and Siebesma, 1986; Paladin and Vulpiani, 1987; Stanley and Meakin, 1988; Tel, 1988). They can be derived from the curve $D_q(q)$ by a geometrical construction.

The numbers $\alpha(q)$ and $f(q)$ give a point $(\alpha(q), f(q))$. The set of these points for various values of q defines a curve $f(\alpha)$, called the multifractal spectrum, that describes the multifractal in yet another way, which is important in relating observations to models of multifractals. In practice, one computes a finite set of points $(\alpha(q), f(q))$ for a limited number of values of $-20 \leq q \leq 20$, and the curve $f(\alpha)$ is determined by a polynomial fit to the points. Both the function $f(\alpha)$ and α have deep geometrical significance: $\alpha(x)$ is the exponent describing the scaling behavior of the field at the position x (or time t), and $f(\alpha_0)$ is the fractal dimension of the set of points at which α has a particular value α_0. A multifractal has an infinite number of parts, each of which is a fractal set, and the union of all these sets fills an interval on the real line.

9.5.2 Magnetic Field Strength Fluctuations at 1 AU

The fluctuations of the magnetic field strength observed by IMP-8 at 1 AU during 1974, when corotating streams were dominant, were analyzed by Burlaga (1992b). The fluctuations were bursty, inhomogeneous, and asymmetric, as one expects for multifractal fluctuations. The fluctuations on the various scales appear to be similar for periods from 2 to 32 hours, again suggesting a multifractal structure.

If the fluctuations have multifractal structure, then equation (9.15) implies that a plot of $\log \langle B_n^q \rangle$ versus $\log(\tau_n)$ should give a straight line with a slope $\tau(q)$ for any value of q in some range. A family of straight lines, parametrized by q, is observed for time averages from 2 to 32 hours. Such lines were obtained for all integer values of q in the range $-14 \leq q \leq 14$. The slopes $\tau(q)$ and the uncertainties in these slopes were determined by least squares fits.

Burlaga (1992b) suggested that the multifractal magnetic field strength fluctuations described above are a consequence of the multifractal speed fluctuations he observed during the same period (Burlaga, 1992a). Several observations are consistent with this hypothesis. The magnetic field strength fluctuations are generally small near the sun and grow with increasing distance from the sun (e.g., Whang, 1991), indicating the formation of magnetic field strength fluctuations in the interplanetary medium. The fluctuations are largest in the interaction regions where the speed is increasing. Finally, the magnetic field fluctuations show multifractal behavior in the same range of periods as the speed fluctuations.

A simple conceptual model illustrates semiquantitatively how multifractal magnetic field fluctuations might be produced kinematically to first approximation by velocity fluctuations (Burlaga, 1992b). Assume that the magnetic field strength is constant across a corotating stream at a certain distance near the sun, having an azimuthal extent ϕ and a corresponding time scale T. As the corotating stream evolves, it redistributes the magnetic flux. The magnetic field strength increases where the speed is increasing and the magnetic field strength decreases where the speed is decreasing. This is basically a kinematic effect (Burlaga and Barouch, 1976). Assume for simplicity that the region of compression has the same extent as the region of rarefaction. (In reality, the compression region is narrower than the rarefaction region.) In this approximation, the magnetic flux is redistributed in two equal intervals, the magnetic flux being removed from the rarefaction region and transferred to the compression region. This process, which is regarded as the first step of a binomial process, gives the compression and rarefaction regions seen in the 32-hour averages of B. Since Burlaga (1992a) showed that the speed fluctuations in recurrent streams have a multifractal structure, one can assume that there are speed fluctuations on a time scale $T/2$ whose time profiles are similar to the original stream profile. Suppose that these fluctuations also compress and expand the magnetic field, redistributing the flux, albeit on a smaller scale ($T/2$) than the corotating stream (T). Assume again that the compression region has the same size as the rarefaction region, and assume that the measure multipliers have the same values at this step as in the preceding step of the binomial process. Continuing in this way, assume that there are speed fluctuations on a time scale of $T/4$ which redistribute magnetic flux and produce fluctuations in the 8-hour averages of B, etc. If the time for the production of the magnetic field fluctuations on one scale is smaller than that on the next larger scale, then the process can evolve in a multiplicative fashion, and a binomial model reflects the basic kinematic processes involved.

One can calculate $D_{q'}$ for the model just described using equation (9.9) for a binomial multiplicative process, from Meneveau and Sreenivasan (1991). The parameter p is the fraction of the measure on one half of the interval at each stage, and $(1-p)$ is the fraction of the measure on the other half of the interval. It is assumed that p is the same at every stage. The theoretical curve $\tau(q)$, derived from equation (9.17) with $D_{q'}$ given by (9.9) and $p_1 = 0.585$, gives a good fit to the data for $-1 \leq q \leq 10$ and it describes the trend for $-10 \leq q < 10$ (Burlaga, 1992b). Thus, the multifractal magnetic field structure is possibly the result of a multiplicative process resembling that in the illustrative model described above.

The model described above assumes that the measure is distributed on two equal intervals. This is probably not the case for magnetic field strength fluctuations generated by streams and the smaller scale fluctuations within the streams. The compression region is narrower than the rarefaction region on the largest scale, and it is likely to be so on the smaller scales as well. The equations for a two-scale binomial multiplicative process were derived by Pirraglia (1993). For a cell of unit length, assume that the first segment has length β and the second segment has length γ, where

$$\beta + \gamma = 1 \tag{9.20}$$

Assume that the measure is distributed on the first segment with density D_β and on the second segment with density D_γ, where

$$D_\beta \beta + D_\gamma \gamma = p_\beta + p_\gamma = 1 \tag{9.21}$$

Pirraglia shows that in the limit as q approaches infinity

$$\alpha = \frac{\ln p_\beta}{\ln \beta} = 1 + \frac{\ln D_\beta}{\ln \beta} \tag{9.22}$$

and in the limit as q approaches minus infinity

$$\beta = \frac{\ln p_\gamma}{\ln \gamma} = 1 + \frac{\ln D_\gamma}{\ln \gamma} \tag{9.23}$$

From the multifractal spectrum of the fluctuations observed at 1 AU during 1984, calculated from equations (9.17), (9.18), and (9.19), one obtains 0.76 and 1.22 for the limiting values of α. Equations (9.20), (9.21), (9.22), and (9.23) then give $\beta = 0.3609$, $\gamma = 0.6391$, $p_\beta = 0.2884$ and $p_\gamma = 0.7116$. These parameters provide a very good fit to the observed multifractal spectrum for the fluctuations (Pirraglia, 1993). Thus, the two-scale binomial multiplicative process provides a satisfactory description of the multifractal magnetic field strength fluctuations observed in corotating streams.

9.5.3 Multifractal Magnetic Field Strength Fluctuations in the Outer Heliosphere

The existence of multifractal structure in the large-scale fluctuations in the solar wind was demonstrated by Burlaga (1991a) and Burlaga et al. (1991). These authors analyzed the magnetic field strength fluctuations observed by Voyager 2 near 25 AU from approximately day 190, 1987, to day 365, 1988. Plots of $\log_{10}(\langle B_n^q \rangle)$ versus $\log_{10}(\langle \tau_n \rangle)$ for $-10 \leq q \leq 10$ were linear in the range of periods from 16 hours to 21.3 days, suggesting the existence of multifractal structure in the intermediate and low frequency ranges, extending to the solar rotation period at the limit of the low frequency range. The plots of D_q versus q and $f(\alpha)$ derived from the slopes of these lines confirm the existence of multifractal structure in the large-scale magnetic field strength fluctuations. The minimum and maximum values of α (≈ 0.8 and 1.22, respectively) are very close to those derived from the intermediate scale (4–32 hours) magnetic field strength fluctuations in corotating streams at 1 AU discussed in the preceding section.

9.5.4 Solar Cycle Variations of Multifractal Magnetic Field Strength Fluctuations in the Outer Heliosphere

The normalized magnetic field strength fluctuations observed by Voyager 2 from 1983.0 to 1989.6 are shown in Fig. 9.11 from Burlaga et al. (1993b). The data have been divided into three intervals, shown in the three panels on the top of Fig. 9.11. The first panel shows the quasi-periodic magnetic field strength fluctuations observed from 1983.0 to 1985.0; these occur with a period near the solar rotation period at a time when corotating streams from coronal holes were observed at 1 AU. The second panel shows low amplitude, highly irregular magnetic field strength fluctuations observed near the minimum of the solar cycle from 1985.0 to 1987.5. The third panel shows data obtained when the effects of increasing solar activity were first

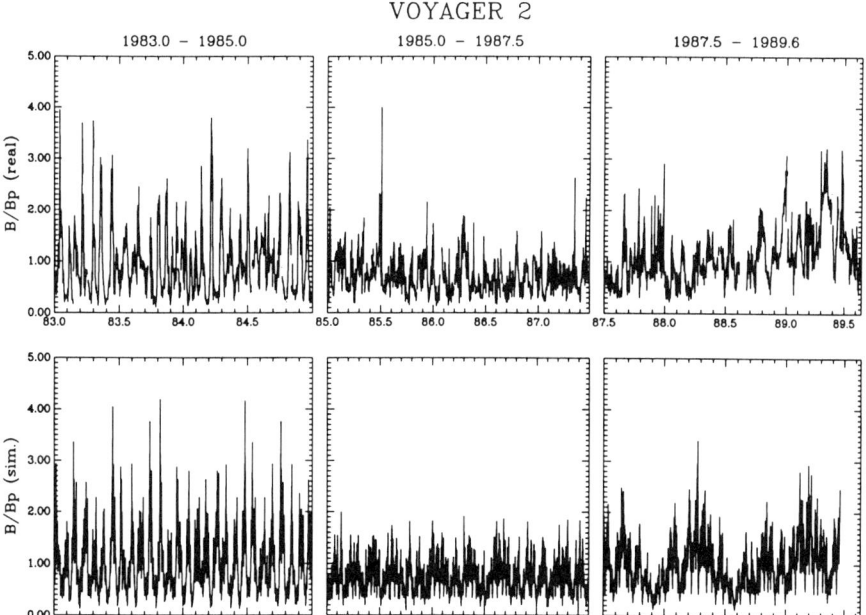

Fig. 9.11. Multifractal large-scale fluctuations of the magnetic field strength in flows of three types. (L.F. Burlaga, J. Perko, and J. Pirraglia, *Astrophys. J.*, **407,** 347, 1993b).

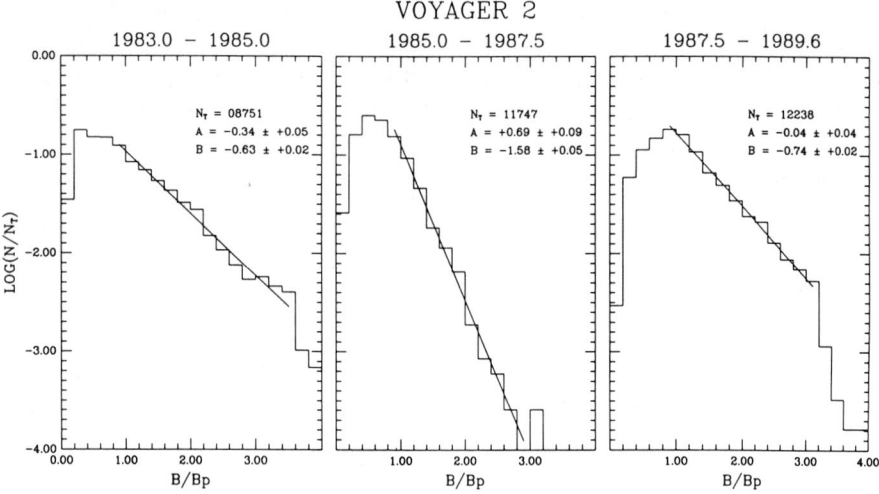

Fig. 9.12. Distribution functions of the magnetic field strength. (L.F. Burlaga, J. Perko, and J. Pirraglia, *Astrophys. J.*, **407**, 347, 1993b.)

observed in the outer heliosphere from 1987.5 to 1989.6. These fluctuations are highly irregular, bursty, inhomogeneous, and of large amplitude. Although the boundaries of these three intervals are not sharply defined, it is clear that there are three different states of the large-scale magnetic field strength fluctuations in the outer heliosphere, corresponding to different phases of the solar cycle: decreasing solar activity, minimum solar activity, and increasing solar activity.

The distribution functions of the hour averages of the magnetic field strengths in each of the three intervals in Fig. 9.11 are shown in Fig. 9.12. All three distributions have exponential tails, but they differ quantitatively. The slope of the tail is much steeper for the period near solar minimum than for the other two intervals, corresponding to the absence of relatively strong fields (MIRs) during this period. The slope for the interval of decreasing solar activity (-0.63 ± 0.02) is comparable to that for the interval of increasing solar activity (-0.7 ± 0.02). However, the amplitude (intercept) for the interval of decreasing solar activity is smaller than that for the interval of increasing solar activity. The greater amplitude in the latter case reflects the presence of stronger and broader MIRs when the sun is more active and ejecta are more numerous.

Plots of $\log_{10}(\langle B_n^q \rangle)$ versus $\log_{10}(\tau_n)$ for $-10 \le q \le 10$ for each of the three intervals discussed above are linear over a certain range, consistent with the existence of multifractal structure. For the period of decreasing solar activity (1983.0–1985.0), the multifractal structure is observed in the range of periods from 16 hours to 128 hours. Near solar minimum (1985.0–1987.5), the multifractal structure is observed at somewhat longer periods, from 32 hours to 256 hours. During the phase of increasing solar

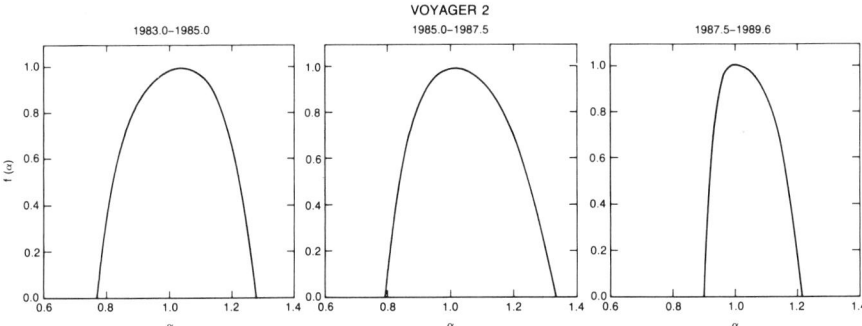

Fig. 9.13. Multifractal spectra of large-scale magnetic field strength fluctuations in flows of three types. (L.F. Burlaga, J. Perko, and J. Pirraglia, *Astrophys. J.*, **407**, 347, 1993b.)

activity (1987.5–1989.6), the multifractal structure is observed over a relatively large range of periods, from 16 hours to 256 hours.

The multifractal spectra for each of the three intervals discussed above are shown in Fig. 9.13. The spectra differ from one another in both their shapes and the range of α. From the minimum and maximum values of $f(\alpha)$ for each of these multifractal spectra, one determines the values of β, γ, p_β, and p_γ using equations (9.20), (9.21), (9.22), and (9.23), which define a two-scale binomial multiplicative process, as described by Pirraglia (1993). Given these parameters and this process, Pirraglia showed how one can generate time series with the same multifractal spectra as those from which these parameters were derived. It is significant that the simulations by Pirraglia show that the multifractal scaling typically extends to periods a factor of 4 lower than and a factor of 4 higher than those derived from spacecraft observations.

The time series for the magnetic field strength, derived from the multifractal spectra in Fig. 9.13 using the method of Pirraglia (1993) are shown in the lower panels of Fig. 9.11. The qualitative correspondence between the magnetic field strength observations in the upper panels of Fig. 9.11 and the model profiles in the lower panels is very good in each of the three intervals, despite the very different characteristics of the time series in these intervals. For the period of declining solar activity (1983.0–1985.0), the fluctuations were quasi-periodic with a period of 26 days (the solar rotation period). The multifractal time series was superimposed on a sine wave with constant amplitude and a period of 26 days. For these quasi-periodic fluctuations, the simulation satisfactorily describes the irregularity of the amplitudes of the CMIRs and the asymmetry between the compression regions and the rarefaction regions. For the period near the minimum of solar activity, the simulation describes the relatively low amplitudes and burstiness of the magnetic field strength fluctuations. For the period of increasing solar activity, the simulation describes the tendency for broad clusters of very intense fields to form among intermittent fluctuations of smaller amplitude. These clusters correspond to the GMIRs and LMIRs that were identified in this period, as discussed in Sections 8.5 and 8.4, respectively. These GMIRs and perhaps also the LMIRS seem to be basically statistical structures, as opposed to CIRs, which are well-localized structures in space with statistically varying amplitudes.

The multifractal structure of the large-scale fluctuations in the heliospheric magnetic field provides a quantitative description of the magnetic field fluctuations as a function of position and time. The binomial multiplicative model and the algorithms of Pirraglia open the way to develop models of the global structure of the large-scale magnetic field strength fluctuations. These models will be of fundamental importance for cosmic ray propagation theories, they will provide global images of the heliospheric magnetic field, and they will lead to a new approach to heliospheric dynamics. We are only at the threshold of these exciting possibilities.

10

Heliospheric Vortex Street

10.1 Observations

10.1.1 North–South Flows

One of the most surprising results from the plasma experiment on Voyager 2 is the discovery of a quasi-periodic meridional flow with a period of about 25.5 days between 20 AU and 25 AU from 1986 to early 1988 (Lazarus and Belcher, 1988; Lazarus et al., 1988; Lazarus and McNutt, 1990). The amplitude of the north–south flow deflections, shown as the $\theta_{NS}(t)$ profile at the top of Fig. 10.1 is about $\pm 5°$ north and south of the heliographic equator. There was no similar deflection of the flow in the east and west directions. The mean speed was approximately 400 km/s.

The bulk speed profile at the bottom of Fig. 10.1 shows small but distinct variations that Lazarus et al. (1988, 1970) attribute to shocks produced by stream–stream interactions close to the sun as described in Chapters 7 and 8. Actually, all but one of the "shocks" occurred in data gaps, so that the identification of the events as shocks is not definitive. In many but not all cases, the transition from northward flow to southward flow is related to increases in speed, density, and pressure. The transitions from southward flow to northward flow are less well correlated with the other parameters.

The magnetic field was pointing toward the sun in at least 77% of the northward flow regions, and it was pointing away from the sun in at least 59% of the regions of southward flow. Lazarus et al. (1988) concluded that the heliospheric current sheet did not separate the regions of different flow, but it was embedded in the flows once per half-period.

10.1.2 Solar Observations

The heliospheric current sheet was close to the ecliptic during 1986 and 1987, as discussed in Section 2.2.4. This configuration is typically observed near solar minimum. Since the solar wind speed is generally lowest near the heliospheric current sheet, one can assume that the speed was nearly a minimum at Voyager 2 during the period in which the north–south flows

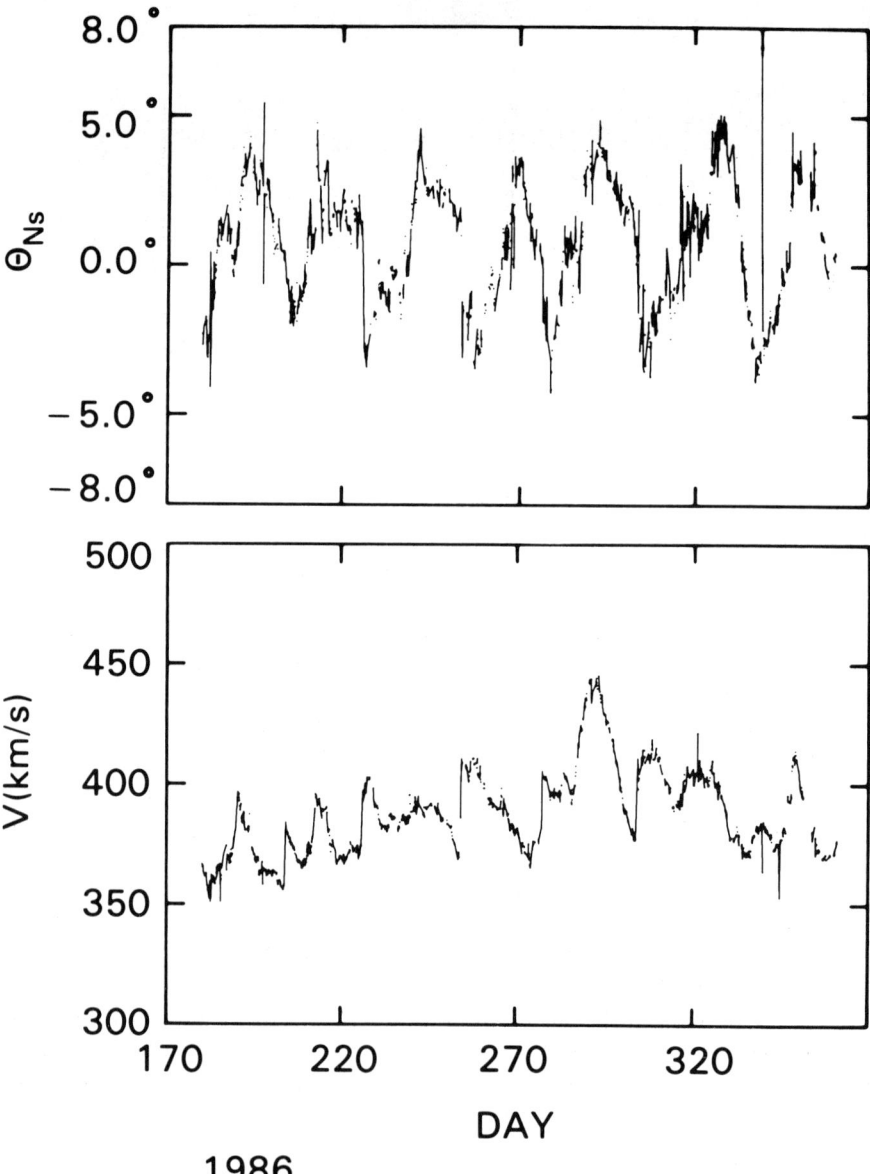

Fig. 10.1. Observations of periodic north–south flows. (A.J. Lazarus, B. Yedidia, L. Villaneuva, R.L. McNutt, Jr., J.W. Belcher, U. Villante, and L.F. Burlaga, *J. Geophys. Res. Lett.*, **15,** 1519, 1988, copyright by the American Geophysical Union.)

were observed. Low speeds are typically observed at the solar equator near solar minimum, when the polar coronal holes do not extend to low latitudes.

The temporal evolution of the latitudinal extent of the polar coronal holes was demonstrated by Gazis et al. (1989). Prior to mid-1985, the polar coronal holes did extend to near-equatorial latitudes. Consequently, the speed observed by Pioneer 11 at a latitude of $\approx 15°$N was nearly the same on average as that observed by Voyager 2 near the ecliptic. However, throughout 1986, the polar coronal holes were present only at high latitudes ($>60°$). These coronal holes produced fast flows that were observed by Pioneer 11 near 15°N but not by Voyager 2 near the heliographic equator. Thus, a large latitudinal gradient in the solar wind speed was observed during 1986 (Gazis et al., 1989). A large and stable shear layer was present above the heliographic equator and presumably below the equator during 1986 as a result of the coronal hole configuration.

10.2 Conventional Flow Models

The initial attempts to explain the north–south flows were based on the conventional models of streams and interaction regions. Lazarus et al. (1988) suggested that the flow deflections were produced by pressure gradients associated with interaction regions. Pizzo and Goldstein (1987) developed a 3-D MHD model based on a tilted dipole magnetic field configuration, with fast flows over the poles and slow flows near the equator. Their model did predict meridional flows out to 10 AU, but it predicted two north–south flow cycles per solar cycle rather than one as observed. It also predicted large east–west flows, which were not observed. Thus, the model of Pizzo and Goldstein does not explain the observed characteristics of the north–south flows.

Another attempt to explain the north–south flows as the consequence of pressure gradients associated with streams was published by McNutt (1988). He did not obtain solutions for N, T, V, and B from the MHD equations. Using the MHD equations in spherical coordinates, NcNutt argued on dimensional grounds that the dominant term was the one involving the gradient in the total pressure, and he concluded from this that the north–south flows were pressure driven. We shall show below that this is not the only possible explanation allowed by the MHD equations.

The most convincing argument that north–south flows can be driven by pressure gradients associated with corotating streams from high latitude coronal holes was presented by Pizzo (1993), using an MHD model that was constrained to be time independent. Pizzo predicted that the amplitude of the north–south flows is largest near several AU and thereafter decreases with increasing distance from the sun. North–south flows have not been observed within 20 AU. Pizzo predicted a correlation between B, N, and T, which is also not observed. However, one must consider the possibility that at least a part of the north–south flow deflections and some of the enhancements in N, T, and B are under some conditions produced by pressure gradients. The flow computed by Pizzo might be unstable. Thus, a generalization of his model to include time variations is needed.

Meridional flows can also be driven, in principle, by gradients in the magnetic pressure. Since the magnetic field is wound in a spiral in the equatorial plane, the magnetic field strength there falls off as R^{-1} far from the sun, whereas the magnetic field strength falls off as R^{-2} in the polar regions along the solar rotation axis. There is thus a gradient of magnetic pressure that tends to drive meridional flows away from the equator (e.g., see Winge and Coleman, 1974). Suess and Nerney (1975) and Nerney and Suess (1975a,b; 1985) argued that the flow could be large. However, Pizzo and Goldstein (1987) showed that the calculations of Nerney and Suess were based on unrealistic assumptions and that the meridional flows produced by this mechanism are an order of magnitude smaller than predicted by Nerney and Suess. Such flows are too small to explain the north–south flows observed in the outer heliosphere during 1986 and 1987. McNutt (1988) also noted that the model of Suess and Nerney predicts that V_θ should be zero at maximum pressure, in contradiction to the observations.

10.3 Vortex Street Models

10.3.1 Hydrodynamic Vortex Street Model

The north–south flows may be explained by a vortex street described analytically by a series of singular line vortices in an incompressible fluid (Burlaga, 1990a). The heliospheric vortex street, driven by two shear layers, above and below the heliographic equator, is illustrated in Fig. 10.2. One row of vortices is largely above the heliographic equator, and the other row of vortices is largely below the heliographic equator. The separation of the vortices in each row is $a = 5$ or 6 AU. The meridional separation between

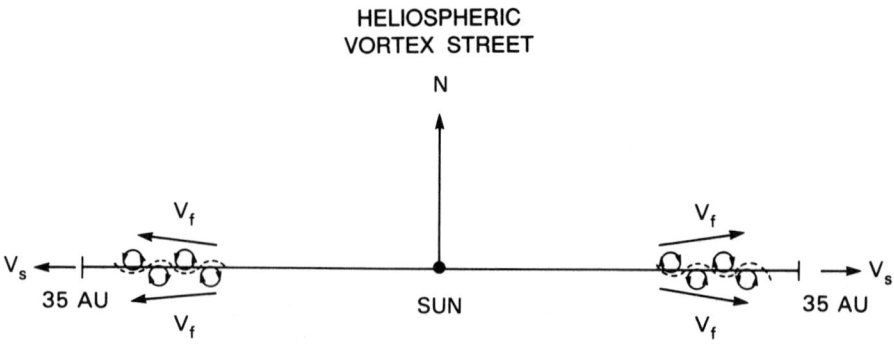

Fig. 10.2. Illustration of the vortex street model. (L.F. Burlaga, *J. Geophys. Res. Lett.*, **95**, 2229, 1990a, published by the American Geophysical Union.)

the rows of vortices is $b = 0.281a$, which is the condition derived by Lamb (1945) for stability of a vortex street formed by a series of singular lines in an incompressible fluid.

Two shear layers existed near the ecliptic during 1986 and 1987. The speed observed by Pioneer 11 near 15°N was 200 km/s greater than that observed by Voyager 2 near the ecliptic (Gazis et al., 1989). A speed gradient of 13 km/s/deg, increasing from the ecliptic to the northern hemisphere, was present. Such a shear across a vortex of diameter 3 AU at 25 AU implies a change of 90 km/s across the vortex, which gives a north–south deflection of ±5° for a 400 km/s solar wind speed, consistent with the observations.

Additional evidence for a speed gradient near the ecliptic at this time, when the heliospheric current sheet was near the ecliptic (see Chapter 2), was presented in the IMP-8 data (Lazarus et al., 1988). IMP-8 observed one stream per solar rotation when it was a few degrees above the heliographic equator and another stream per solar rotation 180° in longitude from the first when it was a few degrees below the heliographic equator. This is consistent with the emission of high speed flows from the polar coronal holes that were presented (Gazis et al., 1989) and the near-equatorial heliospheric current sheet, which is generally associated with slow flows. The heliospheric current sheet was tilted approximately 5–20° during this period, allowing IMP-8 to sample the flow from the northern hemisphere and the southern hemisphere alternately during one solar rotation.

The flows from the north and south polar coronal holes and the constraint of low speeds near the heliospheric current sheet were responsible for the two shear layers that produced the heliospheric vortex street. Fast streams from the northern hemisphere caused a clockwise rotation of the flow in the northern hemisphere on the right of Fig. 10.2. The fast flow from the southern hemisphere caused a counterclockwise rotation of the flow in the southern hemisphere on the right of Fig. 10.2. The fast flows alternately from the north and the south, 180° out of phase, drive the vortices that are seen alternately above and below the heliographic equator (Burlaga, 1990a).

The solution of Lamb (1945) for a shear-driven vortex street consisting of singular line vortices in an incompressible fluid, and planar geometry was used by Burlaga (1990a) to model the heliospheric vortex street locally. He argued that the incompressible fluid approximation is a reasonable first approximation, because the difference in velocity across a vortex, approximately 90 km/s, is smaller than the magnetoacoustic speed. He argued that the planar symmetry provides a reasonable approximation to the local flow configuration observed near the spacecraft, since the size of the vortices (≈ 3 AU) is much smaller than the scale of the system (≈ 25 AU). The magnetic field is negligible to first approximation, because it is along the vortex axis and transverse to the flow so that it does not impede roll-up of the vortices. The magnetic field does stabilize the structure along the vortex axis. The solution for $a = 6.24$ AU, the convective speed $V_\theta = 425$ km/s,

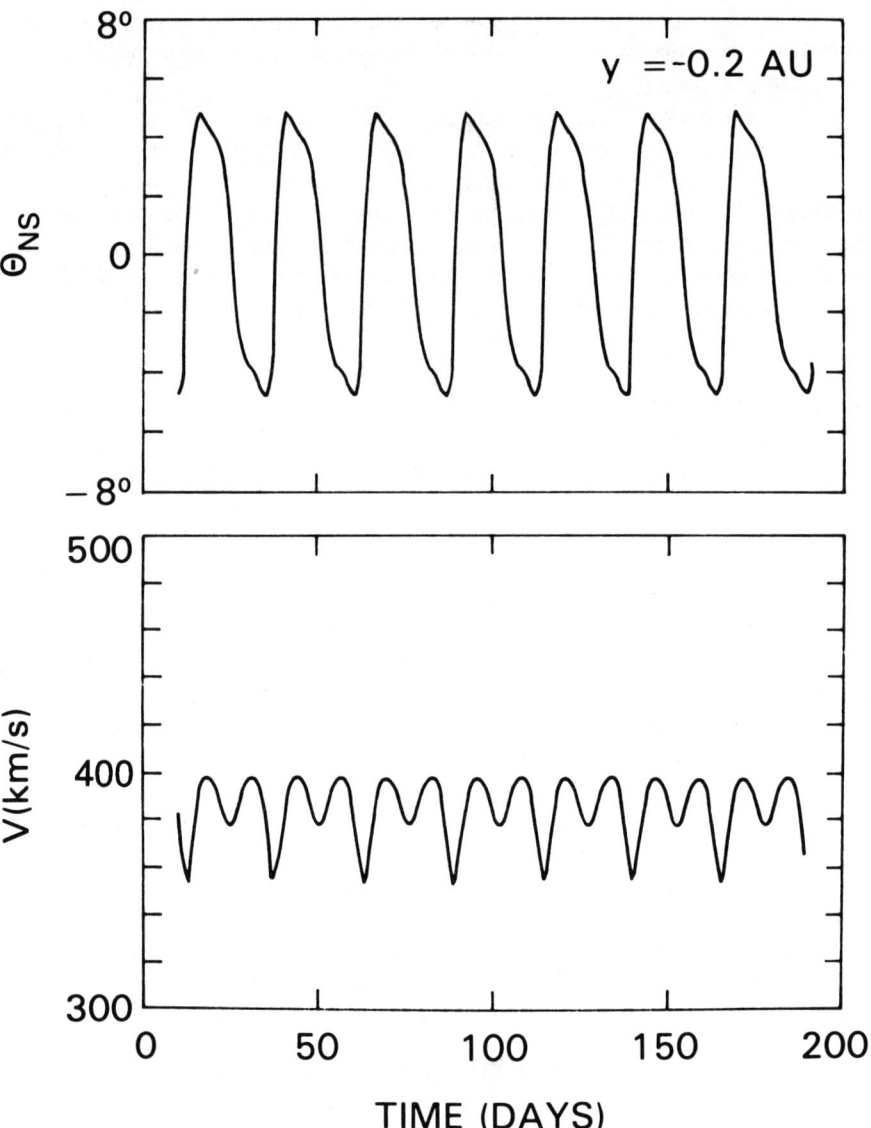

Fig. 10.3. Model velocity profiles for a heliospheric vortex street. (L.F. Burlaga, *J. Geophys. Res. Lett.*, **95**, 2229, 1990a, published by the American Geophysical Union.)

$k = 20$ km/s, and a spacecraft located a distance $y = -0.2$ AU below the plane of symmetry of the vortex street is shown in Fig. 10.3. The top panel shows that north–south flows with an amplitude of $\pm 5°$ and a period of 25 days are described by the solution for a convected vortex street.

The vortex street model predicts two maxima in the solar wind speed on each solar rotation, during one period of the north–south flow oscillation, as shown in the bottom panel of Fig. 10.3. This prediction is consistent with the observations of Lazarus et al. (1988) shown in Fig. 10.1, although the double peak in the speed profile was not discussed in the observational papers on the north–south flows.

The vortex street model makes another surprising prediction. Generally the solar wind speed is independent of the distance from the sun, as discussed in Section 3.5. However, during 1986 and 1987 the speed at 20–25 AU was about 20% lower than the speed at 1 AU (Fig. 10.4). This lower bulk speed at large distances from the sun might be caused by the heliospheric vortex street (Burlaga and Ness, 1993b). Figure 10.2 shows that the vortices produce a sunward component of velocity near the heliographic equator of up to approximately 90 km/s and a mean component of the order of 50 km/s. Given a mean solar wind speed of approximately 400 km/s (Fig. 10.1), one expects the velocity at 20–25 AU to be approximately 13% less than that at 1 AU, consistent with the observations in Fig. 10.4.

10.3.2 Linear Model of Vortex Street Formation

Veselovsky (1990) independently suggested the possibility of vortex perturbations related to the north–south flows observed near 25 AU. In particular, he considered the development of vortex perturbations of the acoustic type in the presence of velocity shears relative to the heliographic equator. His model describes the fields in terms of a linear perturbation solution $\rho_1, \mathbf{v}_1, p_1 \propto \exp[kr - \omega t]$ of the axially symmetric gas dynamic equations for a compressible, polytropic fluid in the presence of a thin, double shear layer.

Veselovsky obtains the dispersion equation for sound waves, indicating that the perturbations are basically sound waves. The solution shows a sequence of "o" points alternately above and below the heliographic equator, which he interprets as the centers of vortices. Note that the velocity is zero at his o-points, whereas it is infinite at the centers of the vortices considered by Burlaga (1990a). The distance of the o-points above and below the heliographic equator is proportional to the amplitude of the perturbation, rather than being fixed by a stability condition as in the solution applied by Burlaga (1990a). The pressure and density all have minima at the center of the vortices, for vortices propagating away from the sun.

The solution of Veselovsky cannot be a precise representation of the observations, since a vortex street is a nonlinear phenomenon whereas his solution is a linear perturbation approximation. The relative phases of the perturbations in the linear theory are not in agreement with the observations. Moreover, the positions of the vortices relative to the equator depend on the amplitude of the perturbation, rather than on the separation of the vortices as in the nonlinear solution discussed above. Nevertheless, Veselovsky's consideration of linear instability, the axial symmetry, and the

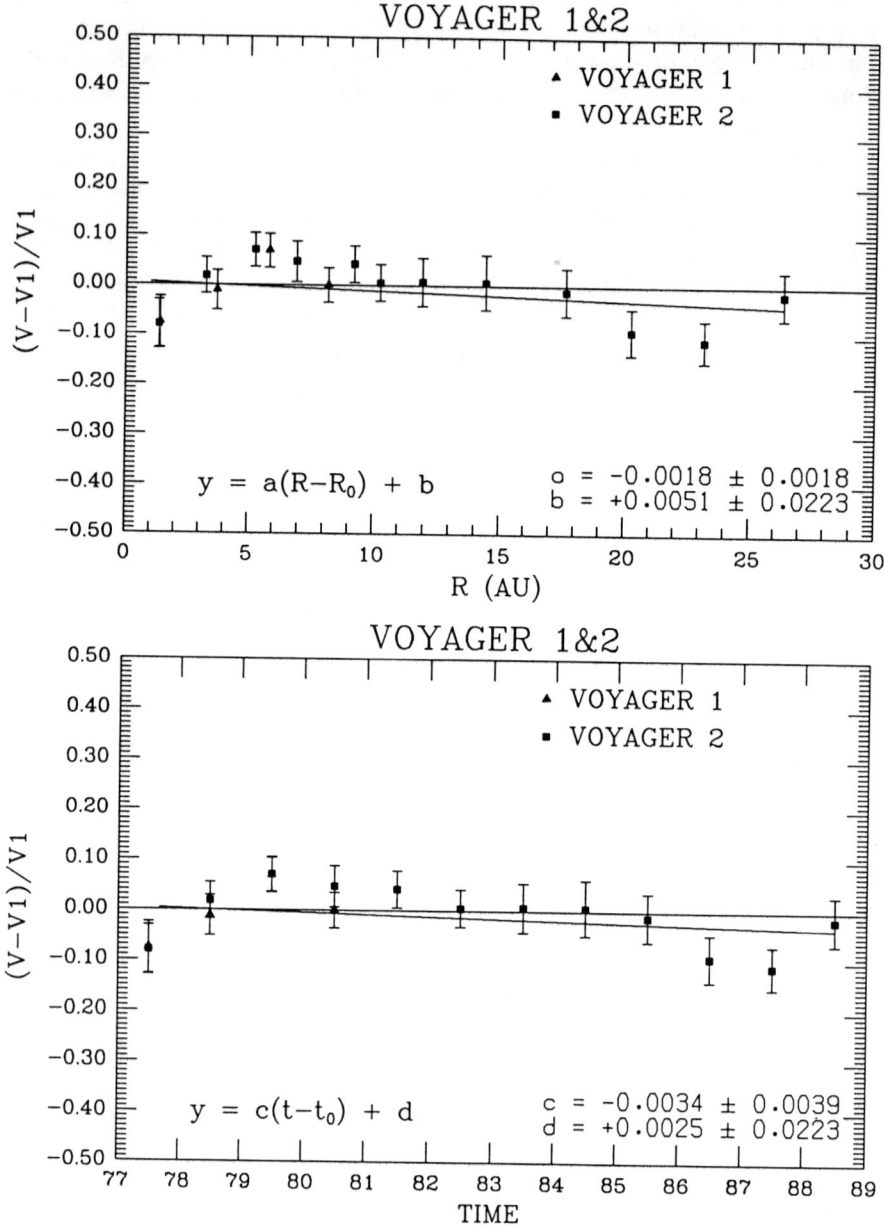

Fig. 10.4. The relative difference between the solar wind speed in the outer heliosphere and at 1 AU when the north–south flows were observed. (L.F. Burlaga and N.F. Ness, *J. Geophys. Res. Lett.*, **98**, 3539, 1993b, copyright by the American Geophysical Union.)

effects of compressibility complement the approach of Burlaga (1990a), which was aimed at a quantitative explanation of the observed velocity profiles.

10.3.3 Kelvin–Helmholtz Instability, Nonlinear Models

A nonlinear model of the growth of the Kelvin–Helmholtz instability and the formation of a vortex street from two shear layers is given in Siregar et al. (1992, 1993). The basic equations are the single-fluid MHD equations (Chapter 1), including viscous stress. They assume that the plasma obeys a polytropic law. The equations are solved with a spectral code. The initial condition is two plane, equilibrium shear layers, corresponding to the local shear layers above and below the solar heliographic equator. The form of the initial perturbation, which is an important aspect of the model, is a small sinusoidal perturbation that leads to a staggered vorticity distribution. Such a perturbation is produced by the streams alternately from the north and south polar coronal holes, as discussed by Burlaga (1990a).

The solution of Siregar et al. (1992, 1993) shows that the Kelvin–Helmholtz instability produces certain growing modes that induce an interaction between the two shear layers, leading to the formation of a Karman vortex street (Fig. 10.5). The density and pressure are low in the vortices, consistent with the linear compressible result of Veselovsky. Representative variations of the total speed U, azimuthal flow angle $\theta = \arctan(U_x/U)$, and density along two radial trajectories in the neighborhood of the current sheet separating the two shear layers are shown by the solid and dotted curves in Fig. 10.6. The quasi-periodic north–south flow of several degrees north and south, corresponding to the vortex motion, is evident in the second panel from the top of Fig. 10.6. The two maxima in the speed profile during one cycle of the north–south flow, which were predicted by the solution of Burlaga (1990a), are also predicted by the nonlinear numerical model, as shown in the top panel of Fig. 10.6. The details of the profiles depend on the position of the cut, as was also demonstrated by Burlaga (1990a). The numerical model also shows that the neglect of the magnetic forces is justified in the vortex street model, insofar as the basic structure of the vortex street is concerned.

The model of Siregar et al. (1992, 1993) confirms the basic features of the vortex street model of Burlaga (1990a), but it also provides significant new results. The numerical model shows that the vortex street will evolve from an equilibrium configuration of two shear layers for parameters representative of the solar wind. In particular, it suggests that the flow considered by Pizzo (1994) cannot be assumed to be steady. The model of Siregar and coworkers predicts finite speeds at the centers of the vortices, and low densities and pressures there. It also predicts a strong correlation of the pressure–density maxima with the increase in total velocity (Fig. 10.6), consistent with the observations of Lazarus et al. (1988). However, the density maxima are smaller than those observed, suggesting that the

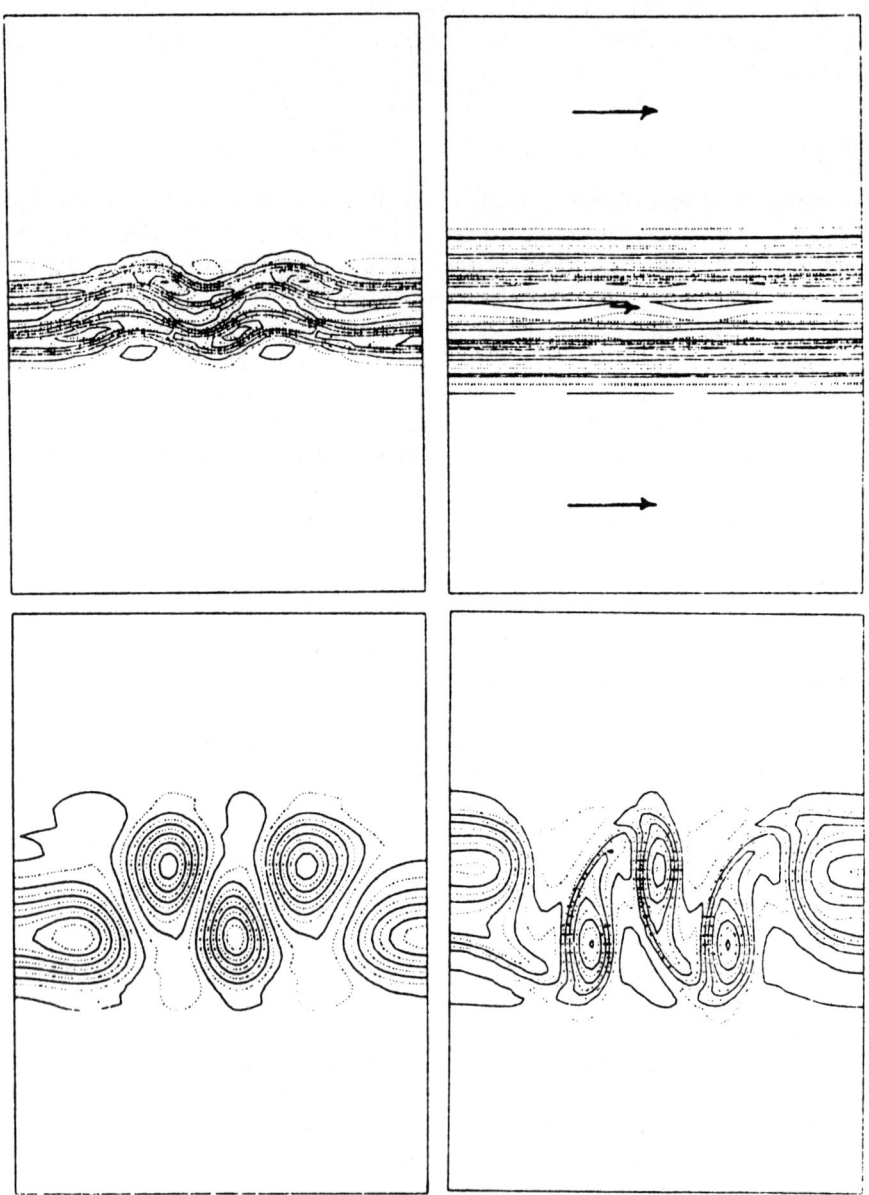

Fig. 10.5. Nonlinear vortex street model vorticity contours. The initial configuration is in the upper right panel, and the evolution of the flow is shown in successive panels in the counterclockwise direction. (E.D. Siregar, D. Aaron Roberts, and M.L. Goldstein, *Geophys. Res. Lett.,* **19,** 1427–1992, copyright by the American Geophysical Union.)

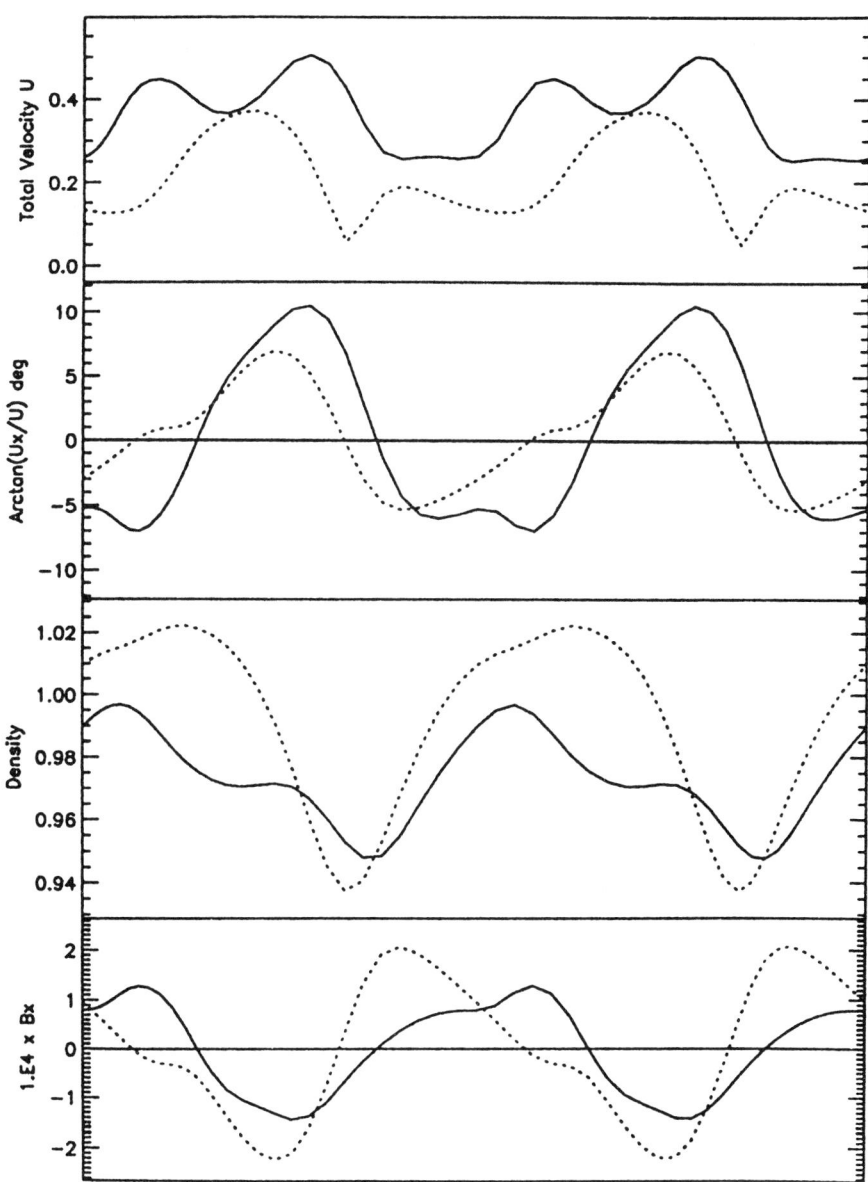

Fig. 10.6. Nonlinear vortex street model profile. (E. Siregar, D. Aaron Roberts, and M.L. Goldstein, *J. Geophys. Res. Lett.,* **98,** 13233, 1993, copyright by the American Geophysical Union.)

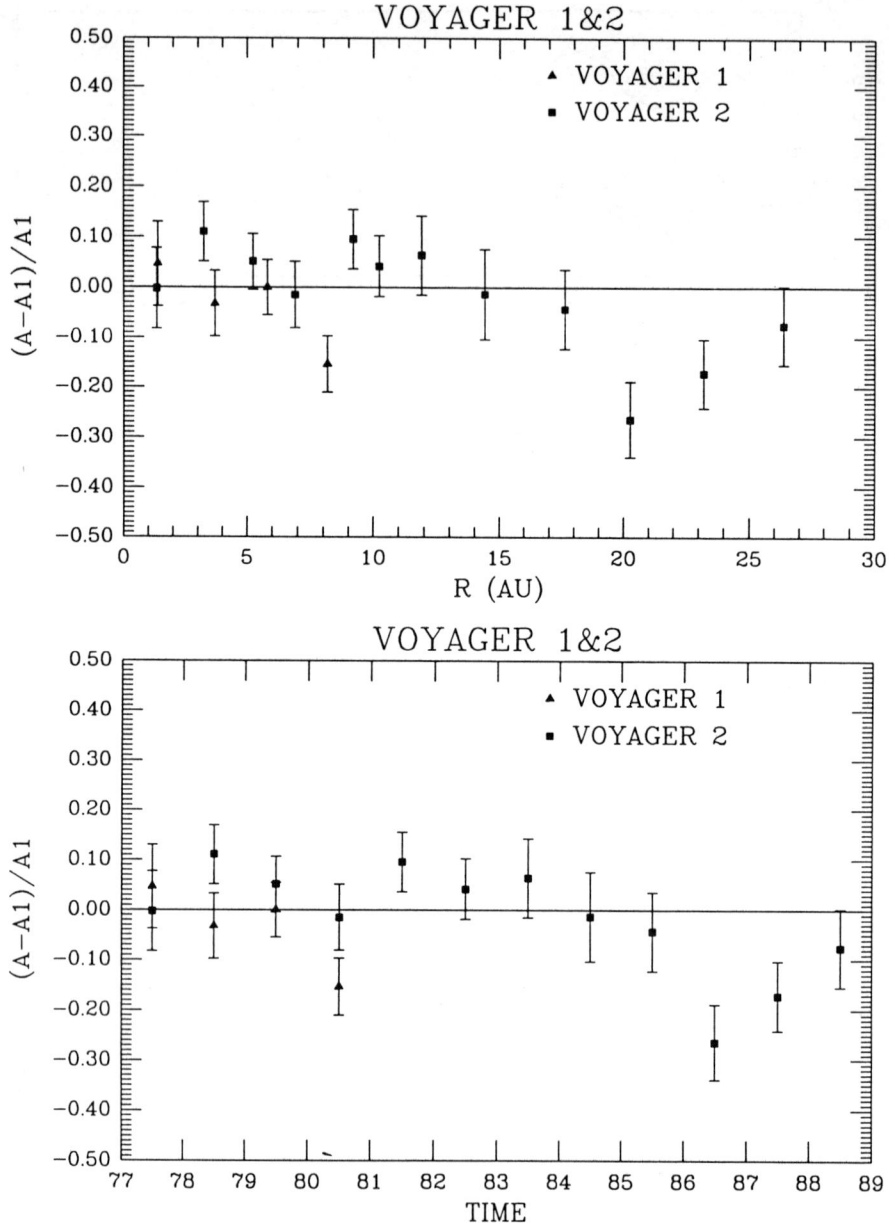

Fig. 10.7. The relative difference between the magnetic field strength in the outer heliosphere and at 1 AU when the north–south flows were observed during 1986–87. (L.F. Burlaga, and N.F. Ness, *J. Geophys. Res. Lett.,* **98,** 3539, 1993b, copyright by the American Geophysical Union.)

interaction regions might also be present in the observations of Lazarus et al. (1988). In other words, the interaction regions might be interacting with the vortex street, but they are not primarily responsible for either the north–south flows or the double peak in the speed profile.

Another prediction of the model of Siregar et al. (1993) is that the magnetic field at 20–25 AU, where the vortex street is present, should be less than that given by Parker's model. In other words, a flux deficit should develop when the vortex street is present. A flux deficit of about 10% at 20–25 AU was observed by the Voyager 2 magnetometer when the north–south flows were observed (Burlaga and Ness, 1993b), as shown in Fig. 10.7. Recent 64^3 3-D simulations confirm the basic picture of a vortex street. They show that the Parker field, which is observed to be transverse to the flow, is actually required in order to suppress 3-D instabilities that would disrupt the flow and produce fully three-dimensional turbulence. The Parker field keeps the flow very nearly two-dimensional. Further studies of the heliospheric vortex street based on MHD models are needed to elucidate the mechanism that produces the flux deficit.

REFERENCES

Abraham-Shrauner, B., *Determination of magnetohydrodynamic shock normals*, J. Geophys. Res., **77,** 736, 1972.

Abraham-Shrauner, B., and S.H. Yun, Interplanetary shocks seen by Ames plasma probe on Pioneer 6 and 7, J. Geophys. Res., **81,** 2097, 1976.

Acuna, M.H., and R.P. Lepping, Modification to shock fitting program, J. Geophys. Res., **89,** 11004, 1984.

Acuna, M.H., L.F. Burlaga, R.P. Lepping, and N.F. Ness, Initial results from Voyagers 1, 2 magnetic field experiment, in *Solar Wind Four,* edited by H. Rosenbauer, p., 143, Rep. MPAE-100-81-31, Max Planck Institute, Lindau, Germany, 1981.

Akasofu, S.-I., Reply, in *Solar Wind Five,* edited by M. Neugebauer, p. 484, NASA Conf. Publ. 2280, Washington, DC, 1983.

Akasofu, S.-I., and S. Chapman, *Solar–Terrestrial Physics,* p. 506, Clarendon Press, Oxford, 1972.

Akasofu, S.-I., and C.D. Fry, Heliospheric current sheet and its solar cycle variations, J. Geophys. Res., **91,** 13679, 1986.

Akasofu, S.-I., and K. Hakamada, Solar wind disturbances in outer heliosphere caused by successive solar flares from the same active region, in *Solar Wind Five,* edited by M. Neugebauer, p. 484, NASA Conf. Publ. 2280, Washington, DC, 1983a.

Akasofu, S.-I., and K. Hakamada, Solar wind disturbances in outer heliosphere caused by six successive solar flares from the same active region, Geophys. Res. Lett., **10,** 577, 1983b.

Akasofu, S.-I., K. Hakamada, and C. Fry, Solar wind disturbances caused by solar flares: Equatorial plane, *Planet. Space Sci.,* **31,** 1435, 1983.

Akasofu, S.-I., C. Olmsted, and J.A. Lockwood, Solar activity and modulation of the cosmic ray intensity, J. Geophys. Res., **90,** 4439, 1985a.

Akasofu, S.-I., W. Fillius, W. Sun, C. Fry, and M. Dryer, A simulation study of two major events in the heliosphere during the present sunspot cycle, J. Geophys. Res., **90,** 8193, 1985b.

Alfvén, H., *Cosmical Electrodynamics,* Clarendon Press, Oxford, 1950.

Alfvén, H., Electrical currents in cosmic plasmas, Rev. Geophys. Space Phys., **15,** 22, 1977.

Altschuler, M.D., and G. Newkirk, Jr., Magnetic fields and the structure of the solar corona, *Solar Phys.,* **9,** 131, 1969.

Anderson, J.E., *Magnetohydrodynamic Shock Waves,* MIT Press, Cambridge, MA, 1963.

Anselmet, F., Y. Gagne, E.J. Hopfinger, and R.A. Antonia, High-order velocity structure functions in turbulent shear flows, J. Fluid Mech., **140,** 63, 1984.

Arnold, V.I., Lectures on bifurcations in versal families, *Russ. Math. Surv.,* **27**:5, 54, 1972.
Arnold, V.I., *Catastrophe Theory,* Springer-Verlag, Berlin-Heidelberg, 1986.
Bame, S., J. Asbridge, and W. Feldman, Bidirectional streaming of solar wind electrons >80 MeV, ISEE evidence for a closed-field structure within the driver gas of an interplanetary shock. *Geophys. Res. Lett.,* **8,** 173, 1981.
Bame, S.J., J.L. Phillips, D. J. McComas, J.T. Gosling, and B.E. Goldstein, The Ulysses plasma investigation: Experimental description and initial in-ecliptic results, in *Solar Wind Seven,* edited by E. Marsch and R. Schwenn, p. 139, Pergamon Press, New York, 1992.
Barnes, A., Distant solar wind plasma—View from the Pioneers, in *Physics of the Outer Heliosphere,* edited by S. Grzedzielski and D.E. Page, p. 235, Pergamon Press, New York, 1990.
Barnes, A., J.D. Mihalov, P.R. Gazis, A.J. Lazarus, J.W. Belcher, G.S. Gordon, Jr., and R.L. McNutt, Jr., Global properties of the plasma in the outer heliosphere: 1. Large-scale structure and evolution, in *Solar Wind Seven,* edited by E. Marsch and R. Schwenn, p. 143, Pergamon Press, New York, 1992.
Barouch, E., and L.F. Burlaga, Causes of Forbush decreases and other cosmic ray variations, *J. Geophys. Res.,* **80,** 449, 1975.
Barouch, E., and L.F. Burlaga, Three-dimensional interplanetary stream magnetism and energetic particle motion, *J. Geophys. Res.,* **81,** 2103, 1976.
Bavassano, B., M. Dobrowolny, and F. Mariani, Evidence of magnetic field line merging in the solar wind, *J. Geophys. Res.,* **81,** 1, 1976.
Bavassano, B., M. Dobrowolny, G. Ganfani, F. Mariani, and N.F. Ness, Statistical properties of MHD fluctuations associated with high speed streams from Helios-2 observations, *Solar Phys.,* **78,** 373, 1982.
Bavassano-Cattaneo, M.B., B.T. Tsurutani, E.J. Smith, and R.P. Lin, Subcritical and supercritical shocks: Magnetic field and energetic particle observations, *J. Geophys. Res.,* **91,** 11929, 1986.
Behannon, K.W., Heliocentric distance dependence of the interplanetary magnetic field, *Rev. Geophys. Space Res.,* **16,** 125, 1978.
Behannon, K.W., and L.F. Burlaga, Alfvén waves and Alfvénic fluctuations in the solar wind, in *Solar Wind Four,* edited by H. Rosenbauer, p. 375, Rep. MPAE-100-81-31, Max-Planck Institute, Lindau, Germany, 1981.
Behannon, K.W., F.M. Neubauer, and H. Barnsdorf, Fine-scale characteristics of interplanetary sector boundaries, *J. Geophys. Res.,* **86,** 3273, 1981.
Behannon, K.W., L.F. Burlaga, and A.J. Hundhausen, A comparison of coronal and interplanetary current sheet inclinations, *J. Geophys. Res.,* **88,** 7837, 1983.
Behannon, K.W., L.F. Burlaga, J.T. Hoeksema, and L.W. Klein, Spatial variation and evolution of heliosphere sector structure, *J. Geophys. Res.,* **94,** 1245, 1989.
Behannon, K.W., L.F. Burlaga, and A. Hewish, Structure and evolution of compound streams at 1 AU, *J. Geophys. Res.,* **96,** 21213, 1991.
Belcher, J.W., and L. Davis, Jr., Large-amplitude Alfvén waves in the interplanetary medium, *J. Geophys. Res.,* **76,** 3534, 1971.
Belcher, J.W., and C.V. Solodyna, Alfvén waves and directional discontinuities in the interplanetary medium, *J. Geophys. Res.,* **80,** 181, 1975.

Bieber, J.W., J. Chen, W.H. Matthaeus, C.W. Smith, and M.A. Pomerantz, Long-term variation of interplanetary magnetic field spectra with implications for cosmic ray modulation, *J. Geophys. Res.*, **98**, 3585, 1993.

Billings, D.E., Distribution of matter with temperature in the emission corona, *Astrophys. J.*, **130**, 961, 1959.

Billings, D.E., and W.O. Roberts, The origin of M-region geomagnetic storms, *Astrophys. Norv.*, **9**, 1947, 1964.

Borgas, M.S., A comparison of intermittency models in turbulence, *Phys. Fluids A*, **4**, 2055, 1992.

Borrini, G., J.T. Gosling, S.J. Bame, W.C. Feldman, and J.M. Wilcox, Solar wind helium and hydrogen structure near the heliospheric current sheet—A signal of coronal streamers at 1 AU, *J. Geophys. Res.*, **86**, 4565, 1981.

Borrini, G., J.T. Gosling, S. J. Bame, and W.C. Feldman, An analysis of shock wave disturbances observed at 1 AU from 1971 through 1978, *J. Geophys. Res.*, **87**, 65, 1982.

Boyd, T.J., *Plasma Dynamics,* Barnes and Noble, New York, 1969.

Bruno, R., L.F. Burlaga, and A.J. Hundhausen, Quadrupole distortions of the heliospheric current sheet in 1976 and 1977, *J. Geophys. Res.*, **87**, 10339, 1982.

Bruno, R., L.F. Burlaga, and A.J. Hundhausen, K-Coronameter observations and potential field model comparison in 1976 and 1977, *J. Geophys. Res.*, **89**, 5381, 1984.

Buck, G.J., Force-free magnetic field solution in toroidal coordinates, *J. Appl. Phys.*, **36**, 2231, 1965.

Burlaga, L.F., Micro-scale structures in the interplanetary medium, *Solar Phys.*, **4**, 67, 1968.

Burlaga, L.F., Directional discontinuities in the interplanetary magnetic field, *Solar Phys.*, **7**, 54, 1969a.

Burlaga, L.F., Large velocity discontinuities in the interplanetary magnetic field, *Solar Phys.*, **7**, 72, 1969b.

Burlaga, L.F., A reverse hydromagnetic shock in the solar wind, *Cosmic Electrodyn.*, **1**, 233, 1970.

Burlaga, L.F., Hydrodynamic waves and discontinuities in the solar wind, *Space Sci. Rev.*, **5**, 600, 1971a.

Burlaga, L.F., Nature and origin of directional discontinuities, *J. Geophys. Res.*, **76**, 4360, 1971b.

Burlaga, L.F., Microstructure of the interplanetary medium, in *Solar Wind,* edited by C.P. Sonett, P.J. Coleman, Jr., and J.M. Wilcox, p. 309, NASA Spec. Publ., 308, Washington, DC, 1972.

Burlaga, L.F., Interplanetary stream interfaces, *J. Geophys. Res.*, **79**, 3717, 1974.

Burlaga, L.F., Interplanetary streams and their interaction with the earth, *Space Sci. Rev.*, **17**, 327, 1975.

Burlaga, L.F., Magnetic fields, plasmas, and coronal holes: The inner solar system, *Space Sci. Rev.*, **23**, 201, 1979.

Burlaga, L.F., Corotating pressure waves without fast streams in the solar wind, *J. Geophys. Res.*, **88**, 6085, 1983.

Burlaga, L.F., MHD processes in the outer heliosphere, *Space Sci. Rev.*, **39**, 255, 1984.

Burlaga, L.F., Interaction regions in the distant heliosphere, in *Solar Wind Six,* edited by V.J. Pizzo, T.E. Holzer, and D.G. Sime, p. 547, NCAR Technical Note 306 + Proc, p. 547, Boulder, CO, 1987.

Burlaga, L.F., Magnetic clouds: Constant alpha force-free configurations, *J. Geophys. Res.*, **93**, 7217, 1988a.

Burlaga, L.F., Period doubling in the outer heliosphere, *J. Geophys. Res.*, **93**, 4103, 1988b.

Burlaga, L.F., A heliospheric vortex street?, *J. Geophys. Res.*, **95**, 2229, 1990a.

Burlaga, L.F., Heliospheric shocks and catastrophe theory, *Geophys. Res. Lett.*, **17**, 1633, 1990b.

Burlaga, L.F., Multifractal structure of the interplanetary magnetic field near 25 AU, *J. Geophys. Res.*, **18**, 69, 1991a.

Burlaga, L.F., Intermittent turbulence in the solar wind, *J. Geophys. Res.*, **96**, 5847, 1991b.

Burlaga, L.F., Magnetic clouds, Chapter 5 in *Physics of the Inner Heliosphere*, Vol. 2, edited by R. Schwenn and E. Marsch, p. 1, Springer-Verlag, Berlin-Heidelberg, 1991c.

Burlaga, L.F., Multifractal structure in recurrent streams at 1 AU and near 6 AU, *J. Geophys. Res.*, **18**, 1651, 1992a.

Burlaga, L.F., Multifractal structure of the interplanetary magnetic field and plasma in recurrent streams at 1 AU, *J. Geophys. Res.*, **97**, 4283, 1992b.

Burlaga, L.F., Intermittent turbulence in large-scale speed fluctuations at 1 AU near solar maximum, *J. Geophys. Res.*, **98**, 17467, 1993.

Burlaga, L.F., and E. Barouch, Interplanetary stream magnetism: Kinematic effects, *Astrophys. J.*, **203**, 257, 1976.

Burlaga, L.F., and K.W. Behannon, Magnetic clouds: Voyager observations between 2 and 4 AU, *Solar Phys.*, **81**, 181, 1982.

Burlaga, L.F., and J.K. Chao, Reverse and forward slow shocks in the solar wind, *J. Geophys. Res.*, **76**, 7516, 1971.

Burlaga, L.F., and M.L. Goldstein, Radial variations of large-scale magnetohydrodynamic fluctuations in the solar wind, *J. Geophys. Res.*, **89**, 6813, 1984.

Burlaga, L.F., and J.H. King, Intense interplanetary magnetic fields observed by geocentric spacecraft during 1963–1975, *J. Geophys. Res.*, **84**, 6633, 1979.

Burlaga, L.F., and L.W. Klein, Configurations of corotating shocks in the outer heliosphere, *J. Geophys. Res.*, **91**, 8975, 1986a.

Burlaga, L.F., and L.W. Klein, Fractal structure of the interplanetary magnetic field, *J. Geophys. Res.*, **91**, 347, 1986b.

Burlaga, L.F., and J.F. Lemaire, Interplanetary magnetic holes: Theory, *J. Geophys. Res.*, **83**, 5127, 1978.

Burlaga, L.F., and W.H. Mish, Large-scale fluctuations in the interplanetary medium, *J. Geophys. Res.*, **92**, 1261, 1987.

Burlaga, L.F., and N.F. Ness, Macro- and microstructure of the interplanetary magnetic field, *Can. J. Phys.*, **46**, S962, 1968.

Burlaga, L.F., and N.F. Ness, Tangential discontinuities in the solar wind at 1 AU, *Solar Phys.*, **9**, 467, 1969.

Burlaga, L.F., and N.F. Ness, Large-scale distant heliospheric magnetic field: Voyager 1 and 2 observations from 1986 through 1989, *J. Geophys. Res.*, **98**, 17451, 1993a.

Burlaga, L.F., and N.F. Ness, Radial and latitudinal variations of the magnetic field strength in the outer heliosphere, *J. Geophys. Res.*, **98**, 3539, 1993b.

Burlaga, L.F., and K.W. Ogilvie, Causes of sudden commencements and sudden impulses, *J. Geophys. Res.*, **74**, 2815, 1969.

Burlaga, L.F., and K.W. Ogilvie, Heating of the solar wind, *Astrophys. J.*, **159**, 659, 1970a.

Burlaga, L.F., and K.W. Ogilvie, Magnetic and thermal pressures in the solar wind, *Solar Phys.*, **15**, 61, 1970b.

Burlaga, L.F., and K.W. Ogilvie, Solar wind temperature and speed, *J. Geophys. Res.*, **78**, 2028, 1973.

Burlaga, L.F., and J.D. Scudder, Motion of shocks through interplanetary streams, *J. Geophys. Res.*, **80**, 4044, 1975.

Burlaga, L.F., K.W. Ogilvie, and D.H. Fairfield, Microscale fluctuations in the interplanetary magnetic field, *Astrophys. J.*, **155**, L171, 1969.

Burlaga, L.F., K. Ogilvie, D. Fairfield, M. Montgomery, and S. Bame, Energy transfer at colliding streams in the solar wind, *Astrophys. J.*, **164**, 137, 1971.

Burlaga, L.F., J.F. Lemaire, and J.M. Turner, Interplanetary current sheets at 1 AU, *J. Geophys. Res.*, **82**, 3191, 1977.

Burlaga, L.F., K.W. Behannon, S.F. Hansen, G.W. Pneuman, and W.C. Feldman, Sources of magnetic fields in recurrent interplanetary streams, *J. Geophys. Res.*, **83**, 4177, 1978a.

Burlaga, L.F., N.F. Ness, F. Mariani, B. Bavassano, U. Villante, H. Rosenbauer, R. Schwenn, and J. Harvey, Magnetic field and flows between 1 and 0.3 AU during the primary mission of Helios 1, *J. Geophys. Res.*, **83**, 5167, 1978b.

Burlaga, L.F., R. Lepping, R. Weber, T. Armstrong, C. Goodrich, J. Sullivan, D. Gurnett, P. Kellogg, E. Keppler, F. Mariani, F. Neubauer, H. Rosenbauer, and R. Schwenn, Interplanetary particles and fields, November 22 to December 6, 1977: Helios, Voyager and IMP observations between 0.6 AU and 1.6 AU, *J. Geophys. Res.*, **85**, 2227, 1980.

Burlaga, L.F., A.J. Hundhausen, and X.-P. Zhao, The coronal and interplanetary current sheet in early 1976, *J. Geophys. Res.*, **86**, 8893, 1981a.

Burlaga, L.F., E. Sittler, F. Mariani, and R. Schwenn, Magnetic loop behind an interplanetary shock: Voyager, Helios and IMP-8 observations, *J. Geophys. Res.*, **86**, 6673, 1981b.

Burlaga, L.F., L.W. Klein, N.R. Sheeley, Jr., D.J. Michels, R.A. Howard, M.J. Koomen, R. Schwenn, and H. Rosenbauer, A magnetic cloud and coronal mass ejection, *Geophys. Res. Lett.*, **9**, 1317, 1982a.

Burlaga, L.F., R. Lepping, K.W. Behannon, L.W. Klein, and F.M. Neubauer, Large-scale variations in the IMF: Voyager 1 and 2 observations between 1 and 5 AU, *J. Geophys. Res.*, **87**, 4345, 1982b.

Burlaga, L.F., R. Schwenn, and H. Rosenbauer, Dynamical evolution of interplanetary magnetic fields and flows between 0.3 AU and 8.5 AU: Entrainment, *Geophys. Res. Lett.*, **10**, 413, 1983.

Burlaga, L.F., L. Klein, R.P. Lepping, and K.W. Behannon, Large-scale interplanetary magnetic fields: Voyager 1 and 2 observations between 1 AU and 9.5 AU, *J. Geophys. Res.*, **89**, 10659, 1984a.

Burlaga, L.F., F.B. McDonald, N.F. Ness, R. Schwenn, A.J. Lazarus, and F. Mariani, Interplanetary flow systems associated with cosmic ray modulation in 1977–1978, *J. Geophys. Res.*, **89**, 6579, 1984b.

Burlaga, L.F., F.B. McDonald, M.L. Goldstein, and A.J. Lazarus, Cosmic ray modulation and turbulent interaction regions near 11 AU, *J. Geophys. Res.*, **90**, 12127, 1985a.

Burlaga, L.F., V. Pizzo, A. Lazarus, and P. Gazis, Stream dynamics between 1 AU and 2 AU: A comparison of observations and theory, *J. Geophys. Res.,* **90,** 7377, 1985b.

Burlaga, L.F., F.B. McDonald, and R. Schwenn, Formation of a compound stream between 0.85 and 6.2 AU and its effects on solar energetic particles and galactic cosmic rays, *J. Geophys. Res.,* **91,** 13331, 1986.

Burlaga, L.F., K.W. Behannon, and L.W. Klein, Compound streams, magnetic clouds and major magnetic storms, *J. Geophys. Res.,* **92,** 5725, 1987a.

Burlaga, L.F., N.F. Ness, and F.B. McDonald, Large-scale fluctuations in B between 13 AU and 22 AU and their effects on cosmic rays, *J. Geophys. Res.,* **92,** 13647, 1987b.

Burlaga, L.F., W.H. Mish, and D.A. Roberts, Large-scale fluctuations at 1 AU: 1978–1982, *J. Geophys. Res.,* **94,** 177, 1989.

Burlaga, L.F., R. Lepping, and J. Jones, Global configuration of a magnetic cloud, in *Physics of Flux Ropes,* edited by C.T. Russell, E.R. Priest, and L.C. Lee, p. 373, AGU Geophysical Monograph 58, American Geophysical Union, Washington, DC, 1990a.

Burlaga, L.F., W.H. Mish, and Y.C. Whang, Coalescence of recurrent streams of different sizes and amplitudes, *J. Geophys. Res.,* **95,** 4247, 1990b.

Burlaga, L.F., J.D. Scudder, L.W. Klein, and P.A. Isenberg, Pressure balanced structures between 1 AU and 24 AU and their implications for solar wind electrons and interstellar pickup ions, *J. Geophys. Res.,* **95,** 2229, 1990c.

Burlaga, L.F., F.B. McDonald, N.F. Ness, and A.J. Lazarus, Multifractal structure of the interplanetary magnetic field: Voyager 2 observations near 25 AU, 1987–1988, *Geophys. Res. Lett.,* **18,** 69, 1991.

Burlaga, L.F., F.B. McDonald, and N.F. Ness, Cosmic ray modulation and the distant heliospheric magnetic field: Voyager 1 and 2 observations from 1986 through 1989, *J. Geophys. Res.,* **98,** 1, 1993a.

Burlaga, L.F., J. Perko, and J. Pirraglia, Cosmic ray modulation, merged interaction regions and multifractals, *Astrophys. J.,* **407,** 347, 1993b.

Burlaga, L.F., N.F. Ness, J. Belcher, and A. Szabo, Pickup protons and pressure balanced structures in MIRs: Voyager 2 observations at 35 AU, *J. Geophys. Res.,* **98,** 21511, 1994.

Cane, H.V., The evolution of interplanetary shocks, *J. Geophysical Res.,* **87,** 191, 1985.

Cane, H.V., Cosmic ray decreases and magnetic clouds, *J. Geophys. Res.,* **98,** 3509, 1993.

Chandrasekhar, S., and P.C. Kendall, On force-free magnetic fields, *Astrophys. J.,* **126,** 457, 1957.

Chao, J.K., and R.P. Lepping, A correlative study of SSC's, interplanetary shocks, and solar activity, *J. Geophys. Res.,* **79,** 1799, 1974.

Chao, J.K., and S. Olbert, Observations of slow shocks in interplanetary space, *J. Geophys. Res.,* **75,** 639, 1970.

Chao, J.K., R.L. Lepping, and J. Binsak, A reverse shock associated with a stream–stream interaction: The February 29, 1968, event. *J. Geophys. Res.,* **79,** 2767, 1974.

Chapman, S., and J. Bartels, *Geomagnetism,* Chapter 12, Oxford University Press, New York, 1940.

Chapman, S., and V.C.A. Ferraro, Solar streams of corpuscles, their geometry, absorption of light and penetration, *Mon. Not. R. Astron. Soc.,* **89,** 470, 1929.

Chapman, S., and V.C.A. Ferraro, *Terr. Magn. Atmos. Electr.*, **45**, 245, 1940.
Chen, J., Effects of toroidal forces in current loops embedded in a background plasma, *Astrophys. J.*, **338**, 453, 1989.
Chen, J., Dynamics, catastrophe and magnetic energy release of toroidal solar current loops, in *Physics of Magnetic Flux Ropes*, edited by C.T. Russell, E.R. Priest, and L.C. Lee, p. 269, Geophysical Monograph 58, American Geophysical Union, Washington, DC, 1990.
Chen, J., A current loop model of magnetic clouds, in *Proceedings of the First SOLTIP Symposium, Vol. 1*, edited by S. Fisher and M. Vandas, p. 79, Astronomical Institute of the Czechoslovak Academy of Sciences, Prague, 1992.
Chen, J., and D.A. Garren, Interplanetary magnetic clouds: Topology and driving mechanism, *Geophys. Res. Lett.*, **20**, 2319, 1993.
Cliver, E.W., J.D. Mihalov, N.R. Sheeley, Jr., R.A. Howard, M.J. Koomen, and R. Schwenn, Solar activity and heliosphere-wide cosmic ray modulation in mid-1982, *J. Geophys. Res.*, **92**, 8487, 1987.
Cocconi, G., T. Gold, K. Greisen, S. Hayakawa, and J.P. Morrison, The cosmic ray flare effect, *Nuovo Cimento*, **8**, 161, 1958.
Colburn, D.S., and C.P. Sonett, Discontinuities in the solar wind, *Space Sci. Rev.*, **9**, 467, 1966.
Coleman, P.J., Jr., Turbulence, viscosity and dissipation in the solar wind plasma, *Astrophys. J.*, **153**, 371, 1968.
Coles, W.A. and S. Maagoe, Solar-wind velocity from IPS observations, *J. Geophys. Res.*, **77**, 5622, 1972.
Coles, W.A., and B.J. Rickett, IPS observations of the solar wind speed out of the ecliptic, *J. Geophys. Res.*, **81**, 4797, 1976.
Coles, W.A., and B.J. Rickett, Interplanetary scintillation observations of the solar wind at high latitudes, in *The Sun and Heliosphere in Three Dimensions*, edited by R.G. Marsden, p. 143, D. Reidel, Dordrecht, 1986.
Collard, H.R., and J.H. Wolfe, Radial gradient of solar wind velocity from 1 to 15 AU, in *Solar Wind Three*, edited by C.T. Russell, p. 281, University of California Press, Los Angeles, 1974.
Collard, H.R., J.D. Mihalov, and J.H. Wolfe, Radial variation of the solar wind speed between 1 and 15 AU, *J. Geophys. Res.*, **87**, 2203, 1982.
Coniglio, A., Multifractal structure of clusters and growing aggregates, *Physica*, **140A**, 51, 1986.
Crooker, N.U., J.T. Gosling, E.J. Smith, and C.T. Russell, A bubblelike coronal mass ejection flux rope in the solar wind, in *Physics of Magnetic Flux Ropes*, edited by C.T. Russell, E.R. Priest, and L.C. Lee, pp. 365–372, AGU Geophysical Monograph 58, American Geophysical Union, Washington, DC, 1990.
Decker, R.B., Particle acceleration at shocks with surface ripples, *J. Geophys. Res.*, **95**, 11993, 1990.
Denskat, K.U., and L.F. Burlaga, Multispacecraft observations of microscale fluctuations in the solar wind, *J. Geophys. Res.*, **82**, 2693, 1977.
Denskat, K.U., and F.M. Neubauer, Statistical properties of low-frequency magnetic field fluctuations in the solar wind from 0.29 to 1.0 AU during solar minimum conditions: Helios 1 and Helios 2, *J. Geophys. Res.*, **87**, 2215, 1982.
Denskat, K.U., and F.M. Neubauer, Observations of hydromagnetic turbulence in

the solar wind, in *Solar Wind Five,* edited by M. Neugebauer, p. 81, NASA Conf. Publ. 2280, Washington, DC, 1983.

Detman, T.R., M. Dryer, T. Yeh, S.M. Han, S.T. Wu, and D.J. McComas, A time-dependent, three-dimensional MHD numerical study of interplanetary magnetic field draping around plasmoids in the solar wind, *J. Geophys. Res.,* **96,** 9531, 1991.

Dioddato, L., G. Moreno, C. Signorini, and K.W. Ogilvie, Long-term variations of the solar wind parameters, *J. Geophys. Res.,* **79,** 5095, 1974.

Dryer, M., Interplanetary shock waves: Recent developments, *Space Sci. Rev.,* **17,** 277, 1975.

Dryer, M., Coronal transient phenomena, *Space Sci. Rev.,* **33,** 233, 1982.

Dryer, M., and D.F. Smart, Dynamical models of coronal transients and interplanetary disturbances, *Adv. Space Res.,* **4,** 291, 1984.

Dryer, M., and R.S. Steinolfson, MHD solutions of interplanetary disturbances generated by simulated velocity perturbations, *J. Geophys. Res.,* **81,** 5413, 1976.

Dryer, M., C. Candelaria, Z.K. Smith, R.S. Steinolfson, E.J. Smith, J.H. Wolfe, J.D. Mihalov, and P. Rosenau, Dynamic MHD modeling of the solar wind disturbances during the August 1972 events, *J. Geophys. Res.,* **83,** 532, 1978a.

Dryer, M., Z.K. Smith, J.D. Mihalov, J.H. Wolfe, R.S. Steinolfson, and S.T. Wu, Dynamic MHD modeling of solar wind corotating stream interaction regions observed by Pioneer 10 and 11, *J. Geophys. Res.,* **83,** 4347, 1978b.

Dryer, M.S., S.T. Wu, G. Gislason, S.M. Han, Z.K. Smith, J.F. Whang, D.F. Smart, and M.A. Shea, Magnetohydrodynamic modeling of interplanetary disturbances between the Sun and Earth, *Astrophys. Space Sci.,* **105,** 187, 1984.

Dryer, M.S., S.T. Wu, and S.T. Han, Three-dimensional, time-dependent, MHD model of a solar flare-generated interplanetary shock wave, in *The Sun and the Heliosphere in Three Dimensions,* edited by R.G. Marsden, p. 135, D. Reidel, D. Dordrecht, 1986.

Farrugia, C.J., and L.F. Burlaga, A fast-moving magnetic cloud and features of its interaction with the dayside magnetosheath, in *The Solar Wind–Magnetosphere System,* edited by S. Bauer and H.K. Biernat, Austrian Academy of Sciences Press, Vienna, p. 33, 1994.

Farrugia, C.J., M.W. Dunlop, F. Geurts, A. Balogh, D.J. Southwood, D.A. Bryant, M. Neugebauer, and A. Etemadi, An interplanetary magnetic field structure oriented at a large (80°) angle to the Parker spiral, *Geophys. Res. Lett.,* **17,** 1025–1028, 1990.

Farrugia, C.J., M.W. Dunlop, S. Ellion, M.P. Freeman, A. Balough, S.W.H. Cowley, R.P. Lepping, and D.G. Sibeck, Multipoint observations of planar interplanetary magnetic field structures, *J. Atmos. Terr. Phys.,* **53,** 1039, 1991.

Farrugia, C.J., L.F. Burlaga, V. Osherovich, and R.P. Lepping, A comparative study of expanding force-free constant alpha magnetic configurations with application to magnetic clouds, in *Solar Wind Seven,* edited by E. Marsch and R. Schwenn, p. 611, Pergamon Press, New York, 1992a.

Farrugia, C.J., V.A. Osherovich, and L.F. Burlaga, Model of a rotating magnetic cloud, in *Study of the Solar–Terrestrial System,* Proceedings of the 26th ESLAB Symposium ESA SP-346, p. 231, 1992b.

Farrugia, C.J., V.A. Osherovich, L.F. Burlaga, R.P. Lepping, and M.P. Freeman,

Radial expansion of an ideal MHD configuration and the temporal development of the magnetic field, in *Solar Wind Seven,* edited by E. Marsch and R. Schwenn, p. 615, Pergamon Press, New York, 1992c.

Farrugia, C.J., L.F. Burlaga, V.A. Osherovich, I.G. Richardson, M.P. Freeman, R.P. Lepping, and A.J. Lazarus, A study of an expanding interplanetary magnetic cloud and its interaction with the earth's magnetosphere: The interplanetary aspect, *J. Geophys. Res.,* **98,** 7621, 1993a.

Farrugia, C.J., I.G. Richardson, L.F. Burlaga, R.P. Lepping, and V.A. Osherovich, Simultaneous observations of solar MeV particles in a magnetic cloud and in the earth's northern tail lobe: Implications for the global field line topology of magnetic clouds, and for the entry of solar particles into the magnetosphere during cloud passage, *J. Geophys. Res.,* **98,** 15497, 1993b.

Farrugia, C.J., R.J. Fitzenreiter, L.F. Burlaga, N.V. Erkaev, V.A. Osherovich, H.K. Biernat, and A. Fazakerly, Observations in the sheath region ahead of a magnetic cloud and in the dayside magnetosheath during magnetic cloud passage, *Adv. Space Res.,* **14**:7, 105, 1993c.

Farrugia, C.J., L.F. Burlaga, V.A. Osherovich, and R.P. Lepping, The magnetic flux rope versus the spheromak as models for interplanetary magnetic clouds, submitted to *J. Geophys. Res.,* 1994.

Feldman, W.C., J.R. Asbridge, S.J. Bame, M.D. Montgomery, and S.P. Gary, Solar wind electrons, *J. Geophys. Res.,* **80,** 4141, 1975.

Feldman, W.C., J.R. Asbridge, S.J. Bame, and J.T. Gosling, High-speed solar wind flow parameters at 1 AU, *J. Geophys. Res.,* **81,** 5054, 1976.

Feldman, W.C., J.R. Asbridge, S.J. Bame, and J.T. Gosling, Long-term variations of selected solar wind properties, *J. Geophys. Res.,* **83,** 2177, 1978.

Feldman, W.C., J.R. Asbridge, S.J. Bame, and D.S. Lemons, The core electron temperature profile between 0.5 and 1.0 AU in the steady-state high speed solar wind, *J. Geophys. Res.,* **84,** 4463, 1979.

Feldman, W.C., J.R. Asbridge, S.J. Bame, and J.T. Gosling, *J. Geophys. Res.,* **86,** 5408, 1981.

Ferraro, V.C.A., and C. Plumpton, *An Introduction to Magneto-fluid Dynamics,* Clarendon Press, Oxford, 1966.

Fitzenreiter, R.J., and L.F. Burlaga, Structure of current sheets in magnetic holes at 1 AU, *J. Geophys. Res.,* **83,** 5579, 1978.

Forbush, S.E., Jr., Worldwide cosmic ray variations, 1936–1952, *J. Geophys. Res.,* **59,** 525, 1954.

Forbush, S.E., Jr., Cosmic ray intensity variations during two solar cycles, *J. Geophys. Res.,* **63,** 651, 1958.

Formisano, V., F. Moreno, and E. Amata, Relationships among the interplanetary plasma parameters: Heos 1, December 1968 to December 1969, *J. Geophys. Res.,* **79,** 5109, 1974.

Friere, G.F., Force-free magnetic field problem, *Am. J. Phys.,* **34,** 567, 1966.

Frisch, U., P-L. Sulem, and M. Nelkin, A simple dynamical model of intermittent fully developed turbulence, *J. Fluid Mech.,* **87,** 719, 1978.

Fry, C.D., and S.-I. Akasofu, Three-dimensional structure of the heliospheric current sheet, in *The Sun and the Heliosphere in Three Dimensions,* edited by R.G. Marsden, p. 287, D. Reidel, Dordrecht, 1986.

Gazis, P.R., Observations of plasma bulk parameters and the energy balance between 1 and 10 AU, *J. Geophys. Res.,* **89,** 775, 1984.

Gazis, P.R., Solar wind stream structure at large heliocentric distances: Pioneer observations, *J. Geophys. Res.*, **92**, 2231, 1987.

Gazis, P.R., and A.J. Lazarus, Voyager observations of solar wind temperature: 1–10 AU, *Geophys. Res. Lett.*, **9**, 431, 1982.

Gazis, P.R., and A.J. Lazarus, The radial evolution of the solar wind, 1–10 AU, in *Solar Wind Five*, edited by M. Neugebauer, p. 509, NASA Conf. Publ. 2280, Washington, DC, 1983.

Gazis, P.R., A.J. Lazarus, and K. Hester, Shock evolution in the outer heliosphere: Voyager and Pioneer observations, *J. Geophys. Res.*, **90**, 9454, 1985.

Gazis, P.R., A. Barnes, and A.J. Lazarus, Intercomparison of Voyager and Pioneer plasma observations, in *Solar Wind Six*, edited by V.J. Pizzo, T.E. Holzer, and D.G. Sime, p. 533, NCAR Technical Note 306 + Proc, Boulder, CO, 1988.

Gazis, P.R., J.D. Mihalov, A. Barnes, A.J. Lazarus, and E.J. Smith, Pioneer and Voyager observations of the solar wind at large heliospheric distances and latitudes, *Geophys. Res. Lett.*, **16**, 223, 1989.

Gazis, P.R., A. Barnes, J.D. Mihalov, and A.J. Lazarus, Solar wind temperature observations in the outer heliosphere, in *Solar Wind Seven*, edited by E. Marsch and R. Schwenn, p. 179, Pergamon Press, New York, 1992.

Geranios, A., A search for the origin of very low electron temperatures, *Planet. Space Sci.*, **26**, 571, 1978.

Geranios, A., Plasma temperature depressions and the open magnetic field lines model, *Astrophys. Space Sci.*, **77**, 167, 1981.

Geranios, A., Magnetically closed regions in the solar wind, *Astrophys. Space Sci.*, **81**, 103, 1982.

Geranios, A., Statistical analysis of magnetically closed structures, *Planet. Space Sci.*, **35**, 727, 1987.

Gibson, C.H., G.R. Stegen, and S. McConnell, Measurements of the universal constant in Kolmogoroff's third hypothesis for high Reynolds number turbulence, *Phys. Fluids*, **13**, 2448, 1970.

Gilmore, Robert, *Catastrophe Theory for Scientists and Engineers*, John Wiley & Sons, New York, 1981.

Gold, T., Contribution to discussion, in *Gas Dynamics of Cosmic Clouds*, edited by H.C. van de Hulst and J.M. Burgers, p. 103, North-Holland, Amsterdam, 1955.

Gold, T., Magnetic field in the solar system, *Nuovo Cimento, Suppl.*, **13**, Ser. X, 318, 1959.

Gold, T., Magnetic storms, *Space Sci. Rev.*, **1**, 100, 1962.

Gold, T., and F. Hoyle, On the origin of solar flares, *Mon. Not. R. Astron. Soc.* **120**:2, 89, 1960.

Goldstein, H., On the field configuration in magnetic clouds, in *Solar Wind Five*, edited by M. Neugebauer, p. 731, NASA Conf. Publ. 2280, Washington, DC, 1983.

Goldstein, B.E., and J.R. Jokipii, Effects of stream-associated fluctuations upon radial variations of average solar wind parameters, *J. Geophys. Res.*, **82**, 1095, 1077.

Goldstein, M.L., L.F. Burlaga, and W.H. Matthaeus, Power spectral signatures of interplanetary corotating and transient flows, *J. Geophys. Res.*, **89**, 3747, 1984.

Gosling, J.T., Coronal mass ejections and magnetic flux ropes in interplanetary space, in *Physics of Magnetic Flux Ropes*, edited by C.T. Russell, E.R. Priest,

and L.C. Lee, p. 344, AGU Geophysical Monograph 58, American Geophysical Union, Washington DC, 1990.
Gosling, J.T., and S.J. Bame, Solar wind variations 1964–1967: An autocorrelation analysis, *J. Geophys. Res.*, **77**, 12, 1972.
Gosling, J.T., and D.J. McComas, Field line draping about fast coronal mass ejecta. A source of strong out-of-ecliptic interplanetary magnetic fields, *Geophys. Res. Lett.*, **14**, 355, 1987.
Gosling, J.T., J.R. Asbridge, S.J. Bame, A.J. Hundhausen, and I.B. Strong, Satellite observations of interplanetary shock waves, *J. Geophys. Res.*, **73**, 43, 1968.
Gosling, J.T., R.T. Hansen, and S.J. Bame, Solar wind speed distributions: 1962–1970, *J. Geophys. Res.*, **76**, 1811, 1971.
Gosling, J.T., J.R. Asbridge, S.J. Bame, and W.C. Feldman, Solar wind speed variations: 1962–1974, *J. Geophys. Res.*, **81**, 5061, 1976a.
Gosling, J.T., A.J. Hundhausen, and S.J. Bame, Solar wind stream evolution at large heliocentric distances: Experimental demonstration and the test of a model, *J. Geophys. Res.*, **81**, 2111, 1976b.
Gosling, J.T., J.R. Asbridge, S.J. Bame, and W.C. Feldman, Solar wind stream interfaces, *J. Geophys. Res.*, **83**, 1401, 1978.
Gosling, J., T.G. Borrini, J.R. Asbridge, S.J. Bame, W.C. Feldman, and R.T. Hansen, Coronal streamers in the solar wind at 1 AU, *J. Geophys. Res.*, **86**, 5438, 1981.
Gosling, J.T., J.R. Asbridge, S.J. Bame, W.C. Feldman, R.D. Zwickl, G. Paschmann, N. Sckopke, and C.T. Russell, A sub-Alfvénic solar wind: Interplanetary and magnetosheath observations, *J. Geophys. Res.*, **87**, 239, 1982.
Gosling, J.T., D.N. Baker, S.J. Bame, W.C. Feldman, and R. Zwickl, Bidirectional solar wind electron heat flux events, *J. Geophys. Res.*, **92**, 8519, 1987.
Gosling, J.T., S.J. Bame, E. J. Smith, and M.E. Burton, Forward–reverse shock pairs associated with transient disturbances in the solar wind at 1 AU, *J. Geophys. Res.*, **93**, 8741, 1988.
Gurnett, D.A., Waves and instabilities, Chapter 9 in *Physics of the Inner Heliosphere II*, edited by R. Schwenn and E. Marsch, p. 135, Springer-Verlag, Berlin-Heidelberg, 1991.
Hada, T., and C.F. Kennel, Nonlinear evolution of slow waves in the solar wind, *J. Geophys. Res.*, **90**, 531, 1985.
Hakamada, K., and S.-I. Akasofu, A cause of solar wind speed variations observed at 1 AU, *J. Geophys. Res.*, **86**, 1290, 1981.
Hakamada, K., and S.-I. Akasofu, Simulation of three-dimensional solar wind disturbances as a result of geomagnetic storms, *Space Sci. Rev.*, **31**, 3, 1982.
Hakamada, K., and L. Munakata, A cause of the solar wind speed variations: An update, *J. Geophys. Res.*, **89**, 357, 1984.
Halsey, T.C., M.H. Jensen, L.P. Kadanoff, I. Procaccia, and B.I. Shraiman, Fractal measures and their singularities: The characterization of strange sets, *Phys. Rev. A*, **33**:2, 1141, 1986.
Hansen, R.T., S.F. Sawyer, and R.T. Hansen, K-Corona and magnetic sector boundaries, *Geophys. Res. Lett.*, **1**, 13, 1974.
Harvey, K.L., N.R. Sheeley, Jr., and J.W. Harvey, Magnetic measurements of coronal holes during 1975–1980, *Solar Phys.*, **74**, 131, 1981.
Heineman, M.A., and G.L. Siscoe, Shape of strong shock fronts in an inhomogeneous solar wind, *J. Geophys. Res.*, **79**, 1349, 1974.

Hentschel, H.G.E., and I. Procaccia, The infinite number of generalized dimensions of fractals and strange attractors, *Physica,* **8D,** 435, 1983.

Hewish, A., and S. Bravo, The sources of large-scale heliospheric disturbances, *Solar Phys.,* **106,** 185, 1986.

Hewish, A., and P.A. Dennison, Measurements of the solar wind and the small-scale structure of the interplanetary medium, *J. Geophys. Res.,* **72,** 1977, 1967.

Hewish, A., and M.D. Symonds, Radio investigation of solar plasma, *Planet. Space Sci.,* **17,** 313, 1969.

Hewish, A., S.J. Tappin, and G.R. Gapper, Origin of interplanetary shocks, *Nature,* **314,** 137, 1985.

Hida, K., An approximate study on the detached shock wave in front of a circular cylinder and a sphere, *J. Phys. Soc. Japan,* **8**:6, 740, 1953.

Hirshberg, J., The transport of flare plasma from the sun to the earth, *Planet. Space Sci.,* **16,** 309, 1968.

Hirshberg, J., Y. Nakagawa, and R.E. Wellck, Propagation of sudden disturbances through a nonhomogeneous solar wind, *J. Geophys. Res.,* **79,** 3726, 1974.

Hoeksema, J.T., The relationship of the large-scale solar field to the interplanetary magnetic field: What will Ulysses find? in *The Sun and the Heliosphere in Three Dimensions,* edited by R.G. Marsden, p. 241, D. Reidel, Dordrecht, 1986.

Hoeksema, J.T., Extending the sun's magnetic field through the three-dimensional heliosphere, *Adv. Space Res.,* **9**:4, 141–152, 1989.

Hoeksema, J.T., Large-scale structure of the heliospheric magnetic field: 1976–1991, in *Solar Wind Seven,* edited by E. Marsch and R. Schwenn, p. 191, Pergamon Press, New York, 1992.

Hoeksema, J.T., and S.T. Suess, The outer magnetic field, in *Solar Magnetic Fields,* edited by G. Poletto, p. 2, *Memorie della Societá Astronomica Italiana,* Vol. 62, 1990.

Hoeksema, J.T., J.M. Wilcox, and P.H. Scherrer, Structure of the heliospheric current sheet in the early portion of sunspot cycle 21, *J. Geophys. Res.,* **87,** 10331, 1982.

Hoeksema, J.T., J.M. Wilcox, and P.H. Scherrer, The structure of the heliospheric current sheet: 1978–1982, *J. Geophys. Res.,* **88,** 9910, 1983.

Hollweg, J.V., Surface waves on solar wind tangential discontinuities, *J. Geophys. Res.,* **87,** 80655, 1982.

Holm, D., and B.A. Kupershmidt, Noncanonical Hamiltonian formulation of ideal magnetohydrodynamics, *Physica,* **7D,** 330, 1980.

Holzer, T.E., The solar wind and astrophysical phenomena, in *Solar System Plasma Physics,* Vol. 1, edited by E.N. Parker, C.F. Kennell, and L.J. Lanzerotti, p. 101, North-Holland Publishing Co., New York, 1979.

Howard, R.A., and M.J. Koomen, Observations of sectored structure in the outer solar corona: Correlation with interplanetary magnetic field, *Solar Phys.* **37,** 469, 1974.

Hsieh, K.C., and A.K. Richter, The importance of being earnest about shock fitting, *J. Geophys. Res.,* **91,** 4157, 1986.

Hu, Y.Q., and S.R. Habbal, Double shock pairs in the solar wind, *J. Geophys. Res.,* **98,** 3551, 1993.

Hudson, P.D., Discontinuities in anisotropic plasma and their identification in the solar wind, *Planet. Space Sci.,* **18,** 1611, 1970.

Hudson, P.D., Rotational discontinuities in an anisotropic plasma, *Planet. Space Sci.*, **19,** 1693, 1971.
Hundhausen, A.J., *Coronal Expansion and Solar Wind,* Springer-Verlag, New York, 1972.
Hundhausen, A.J., An interplanetary view of coronal holes, in *Coronal Holes and High Speed Streams,* edited by J.B. Zirker, p. 225, Colorado Associated University Press, Boulder, 1977.
Hundhausen, A.J., Solar activity and the solar wind, *Rev. Geophys. Space Sci.,* **17,** 2034, 1979.
Hundhausen, A.J., The origin and propagation of coronal mass ejections, in *Solar Wind Six,* edited by V.J. Pizzo, T.E. Holzer, and D.G. Sime, p. 181, NCAR Technical Note 306 + Proc, Boulder, CO, 1988.
Hundhausen, A.J., and L.F. Burlaga, A model for the origin of solar wind stream interfaces, *J. Geophys. Res.,* **80,** 1845, 1975.
Hundhausen, A.J., and J.T. Gosling, Solar wind structure at large heliocentric distances: An interpretation of Pioneer 10 observations, *J. Geophys. Res.,* **81,** 1845, 1976.
Hundhausen, A.J., and M.D. Montgomery, Heat conduction and non-steady phenomena in the solar wind, *J. Geophys. Res.,* **76,** 2236, 1971.
Hundhausen, A.J., S.J. Bame, J.R. Asbridge, and S.J. Sydoriak, Solar wind observations from July 1965 to June 1967, *J. Geophys. Res.,* **75,** 4643, 1970.
Hundhausen, A.J., S.J. Bame, and M.D. Montgomery, Variations of solar-wind plasma properties: Vela observations of a possible heliographic latitude dependence, *J. Geophys. Res.,* **76,** 5145, 1971.
Hundhausen, A.J., T.E. Holzer, and B.C. Low, Do slow shocks precede some coronal mass ejections?, *J. Geophys. Res.,* **92,** 11173, 1987.
Intriligator, D.S., and M. Neugebauer, A search for solar wind velocity change between 0.7 AU and 1 AU, *J. Geophys. Res.,* **80,** 1332, 1975.
Ivanov, K.G., and A.F. Harshiladze, Interplanetary hydromagnetic clouds as flare generated spheromaks, *Solar Phys.,* **98,** 379, 1985.
Ivanov, K.G., A.F. Harshiladze, E.G. Eroshenko, and V.A. Styazhkin, Configuration, structure and dynamics of magnetic clouds from solar flares in light of measurements on board Vega 1 and Vega 2 in January–February, 1986, *Solar Phys.,* **120,** 407, 1989.
Jackson, B.V., R.A. Howard, N.R. Sheeley, Jr., D.J. Michels, M.J. Kooman, and R.M. Illing, Helios spacecraft and earth perspective observations of three looplike solar mass ejection transients, *J. Geophys. Res.,* **90,** 5075, 1985.
Jeffrey, A., *Magnetohydrodynamics,* Oliver and Boyd, Edinburgh, 1966.
Jeffrey, A., and T. Taniuti, *Nonlinear Wave Propagation,* Academic Press, New York, 1964.
Kahler, S.W., Solar flares and coronal mass ejections, *Annu. Rev. Astron. Astrophys.,* **30,** 113, 1992.
Kahler, S.W., and D.V. Reames, Probing magnetic topologies of magnetic clouds by means of solar energetic particle, *J. Geophys. Res.,* **96,** 9419, 1991.
Kayser, S.E., Persistence of shocks to large distances, *J. Geophys. Res.,* **90,** 3967, 1985.
Kayser, S.E., A. Barnes, and J.D. Mihalov, The far reaches of the solar wind: Pioneer 10 and 11 plasma results, *Astrophys. J.,* **285,** 339, 1984.
Kennel, C.F., F.L. Scarf, F.V. Coroniti, E.J. Smith, and D.A. Gurnett, Nonlocal

plasma turbulence associated with interplanetary shocks, *J. Geophys. Res.*, **87**, 17, 1982.

King, J., Solar cycle variations in IMF intensity, *J. Geophys. Res.*, **84**, 5983, 1979.

Klein, L.W., and L.F. Burlaga, Interplanetary sector boundaries 1971–1973, *J. Geophys. Res.*, **85**, 2269, 1980.

Klein, L.W., and L.F. Burlaga, Interplanetary magnetic clouds at 1 AU, *J. Geophys. Res.*, **87**, 613, 1982.

Klein, L.W., L.F. Burlaga, and N.F. Ness, Radial and latitudinal variations of the interplanetary magnetic field, *J. Geophys. Res.*, **92**, 9885, 1987.

Koenigl, A., and A.R. Choudhuri, Force-free equilibria of magnetized jets, *Astrophys. J.*, **289**, 173, 1985.

Koenigl, A., and A.R. Choudhuri, Errata, *Astrophys. J.*, **30**, 954, 1986.

Kojima, M., and T. Kakinuma, Solar cycle evolution of solar wind speed structure between 1973 and 1985 observed with the interplanetary scintillation method, *J. Geophys. Res.*, **92**, 7269, 1987.

Kolmogorov, A.N., Local structure of turbulence in incompressible fluid, *Dokl. Akad. Nauk SSSR*, **30**, 299, 1941.

Kolmogorov, A.N., A refinement of previous hypothesis concerning the local structure of turbulence in viscous incompressible fluid at high Reynolds number, *J. Fluid Mech.*, **13**, 82, 1962.

Korzhov, N.P., V.V. Mishin, and V.M. Tomozov, On the role of plasma parameters and the Kelvin–Helmholtz instability in a viscous interaction of solar wind streams, *Planet. Space Sci.*, **32**, 1169, 1984.

Krieger, A.S., A.F. Timothy, and E.C. Roelof, A coronal hole and its identification as the source of a high velocity solar wind stream, *Solar Phys.*, **29**, 5005, 1973.

Lamb, Sir Horace, *Hydrodynamics,* 6 ed., p. 224, Dover, New York, 1945.

Landau, L.D., and E.M. Lifshitz, *Electrodynamics of Continuous Media,* Pergamon Press, New York, 1960.

Lazarus, A.J., K.W. Ogilvie, and L.F. Burlaga, Interplanetary shock observations, made by Mariner 2 and Explorer 34, *Solar Phys.*, **13**, 232, 1970.

Lazarus, A.J., and J. Belcher, Large-scale structure of the distant solar wind and heliosphere, in *Proceedings of the Sixth International Solar Wind Conference,* edited by V.J. Pizzo, T.E. Holzer, and D.G. Sime, p. 533, NCAR Technical Note 306 + Proc, Boulder, CO, 1988.

Lazarus, A.J., and R.L. McNutt, Jr., Plasma observations in the distant heliosphere: A view from Voyager, in *Physics of the Outer Heliosphere,* edited by S. Grzedzielski and D.E. Page, p. 229, Pergamon Press, New York, 1990.

Lazarus, A.J., B. Yedidia, L. Villanueva, R.L. McNutt, Jr., J.W. Belcher, U. Villante, and L.F. Burlaga, Meridional plasma flow in the outer heliosphere, *J. Geophys. Res. Lett.* **15**, 1519, 1988.

Lemaire, J., and L.F. Burlaga, Diamagnetic boundary layers: A kinematic model, *Astrophys. Space Sci.*, **43**, 303, 1976.

Lepping, R.L., and P.D. Argentiero, Single spacecraft method of estimating shock normals, *J. Geophys. Res.*, **76**, 4349, 1971.

Lepping, R.P., and K.W. Behannon, Magnetic field directional discontinuities. 1. Minimum variance errors, *J. Geophys. Res.*, **85**, 4695, 1980.

Lepping, R.P., and K.W. Behannon, Magnetic field directional discontinuities: Characteristics between 0.46 and 1.0 AU, *J. Geophys. Res.*, **91**, 8725, 1986.

Lepping, R.P., J.A. Jones, and L.F. Burlaga, Magnetic field structure of interplanetary magnetic clouds at 1 AU, *J. Geophys. Res.*, **95**, 11957 1990.

Lepping, R.P., L.F. Burlaga, B.T. Tsurutani, K.W. Ogilvie, A.J. Lazarus, D.S. Evans, and L.W. Klein, The interaction of a very large interplanetary magnetic cloud with the magnetosphere and with cosmic rays, *J. Geophys. Res.*, **96**, 9425, 1991.

LeRoux, J.S., and M.S. Potgieter, The simulated drift features of heliospheric cosmic ray modulation with a time-dependent drift model. IV. The role of the heliospheric neutral sheet deformation, *Astrophys. J.*, **397**, 686, 1992.

Levine, R.H., and M.D. Altschuler, Representations of coronal magnetic fields including currents, *Solar Phys.*, **36**, 345, 1974.

Levine, R.H., M.D. Altschuler, and J.W. Harvey, Solar sources of the interplanetary magnetic field and solar wind, *J. Geophys. Res.*, **82**, 1061, 1977.

Levine, R.H., M. Schulz, and E.N. Frazier, Simulation of the magnetic structure of the inner heliosphere by means of a non-spherical source surface, *Solar Phys.*, **77**, 363, 1982.

Levy, E.H., The interplanetary magnetic field structure, *Nature*, **261**, 394, 1976.

Lindeman, F.A., *Philos. Mag.*, **38**, 669, 1919.

Lockwood, J.A. On the long-term variation in the cosmic radiation, *J. Geophys. Res.*, **65**, 19, 1960.

Lockwood, J.A., Forbush decreases in the cosmic radiation, *Space Sci. Rev.*, **12**, 658, 1971.

Lopez, R.E., Solar cycle invariance in solar wind proton temperature relationships, *J. Geophys. Res.*, **92**, 11189, 1987.

Lopez, R.E., and J.W. Freeman, The solar wind proton temperature–velocity relation, *J. Geophys. Res.*, **91**, 1701, 1986.

Luhman, J., C.T. Russell, and E.J. Smith, Asymmetries of the interplanetary field inferred from observations at two heliospheric distances, in *Solar Wind Six*, edited by V.J. Pizzo, T.E. Holzer, and D.C. Sime, p. 323, NCAR Technical Note 306 + Proc, Boulder, CO, 1988.

Lundquist, S., Magnetohydrostatic fields, Ark. Fys., **2**, 361, 1950.

Lust, R., and A. Schluter, Force-free magnetic fields, *Z. Astrophys.*, **34**, 353, 1954.

Mandelbrot, B., Possible refinement of the lognormal hypothesis concerning the distribution of energy dissipation in intermittent turbulence, in *Statistical Models and Turbulence*, edited by M. Rosenblatt and C. Van Atta, p. 333, Springer-Verlag, Berlin-Heidelberg 1972.

Mandelbrot, B., Intermittent turbulence in self-similar cascades: Divergence of high moments and dimension of the carrier, *J. Fluid Mech.*, **62**, part 2, 331, 1974.

Mandelbrot, B.B., *The Fractal Geometry of Nature*, W.H. Freeman, New York, 1977, 1982.

Mandelbrot, B., Multifractal measures, especially for the geophysicist, *PAGEOPH*, **131**, 5, 1989.

Mandt, M.E., J.R. Kan, and C.T. Russell, Comparison of magnetic field structures in quasi-parallel interplanetary shocks: Observations versus simulations, *J. Geophys. Res.*, **91**, 8981, 1986.

Mariani, F., and F.M. Neubauer, The interplanetary magnetic field, Chapter 4 in *Physics of the Inner Heliospere II*, Vol. 1, edited by R. Schwenn and E. Marsch, p. 183, Springer-Verlag, Berlin-Heidelberg, 1990.

Mariani, F., B. Bavassano, U. Villante, and N.F. Ness, Variations in the occurrence rate of discontinuities in the interplanetary magnetic field, *J. Geophys. Res.*, **78**, 8011, 1973.

Mariani, F., B. Bavassano, and U. Villante, Comment on "A reexamination of rotational and tangential discontinuities in the solar wind" by M. Neugebauer et al., *J. Geophys. Res.*, **90,** 5363, 1985.

Marsch, E., MHD turbulence in the solar wind, Chapter 6 in *Physics of the Inner Heliosphere,* Vol. 2, edited by R. Schwenn and E. Marsch, p. 159, Springer-Verlag, Berlin-Heidelberg, 1991.

Marsch, E., and S. Liu, Structure functions and intermittency of velocity fluctuations in the inner solar wind, *Ann. Geophys.,* **11,** 227, 1993.

Marsden, J.E., A group theoretic approach to the equations of plasma physics, *Can. Math. Bull.,* **25:**2, 129, 1982.

Marsden, R.G., T.R. Sanderson, C. Tranquille, and K.-P. Wenzel, ISEE-3 observations of low energy proton bidirectional events and their relation to isolated interplanetary magnetic structures, *J. Geophys. Res.,* **92,** 11009, 1987.

Martin, R.N., J.W. Belcher, and A.J. Lazarus, Observation and analysis of abrupt changes in the interplanetary plasma velocity and magnetic field, *J. Geophys. Res.,* **78,** 3653, 1973.

Marubashi, K., Structure of interplanetary magnetic clouds and their origins, *Adv. Space Res.,* **6:**6, 33, 1986.

Matthaeus, W.H., and M. Goldstein, Measurements of the rugged invariants of magnetohydrodynamic turbulence in the solar wind, *J. Geophys. Res.,* **87,** 6011, 1982.

Matthaeus, W.H., and M.L. Goldstein, Magnetohydrodynamic turbulence in the solar wind, in *Solar Wind Five,* edited by M. Neugebauer, p. 73, NASA Conf. Publ. 2280, Washington, DC, 1983.

Matthaeus, W.H., and M.L. Goldstein, Low-frequency $1/f$ noise in the interplanetary magnetic field, The turbulent generation of outward traveling Alfvénic fluctuations in the solar wind, *Phys. Rev. Lett.,* **57,** 495, 1986.

Maunder, E.W., Magnetic disturbances, 1882 to 1903, as recorded at the Royal Observatory, Greenwich, and their association with sunspots, *Mon. Not. R. Astron. Soc. London,* **65,** 2, 1905.

McComas, D.J., J.T. Gosling, D. Winterhalter, and E.J. Smith, Interplanetary magnetic field draping about fast coronal mass ejecta in the outer heliosphere, *J. Geophys. Res.,* **93,** 2519, 1988.

McComas, D.J., J.T. Gosling, J.L. Phillips, S.J. Bame, J.G. Luhman, and E.J. Smith, Electron heat flux dropouts in the solar wind: Evidence for interplanetary magnetic field reconnection? *J. Geophys. Res.,* **94,** 6907, 1989.

McComas, J.T. Gosling, and J.L. Phillips, Interplanetary magnetic flux: Momentum and balance, *J. Geophys. Res.,* **97,** 171, 1992.

McDonald, F.B., N. Lal, J.H. Trainor, M.A.I. Van Hollebeke, and W.R. Webber, The solar modulation of galactic cosmic rays in the outer heliosphere, *Astrophys. J.,* **249,** L71, 1981.

McDonald, F.B., N. Lal, and R.E. McGuire, The role of drifts and merged interaction regions in the long term modulation of cosmic rays, in *Proceedings of the 22nd International Cosmic Ray Conference (Dublin),* Dublin Institute for Advanced Studies, 1991.

McNutt, R.L., Jr., Possible explanation of north–south plasma flow in the outer heliosphere and meridional transport of magnetic flux, *Geophys. Res. Lett.,* **15,** 1523, 1988.

Meneveau, C., and K.R. Sreenivasan, Simple multifractal cascade model for fully developed turbulence, *Phys. Rev. Lett.,* **59,** 1424, 1987a.

Meneveau, C., and K.R. Sreenivasan, The multifractal spectrum of the dissipation field in turbulent flows, *Nuclear Phys. B., Suppl. 2*, **9,** 1987b.

Meneveau, C., and K.R. Sreenivasan, The multifractal nature of turbulent energy dissipation, *J. Fluid Mech.*, **224,** 429, 1991.

Mermin, N.D., The topological theory of defects in ordered media, *Rev. Modern Phys.*, **51,** 591, 1979.

Michel, L., Symmetry defects and broken symmetry configurations. Hidden symmetry, *Rev. Modern Phys.*, **52,** 617, 1980.

Mihalov, J.D., and J.H. Wolfe, Pioneer-10 observation of the solar wind proton temperature heliocentric gradient, *Solar Phys.*, **60,** 399, 1978.

Mihalov, J.D., and J.H. Wolfe, Pioneer 10 studies of interplanetary shocks at large heliocentric distances, *Geophys. Res. Lett.*, **6,** 491, 1979.

Miller, G., and L. Turner, Force-free equilibria in toroidal geometry, *Phys. Fluids*, **24,** 363, 1981.

Mitchel, D.G., E.C. Roelof, and J.H. Wolfe, Latitude dependence of solar wind velocity observed >1 AU, *J. Geophys. Res.*, **86,** 165, 1981.

Monin, A.S., and A.M. Yaglom, *Statistical Fluid Mechanics: Mechanics of Turbulence*, Vol. 2, MIT Press, Cambridge, MA, 1975.

Montgomery, D., Theory of hydromagnetic turbulence, in *Solar Wind Five*, edited by M. Neugebauer, p. 675, NASA Conf. Publ. 2280, 1983.

Montgomery, M.D., S.J. Bame, and A.J. Hundhausen, Solar wind electrons: Vela 4 measurements, *J. Geophys. Res.*, **73,** 4999, 1968.

Morrison, P., Solar-connected variations of the cosmic rays, *Phys. Rev.*, **95,** 646, 1954.

Morrison, P., Solar origin of cosmic ray time variations, *Phys. Rev.*, **101,** 1397, 1956.

Morrison, P.J., and J.M. Greene, Noncanonical Hamiltonian density formulation of hydrodynamics and ideal magnetohydrodynamics, *Phys. Rev. Lett.*, **45,** 790, 1980.

Nakagawa, T., A. Nishida, and T. Saito, Planar magnetic structures in the solar wind, *J. Geophys. Res.*, **94,** 11761, 1989.

Nerney, S.F., and S.T. Suess, Restricted three-dimensional stellar wind modeling, 1. Polytropic case, *Astrophys. J.*, **196,** 837, 1975a.

Nerney, S.F., and S.T. Suess, Corrections to the azimuthal IMF due to meridional flow in the solar wind, *Astrophys. J.*, **200,** 503, 1975b.

Nerney, S.F., and S.T. Suess, Modeling the effects of latitudinal gradients in stellar winds, with application to the solar wind, *Astrophys. J.*, **296,** 259, 1985.

Ness, N.F., and J.M. Wilcox, Solar origin of the interplanetary magnetic field, *Phys. Rev. Lett.*, **13,** 461, 1964.

Ness, N.F., C.S. Scearce, J.B. Seek, and J.M. Wilcox, Summary of results from the IMP-1 magnetic field experiment, *Space Res.* **6,** 581, 1966.

Ness, N.F., A.J. Hundhausen, and S.J. Bame, Observations of the interplanetary medium: Vela 3 and IMP-3, 1965-1967, *J. Geophys. Res.*, **76,** 6643, 1971.

Neubauer, F.M., Nonlinear interaction of discontinuities in the solar wind and the origin of slow shocks, *J. Geophys. Res.*, **81,** 2248, 1976.

Neubauer, F.M., G. Mussmann, and G. Dehemel, Fast magnetic fluctuations in the solar wind, Helios 1, *J. Geophys. Res.*, **82,** 3201, 1977.

Neugebauer, M., Observations of solar wind helium, *Fund. Cosmic Phys.*, **7,** 131, 1981.

Neugebauer, M., Alignment of velocity and field changes across tangential discontinuities in the solar wind, *J. Geophys. Res.*, **90**, 6627, 1985.
Neugebauer, M., and C.W. Snyder, Mariner 2 observations of the solar wind. 2. Average properties, *J. Geophys. Res.*, **71**, 4469, 1966.
Neugebauer, M., D.R. Clay, B.E. Goldstein, B.T. Tsurutani, and R. Zwickl, A re-examination of rotational and tangential discontinuities in the solar wind, *J. Geophys. Phys.*, **89**, 5395, 1984.
Neugebauer, M., C.J. Alexander, R. Schwenn, and A.K. Richter, Tangential discontinuities in the solar wind: Correlated field and velocity changes and the Kelvin–Helmholtz instability, *J. Geophys. Res.*, **79**, 13694, 1986.
Neupert, W.M., and V. Pizzo, Solar coronal holes as sources of recurrent geomagnetic disturbances, *J. Geophys. Res.*, **79**, 3701, 1974.
Noci, G., Energy budget in coronal holes, *Solar Phys.*, **28**, 403, 1973.
Nolte, J.T., A.S. Krieger, A.F. Timothy, R.E. Gold, E.C. Roelof, G. Vaiana, A.J. Lazarus, J.F. Sullivan, and P.S. McIntosh, Coronal holes as sources of solar wind, *Solar Phys.*, **46**, 303, 1976.
Ogilvie, K.W., and L.F. Burlaga, Hydromagnetic shocks in the solar wind, *Solar Phys.*, **8**, 442, 1969.
Ogilvie, K.W., and L.F. Burlaga, A discussion of interplanetary post-shock flows with two examples, *J. Geophys. Res.*, **79**, 2324, 1974.
Ogilvie, K.W., and J.D. Scudder, The radial gradients and collisional properties of solar wind electrons, *J. Geophys. Res.*, **83**, 3776, 1978.
Ogilvie, K.W., J.D. Scudder, and M. Sugiura, Electron energy flux in the solar wind, *J. Geophys. Res.*, **76**, 8165, 1971.
Olmsted, C., and S.-I. Akasofu, One-dimensional kinematics of particle stream flow with application to solar wind simulation, *Planet. Space Sci.*, **33**, 831, 1985.
Osherovich, V.A., C.J. Farrugia, and L.F. Burlaga, The non-linear evolution of magnetic flux ropes: 1. The low beta limit, *J. Geophys. Res.*, **98**, 13225, 1993a.
Osherovich, V.I., C.J. Farrugia, and L.F. Burlaga, Dynamics of aging magmatic clouds, *Adv. Space Res.*, **13**:6(6), 57, 1993b.
Osherovich, V.A., C.J. Farrugia, L.F. Burlaga, R.P. Lepping, J. Fainberg, and R.G. Stone, Polytropic relationships in interplanetary magnetic clouds, *J. Geophys. Res.*, **98**, 15331, 1993c.
Paladin, G., and A. Vulpiani, Anomalous scaling laws in multifractal objects, *Phys. Rep.*, **4**, 147, 1987.
Parker, E.N., The gross dynamics of a hydromagnetic gas cloud, *Astrophys. J. Suppl.*, **25**, 51, 1957.
Parker, E.N., Dynamics of the interplanetary gas and magnetic fields, *Astrophys. J.*, **128**, 664, 1958.
Parker, E.N., The stellar-wind regions, *Astrophys. J.*, **134**, 20, 1961.
Parker, E.N., *Interplanetary Dynamical Processes*, Wiley-Interscience, New York, 1963.
Piddington, J.H., *Phys. Rev.*, **112**, 589, 1958.
Pietronero, L., and A.P. Siebesma, Self-similarity of fluctations in random multiplicative processes, *Phys. Rev. Lett.*, **57**, 1089, 1986.
Pirraglia, J.A., Simulation of the multifractal behavior in the interplanetary magnetic field, Technical Report, Laboraory for Extraterrestrial Physics, Greenbelt, MD, 1993.
Pizzo, V.J., An evaluation of corotating stream models, in *Solar Wind Four*, edited by H. Rosenbauer, p. 153, MPAE-100-81-31, 1981.

Pizzo, V.J., A three-dimensional model of corotating streams in the solar wind. 3. Magnetohydrodynamic streams, *J. Geophys. Res.,* **87,** 4374, 1982.

Pizzo, V.J., Quasi-steady solar wind dynamics, in *Solar Wind Five,* edited by M. Neugebauer, p. 675, NASA Conf. Publ. 2280, Washington, DC, 1983a.

Pizzo, V.J., Comments on "Solar wind disturbances in the outer heliosphere, caused by successive solar flares from the same active region," by Akasofu and Hakamada, in *Solar Wind Five,* edited M. Neugebauer, p. 481, NASA Conf. Publ. 2280, Washington, DC, 1983b.

Pizzo, V.J., Interplanetary shocks on the large scale—A retrospective on the last decade's theoretical efforts, in *Collisionless Shocks in the Heliosphere: Reviews of Current Research,* edited by Tsurutani, B.T., and R.G. Stone, p. 51, AGU Geophysical Monograph 35, American Geophysical Union, Washington, DC, 1985.

Pizzo, V.J., Global quasi-steady dynamics of the distant solar wind. Origin of north–south flows in the outer heliosphere, *J. Geophys. Res.,* **99,** 4173, 1994.

Pizzo, V.J., and B.E. Goldstein, Meridional transport of magnetic flux in the solar wind between 1 and 10 AU: A theoretical analysis, *J. Geophys. Res.,* **92,** 7241, 1987.

Pneuman, G.W., The solar wind and the temperature–density structure of the solar corona, *Solar Phys.,* **13,** 28, 247, 1973.

Pneuman, G.W., Latitude dependence of the solar wind speed: Influence of the coronal magnetic field geometry, *J. Geophys. Res.,* **81,** 5049, 1976.

Pneuman, G.W., and R.A. Kopp, Coronal streamers III: Energy transport in streamers and interstreamer regions, *Solar Phys.,* **13,** 176, 1970.

Pneuman, G.W., and R.A. Kopp, Gas–magnetic field interactions in the solar corona, *Solar Phys.,* **18,** 258, 1971.

Pneuman, G.W., S.F. Hansen, and R.T. Hansen, On the reality of potential magnetic fields in the solar corona, *Solar Phys.,* **59,** 313, 1978.

Raeder, J., F.M. Neubauer, N.F. Ness, and L.F. Burlaga, Macroscopic perturbations of the IMF by P/Halley as seen by the Giotto magnetometer, *Astron. Astrophys.,* **187,** 61, 1987.

Rhodes, E.J., and E.J. Smith, Multi-spacecraft study of the solar wind velocity at interplanetary sector boundaries, *J. Geophys. Res.,* **80,** 917, 1975.

Rhodes, E.J., Jr., and E.J. Smith, Evidence of a large-scale gradient in the solar wind velocity, *J. Geophys. Res.,* **81,** 2123, 1976a.

Rhodes, E.J., Jr., and E.J. Smith, Further evidence of a large-scale gradient in the solar wind velocity, *J. Geophys. Res.,* **81,** 5833, 1976b.

Rhodes, E.J., Jr., and E.J. Smith, Multispacecraft observation of heliographic latitude–longitude structure, *J. Geophys. Res.,* **86,** 8877, 1981.

Richardson, I.G., C.J. Farrugia, and L.F. Burlaga, Energetic ion observations in the magnetic cloud of 14–15 January 1988 and their implications for the magnetic field topology, in *Proceedings of the 22nd International Cosmic Ray Conference (Dublin),* Vol. 3, SH7.8, p. 597, Dublin Institute for Advanced Studies, 1991.

Richardson, L.F., *Weather Prediction by Numerical Processes,* Cambridge University Press (Republished by Dover, New York, 1965), 1922.

Richter, A.K., Interplanetary slow shocks, Chapter 7 in *Physics of the Inner Heliosphere,* Vol. 2, edited by R. Schwenn and E. Marsch, p. 23, Springer-Verlag, Berlin-Heidelberg, 1991.

Richter, A.K., and E. Marsch, Helios observational constraints on the development of interplanetary slow shocks, *Ann. Geophys.*, **6**, 319, 1988.

Richter, A.K., K.C. Hsieh, H. Rosenbauer, and F.M. Neubauer, Parallel fast-forward shock waves within 1 AU: Helios-1 and -2 observations, Rep. MPAE-W-79-84-27, Max-Planck Institute, Lindau, Germany, 1984.

Richter, A.K., K.C. Hsieh, A.H. Lutrell, E. Marsch, and R. Schwenn, Review of interplanetary shock phenomena near and within 1 AU, in *Collisionless Shocks in the Heliosphere: Reviews of Current Research,* edited by R.G. Stone and B.T. Tsurutani, p. 33, AGU Geophysical Monograph 35, American Geophysical Union, Washington, DC, 1985a.

Richter, A.K., H. Rosenbauer, F.M. Neubauer, and N.G. Ptitsyna, Solar wind observations associated with a slow forward shock wave at 0.31 AU, *J. Geophys. Res.*, **90**, 7581, 1985b.

Richter, A.K., K.C. Hsieh, H. Rosenbauer, and F.M. Neubauer, Parallel fast-forward shock waves within 1 AU: Helios-1 and -2 observations, *Ann. Geophys.*, **4**, 1, 1986.

Roberts, D.A., Heliocentric distance and temporal dependence of the interplanetary density–magnetic field magnitude correlation, *J. Geophys. Res.*, **95**, 1087, 1990.

Roberts, D.A., and M.L. Goldstein, Spectral signatures of jumps and turbulence in interplanetary speed and magnetic field data, *J. Geophys. Res.*, **92**, 10105, 1987.

Roberts, D.A., and M.L. Goldstein, Turbulence and waves in the solar wind, *Rev. Geophys., Suppl.*, 932, 1991.

Roberts, D.A., M.L. Goldstein, and L.W. Klein, The amplitudes of interplanetary fluctuations: Stream structure, heliocentric distance, and frequency dependence, *J. Geophys. Res.*, **95**, 4203, 1990.

Rosenau, P., and S.T. Suess, Slow shocks in the interplanetary medium, *J. Geophys. Res.*, **82**, 3649, 1977.

Rosenbauer, H., E. Marsch, B. Meyer, H. Miggenrieder, M. Montgomery, K.H. Mulhauser, W. Pillip, W. Voges, and S.K. Zink, A survey on initial results of the Helios experiment, *J. Geophys. Res.*, **42**, 561, 1977.

Rosenberg, R.L., and P.J. Coleman, Jr., Heliographic latitude dependence of the dominant polarity of the interplanetary magnetic field, *J. Geophys. Res.*, **74**, 5611, 1969.

Rosenberg, R.L., and P.J. Coleman, Jr., Solar cycle-dependent north–south field configurations observed in solar wind interaction regions, *J. Geophys. Res.*, **85**, 3021, 1980.

Rosenberg, R.L., M. Kivelson, and P.C. Hedgecock, Heliographic latitude dependence of the dominant polarity of the interplanetary magnetic field by comparison of simultaneous Pioneer 10 and Heos 1,2 data, *J. Geophys. Res.*, **82**, 1273, 1977.

Rosenbluth, M.N., and M.N. Bussac, MHD stability of spheromak, *Nuclear Fusion*, **19**:4, 489, 1979.

Russell, C.T., and C.J. Alexander, Multiple spacecraft observations of interplanetary shocks: Shock-normal oscillations and their effects, *Adv. Space Res.*, **4**, 277, 1984.

Russell, C.T., M.M. Mellot, E.J. Smith, and J.H. King, Multiple spacecraft observations of interplanetary shocks: Four spacecraft determinations of shock normals, *J. Geophys. Res.*, **88**, 4739, 1983a.

Russell, C.T., J.T. Gosling, R.D. Zwickl, and E.J. Smith, Multiple spacecraft observations of interplanetary shocks: ISEE three-dimensional plasma measurements, *J. Geophys. Res.*, **88**, 9941, 1983b.

Rust, D.M., Spawning and shedding of helical magnetic fields in the heliosphere, *Geophys. Res. Lett.*, **21**, 24, 1994.

Saito, T., T. Oki, and S.-I. Akasofu, and C. Olmstead. The sunspot cycle variations of the neutral line on the source surface, *J. Geophys. Res.*, **94**, 5453, 1989.

Sanderson, T.R., R.G. Marsden, R. Reinhard, K.-P. Wenzel, and E.J. Smith, Correlated particle and magnetic field observations of a large-scale magnetic loop structure behind an interplanetary shocks, *Geophys. Res. Lett.*, **10**, 916, 1983.

Sanderson, T.R., R. Reinhard, P. van Nes, and K.-P. Wenzel, Observations of three-dimensional anisotropies of 35–100 keV protons associated with interplanetary shocks, *J. Geophys. Res.*, **90**, 19, 1985.

Sanderson, T.R., J. Beeck, R.G. Marsden, C. Tranquille, K.-P. Wenzel, R.B. McKibben, and E.J. Smith, Energetic ion and cosmic ray characteristics of a magnetic cloud, in *Physics of Flux Ropes*, edited by C.T. Russell, E.R. Priest, and L.C. Lee, p. 385, AGU Geophysical Monograph 58, American Geophysical Union, Washington, DC, 1990.

Sarabhai, V. Some consequences of non-uniformity of solar wind velocity, *J. Geophys. Res.*, **68**, 1555, 1963.

Sari, J.W., and N.F. Ness, Power spectra of the interplanetary magnetic field, *Solar Phys.*, **8**, 155, 1969.

Sawyer, C., High-speed streams and sector boundaries, *J. Geophys. Res.*, **81**, 2437, 1976.

Schatten, K.H., J.M. Wilcox, and N.F. Ness, A model on interplanetary and coronal magnetic fields, *Solar Phys.*, **6**, 442, 1969.

Schertzer, D., and S. Lovejoy, Nonlinear variability in geophysics: Multifractal simulations and analysis, in *Fractals: Physical Origin and Properties*, p. 49. edited by L. Pietronero, Plenum Press, New York, 1990.

Schulz, M., Interplanetary sector structure and the heliomagnetic equator, *Astrophys. Space Sci.*, **34**, 371, 1973.

Schwenn, R., Relationship of coronal transients to interplanetary shocks: 3D aspects, *Space Sci. Rev.*, **34**, 85, 1983.

Schwenn, R. Large-scale structure of the interplanetary medium, Chapter 1 in *Physics of the Inner Heliosphere*, Vol. 1, edited by R. Schwenn and E. Marsch, p. 99, Springer-Verlag, Berlin-Heidelberg, 1990.

Schwenn, R., and E. Marsch, editors, *Physics of the Inner Heliosphere*, Vol. 1, Springer-Verlag, Berlin-Heidelberg, 1990.

Schwenn, R.M., D. Montgomery, H. Rosenbauer, H. Miggenrieder, K.H. Mulhauser, S.J. Bame, W.C. Feldman, and R.T. Hansen, Direct observation of the latitudinal extent of a high speed stream in the solar wind, *J. Geophys. Res.*, **83**, 1011, 1978.

Scudder, J.D., Why all stars should possess circumstellar temperature inversions, *Astrophys. J.*, **398**, 319, 1992.

Scudder, J.D., D.L. Lin, and K.W. Ogilvie, Electron observations in the solar wind and magnetosheath, *J. Geophys. Res.*, **78**, 6535, 1973.

Scudder, J.D., L.F. Burlaga, and E.W. Greenstadt, Scale lengths in quasi-parallel shocks. *J. Geophys. Res.*, **89**, 7545, 1984.

Sen, A.K., Stability of hydromagnetic Kelvin–Helmholtz discontinuity, *Phys. Fluids,* **6,** 1154, 1963.

Serbu, G.P., Explorer 35 observations of the solar-wind electron density, temperature and anisotropy, *J. Geophys. Res.,* **77,** 1703, 1972.

Sheeley, N.R., Coronal holes and solar wind streams during the sunspot cycle, in *Solar Wind Seven,* edited by E. Marsch and R. Schwenn, p. 263, Pergamon Press, New York, 1992.

Sheeley, N., and J.W. Harvey, Coronal holes, solar wind streams and geomagnetic disturbances during 1978 and 1979, *Solar Phys.,* **70,** 237, 181.

Sheeley, N.R., Jr., J.W. Harvey, and W.C. Feldman, Coronal holes and solar wind streams and recurrent geomagnetic disturbances: 1973–1976, Naval Research Laboratory, Tech. Rep., Washington, DC, 1976.

Sheeley, N.R., Jr., E.T. Swanson, and Y.-M. Wang, Out-of-ecliptic tests of the inverse correlation between solar wind speed and coronal expansion factor, *J. Geophys. Res.,* **96,** 13861–13868, 1991.

Sime, D.G., and B.J. Rickett, The latitude and longitude structure of the solar wind from IPS observations, *J. Geophys. Res.,* **83,** 5757, 1978.

Sime, D.G., and B.J. Rickett, Coronal density and the solar wind speed at all latitudes, *J. Geophys. Res.,* **86,** 8869, 1981.

Siregar, E., D. Aaron Roberts, and M.L. Goldstein, An evolving vortex street model for quasi-periodic solar wind fluctuations, *Geophys. Res. Lett.* **19,** 1427, 1992.

Siregar, E., D. Aaron Roberts, and M.L. Goldstein, Quasi-periodic transverse plasma flow associated with an evolving MHD vortex street in the outer heliosphere, *J. Geophys. Res.,* **98,** 13233, 1993.

Siscoe, G.L., and R.W. Suey, Significance criteria for variance matrix applications, *J. Geophys. Res.,* **77,** 1321, 1972.

Siscoe, G.L., L. Davis, Jr., P.J. Coleman, Jr., E.J. Smith, and D.E. Jones, Power spectra and discontinuities of the interplanetary magnetic field: Mariner 4, *J. Geophys. Res.,* **77,** 61, 1968.

Siscoe, G.L., M. Turner, and A.J. Lazarus, Simultaneous plasma and magnetic field measurements of probable tangential discontinuities in the solar wind, *Solar Phys.,* **6,** 456, 1969.

Sittler, E.C., Jr., and J.D. Scudder, An empirical polytrope law for solar wind thermal electrons between 0.45 and 4.67 AU: Voyager 2 and Mariner 10, *J. Geophys. Res.,* **85,** 5131, 1980.

Sittler, E.C., Jr., J.D. Scudder, and J. Jesson, Radial variation of solar wind thermal electrons between 1.36 and 2.25 AU: Voyager 2, in *Solar Wind Four,* edited by H. Rosenbauer, p. 257, Max Planck Inst for Aeron., Katlenburg-Lindau, Germany, 1981.

Slavin, J.A., and E.J. Smith, Solar cycle variations in the IMF, in *Solar Wind 5,* edited by M. Neugebauer, p. 323, NASA Conf. Publ. 2280, Washington, DC, 1983.

Slavin, J.A., E.J. Smith, and B.T. Thomas, Large-scale temporal and radial gradients in the IMF: ISEE-3, and Pioneer 10, 11, *Geophys. Res. Lett.,* **11,** 279, 1984.

Smith, E.J., Observed properties of interplanetary rotational discontinuities, *J. Geophys. Res.,* **78,** 2088, 1973a.

Smith, E.J., Identification of interplanetary tangential and rotational discontinuities, *J. Geophys. Res.,* **78,** 2054, 1973b.

Smith, E.J., Solar wind magnetic field observations, in *Solar Wind Four,* edited by

H. Rosenbauer, p. 96, Rep. MPAE-100-81-31, Max-Planck Institute, Lindau, Germany, 1981.

Smith, E.J., Observations of interplanetary shocks: Recent progress, *Space Sci. Rev.,* **34,** 101, 1983.

Smith, E.J., Interplanetary shock phenomena beyond 1 AU, in *Collisionless Shocks in the Heliosphere, Reviews of Current Research,* edited by B. Tsurutani and R.G. Stone, p. 69, AGU Monograph, 35, American Geophysical Union, Washington, DC, 1985.

Smith, E.J., Interplanetary magnetic field over two solar cycles and out to 20 AU, *Adv. Space Res.,* **9,** 159, 1989.

Smith, E.J., Magnetic fields in the heliosphere, in *Physics of the Outer Heliosphere,* edited by S. Grzedzielski and D.E. Page, p. 253, Pergamon Press, New York, 1990.

Smith, E.J., Magnetic fields throughout the heliosphere, *Adv. Space Res.,* **13**:6, 5, 1993.

Smith, E.J., and A. Barnes, Spatial dependences in the distant solar wind, in *Solar Wind Five,* edited by M. Neugebauer, p. 251, NASA Conf. Publ. 2280, Washington, DC, 1983.

Smith, E.J., and B.T. Thomas, The heliospheric current sheet: 3-dimensional structure and solar cycle changes, in *The Sun and the Heliosphere in Three Dimensions,* edited by R.G. Marsden, p. 267, D. Reidel, Dordrecht, 1986.

Smith, E.J., and J.H. Wolfe, Observations of interaction regions and corotating shocks between one and five AU: Pioneer 10 and 11, *Geophys. Res. Lett.,* **3,** 137, 1976.

Smith, E.J., and J.H. Wolfe, Pioneer 10 and 11 observations of evolving solar wind streams and shocks beyond 1 AU, in *Study of Travelling Interplanetary Phenomena,* edited by M. Shea et al., p. 227, D. Reidel, Hingham, MA, 1977.

Smith, E.J., and J.H. Wolfe, Fields and plasmas in the outer solar system, *Space Sci. Rev.,* **23,** 217, 1979.

Smith, E.J., B.J. Tsurutani, and R.L. Rosenberg, Observations of the interplanetary sector structure up to heliographic latitudes of 16°, *J. Geophys. Res.,* **83,** 717, 1978.

Smith, E.J., J.A. Slavin, R.D. Zwickl, and S.J. Bame, Shocks and storm sudden commencements, in *Solar Wind–Magnetospheric Coupling,* edited by Y. Kamide and J.A. Slavin, p. 345, Scientific Publishing Company, Tokyo, 1986.

Smith, E.J., J.D. Winterhalter, and J.A. Slavin, Recent Pioneer 11 observations of the distant heliospheric magnetic field, in *Solar Wind Six,* edited by V.J. Pizzo, T.E. Holzer, and D.G. Sime, p. 58, NCAR Technical Note 306 + Proc., Boulder, CO, 1988.

Solodyna, C.V., J.W. Sari, and J.W. Belcher, Plasma field characteristics of directional discontinuities in the interplanetary medium. *J. Geophys. Res.,* **82,** 10, 1977.

Sonett, C.P., and D.S. Colburn, The SI^+ and SI^- pair and interplanetary forward--reverse shock ensembles, *Planet. Space Sci.,* **13,** 675, 1975.

Sonett, C.P., D.S. Colburn, L. Davis, E.J. Smith, and P.J. Coleman, Evidence for a collision-free magnetohydrodynamic shock in interplanetary space, *Phys. Rev. Lett.,* **13,** 153, 1964.

Sonett, C.P., D.S. Colburn, and B.R. Briggs, Evidence for a collision-free hydromagnetic shock in interplanetary space, in *The Solar Wind,* edited by C.P.

Sonett, P.J. Coleman, Jr., and J.M. Wilcox, p. 65, Pergamon Press, New York, 1966.

Sonnerup, B.U.O., and L.J. Cahill, Magnetopause structure and attitude from Explorer 12 observations, *J. Geophys. Res.*, **72**, 171, 1967.

Sreenivasan, K.R., Fractals and multifractals in fluid turbulence, *Annu. Rev. Fluid Mech.*, **23**, 539, 1991.

Sreenivasan, K.R., and P. Kailasnath, An update of the intermittency exponent in turbulence, *Phys. Fluids A*, **5**:2, 512, 1993.

Sreenivasan, K.R., R.R. Prasad, C. Meneveau, and R. Ramshankar, The fractal geometry of interfaces and the multifractal distribution of dissipation in fully turbulent flows, *PAGEOPH*, **131**: 1 and 2, 42, 1989a.

Sreenivasan, K.R., R. Ramshankar, and C. Meneveau, Mixing, entrainment and fractal dimensions of surfaces in turbulent flows, *Proc. R. Soc. London*, **A421**, 79, 1989b.

Stanley, H.E., and P. Meakin, Multifractal phenomena in physics and chemistry, *Nature*, **335**, 405, 1988.

Steinolfson, R.S., and A.J. Hundhausen, Waves in low beta plasmas, *J. Geophys. Res.*, **94**, 1222, 1989.

Steinolfson, R.S., and A.J. Hundhausen, Concave-outward slow shocks in coronal mass ejections, *J. Geophys. Res.*, **95**, 15251, 1990.

Steinolfson, R.S., M. Dryer, and Y. Nakagawa, Interplanetary shock pair disturbances: Comparison of theory with data, *J. Geophys. Res.*, **80**, 1989, 1975a.

Steinolfson, R.S., M. Dryer, and Y. Nakagawa, Numerical MHD simulation of interplanetary shock pairs, *J. Geophys. Res.*, **80**, 1223, 1975b.

Stone, E.C., R.E. Vogt, F.B. McDonald, B.J. Teegarden, J.H. Trainor, J.R. Jokipii, and W.R. Webber, Cosmic ray investigation for the Voyager missions; Energetic particle studies in the outer heliosphere—and beyond, *Space Sci. Rev.*, **21**, 355, 1977.

Suess, S.T., Magnetic clouds and the pinch effect, *J. Geophys. Res.*, **93**, 5437, 1988.

Suesss, S.T., The heliopause, *Rev. Geophys.*, **28**, 97, 1990.

Suess, S.T., and E. Hildner, Deformation of heliospheric current sheet, *J. Geophys. Res.*, **90**, 9461, 1985.

Suess, S.T., and S.F. Nerney, The global solar wind and predictions for Pioneers 10 and 11, *Geophys. Res. Lett.*, **2**, 75 1975.

Suess, S.T., B.T. Thomas, and S.F. Nerney, Theoretical interpretation of the observed interplanetary magnetic field radial variation in the outer solar system, *J. Geophys. Res.*, **90**, 4378, 1985.

Suess, S.T., P.H. Scherrer, and J.T. Hoeksema, Solar wind speed azimuthal variation along the heliospheric current sheet, in *The Sun and the Heliosphere in Three Dimensions*, edited by R.G. Marsden, p. 275, D. Reidel, Dordrecht, 1986.

Svalgaard, L., and J.M. Wilcox, The spiral interplanetary magnetic field: A polarity and sunspot cycle variation, *Science*, **186**, 51, 1974.

Svalgaard, L., and J.M. Wilcox, A view of solar magnetic fields, the solar corona, and the solar wind in three dimensions, *Annu. Rev. Astron. Astrophys.*, **16**, 429, 1978.

Taylor, H.E., Sudden commencement associated discontinuities in the interplanetary magnetic field observed by IMP-3, *Solar Phys.*, **6**, 320, 1969.

Taylor, J.B., Relaxation of toroidal plasma and generation of reverse magnetic fields, *Phys. Rev. Lett.*, **33**, 1974.

Taylor, J.B., Relaxation and magnetic reconnection in plasmas, *Rev. Modern Phys.*, **58**:3, 741, 1986.
Tel, T., Fractals, multifractals, and thermodynamics, *Z. Naturforsch.*, **43a,** 1154, 1988.
Thomas, B.T., and E.J. Smith, The Parker spiral configuration of the interplanetary magnetic field between 1 and 8.5 AU, *J. Geophys. Res.*, **85,** 6861, 1980.
Thomas, B.T., and E.J. Smith, The structure and dynamics of the heliospheric current sheet, *J. Geophys. Res.*, **86,** 105, 1981.
Thomas, B.J., J.A. Slavin, and E.J. Smith, Radial and latitudinal gradients in the interplanetary magnetic field: Evidence for meridional flux transport, *J. Geophys. Res.*, **91,** 6760, 1986.
Tranquille, C.T., R. Sanderson, R.G. Marsden, K.-P. Wenzel, and E.J. Smith, Properties of a large-scale interplanetary loop structure as deduced from low-energy proton anisotropy and magnetic field measurements, *J. Geophys. Res.*, **92,** 6, 1987.
Tsurutani, B.T., and E.J. Smith, Interplanetary discontinuities: Temporal variations and the radial gradient from 1 to 8.5 AU, *J. Geophys. Res.*, **84,** 2773, 1979.
Tsurutani, B.T., and R.G. Stone, editors, *Collisionless Shocks in the Heliosphere*: *Reviews of Current Research,* AGU Monograph 35, American Geophysical Union, Washington, DC, 1985.
Tsurutani, B.T., E.J. Smith, and D.E. Jones, Waves upstream of interplanetary shocks, *J. Geophys. Res.*, **88,** 5645, 1983.
Tsurutani, B.T., W.D. Gonzalez, F. Tang, S.-I. Akasofu, and E.J. Smith, Origin of interplanetary southward magnetic field responsible for major magnetic storms near solar maximum (1978–1989), *J. Geophys. Res.*, **93,** 8519, 1989.
Turner, J.M., L.F. Burlaga, N.F. Ness, and J.F. Lemaire, Magnetic holes in the solar wind, *J. Geophys. Res.*, **82,** 1921, 1977.
Vandas, M., S. Fischer, and A. Geranios, Spherical and cylindrical models of magnetic clouds and their comparison with spacecraft data, *Planet. Space Sci.*, **39,** 1147, 1191.
Vandas, M., S. Fisher, P. Pelat, and A. Geranios, Magnetic clouds: Comparison between spacecraft measurements and theoretical magnetic force-free solutions, in *Solar Wind Seven,* edited by E. Marsch and R. Schwenn, p. 671, Pergamon, New York, 1992.
Vandas, M., S. Fisher, P. Pelat, and A. Geranios, Spheroidal models of magnetic clouds and their comparison with spacecraft measurements, *J. Geophys. Res.*, **98,** 11467, 1993.
van de Hulst, H.C., The chromosphere and the corona, in *The Sun,* edited by G.P. Kuiper, pp. 207–321, University of Chicago Press, Chicago, 1953.
Vellante, M., and A.J. Lazarus, An analysis of solar wind fluctuations between 1 and 10 AU, *J. Geophys. Res.*, **92,** 9893, 1987.
Veselovsky, I.S., Solar wind vortex flow in the outer heliosphere, in *Physics of the Outer Heliosphere,* edited by S. Grzedzielski and D.E. Page, p. 277, Pergamon Press, New York, 1990.
Villante, U., and R. Bruno, Structure of current sheets in the sector boundaries, Helios 2 observations during early 1976, *J. Geophys. Res.*, **87,** 607, 1982.
Villante, U., R. Bruno, F. Mariani, L.F. Burlaga, and N.F. Ness, The shape and location of the sector boundary surface in the inner solar system, *J. Geophys. Res.*, **84,** 6641, 1979.
Villante, U., F. Mariani, and P. Francia, The IMF pattern through the solar

minimum. Two spacecraft observations during 1974–1978, *J. Geophys. Res.*, **87**, 249, 1982.

Vinas, A.F., and J.D. Scudder, Fast and optimal solution to the "Rankine–Hugoniot" problem, *J. Geophys. Res.*, **91**, 39, 1986.

Vinas, A.F., M.L. Goldstein, and M.R. Acuna, Spectral analysis of magnetohydrodynamic fluctuations near interplanetary shocks, *J. Geophys. Res.*, **89**, 3762, 1984.

Volkmer, P.M., and F.M. Neubauer, Statistical properties of fast magnetoacoustic shock waves in the solar wind between 0.3 and 1 AU: Helios-1,2 observations, *Ann. Geophys.*, **3**, 1, 1985.

Watanabe, T., K. Shibasaki, and T. Kakinuma, Latitudinal distribution of solar wind velocity and its relation to solar EUV corona, *J. Geophys. Res.*, **79**, 3841, 1974.

Webber, W.R., and J.A. Lockwood, A study of the long-term variation and radial gradient of cosmic rays out to 23 AU, *J. Geophys. Res.*, **86**, 11,458, 1981.

Whang, Y.C., A magnetohydrodynamic model for corotating interplanetary structures, *J. Geophys. Res.*, **85**, 2285, 1981.

Whang, Y.C., Slow shocks around the sun, *Geophys. Res. Lett.*, **9**, 1081, 1982.

Whang, Y.C., The forward–reverse shock pair at large heliocentric distances, *J. Geophys. Res.*, **89**, 7367, 1984.

Whang, Y.C., Solar wind flow upstream of the coronal slow shock, *Astrophys. J.*, **307**, 838, 1986.

Whang, Y.C., Slow shocks and their transition to fast shocks in the inner solar wind, *J. Geophys. Res.*, **92**, 4349, 1987.

Whang, Y.C., Evolution of interplanetary slow shocks, *J. Geophys. Res.*, **93**, 251, 1988.

Whang, Y.C., Shock interactions in the outer heliosphere, *Space Sci. Rev.*, **57**, 339, 1991.

Whang, Y.C., and L.F. Burlaga, Coalescence of two pressure waves associated with stream interactions, *J. Geophys. Res.*, **90**, 221, 1985a.

Whang, Y.C., and L.F. Burlaga, Evolution and interaction of interplanetary shocks, *J. Geophys. Res.*, **90**, 10765, 1985b.

Whang, Y.C., and L.F. Burlaga, The coalescence of two merged interaction regions between 0.62 and 9.5 AU: September 1979 event, *J. Geophys. Res.*, **91**, 13341, 1986.

Whang, Y.C., and L.F. Burlaga, Simulation of period doubling of recurrent solar wind structures, *J. Geophys. Res.*, **95**, 20663, 1990a.

Whang, Y.C., and L.F. Burlaga, Radial evolution of interaction regions, in *Physics of the Outer Heliosphere*, edited by S. Grzedzielski and D.E. Page, p. 224, Pergamon Press, New York, 1990b.

Whang, Y.C., and T.H. Chien, Magnetohydrodynamic interaction of high-speed streams, *J. Geophys. Res.*, **86**, 3262, 1981.

Whang, Y.-M., and N.R. Sheeley, Jr., Solar wind speed and coronal flux-tube expansion, *Astrophys. J.*, **335**, 726, 1990.

Whang, Y.-M., and N.R. Sheeley, Jr., The relationship between the solar wind speed and the aerial expansion factor, in *Solar Wind Seven*, edited by E. Marsch and R. Schwenn, p. 128, Pergamon Press, New York, 1992.

Whang, Y.C., K.W. Behannon, L.F. Burlaga, and S. Zhang, Thermodynamic properties of the heliospheric plasma, *J. Geophys. Res.*, **94**, 2345, 1989.

Whang, Y.C., S. Liu, and L.F. Burlaga, Shock heating of the solar wind plasma, *J. Geophys. Res.*, **95,** 18769, 1991.

Wilcox, J.M., and A.J. Hundhausen, Comparison of heliospheric current sheet structure obtained from potential magnetic field computations and from observed maximum coronal brightness, *J. Geophys. Res.*, **88,** 8095, 1983.

Wilcox, J.M., and N.F. Ness, Quasi-stationary corotating structure in the interplanetary medium, *J. Geophys. Res.*, **70,** 5793, 1965.

Wilson, R.M., and E. Hildner, Are interplanetary magnetic clouds manifestations of coronal transients at 1 AU? *Solar Phys.*, **91,** 169, 1984.

Wilson, R.M., and E. Hildner, On the association of magnetic clouds with disappearing filaments, *J. Geophys. Res.*, **91,** 5867, 1986.

Winge, C.R., and P.J. Coleman, Jr., First order latitudinal effects in the solar wind, *Planet. Space Sci.*, **22,** 439, 1974.

Winterhalter, D., and E.J. Smith, Observations of large scale spatial gradients in the heliospheric magnetic field, *Adv. Space Res.*, **9,** 171, 1989.

Winterhalter, D., E.J. Smith, and J.A. Slavin, The radial gradient in the interplanetary magnetic field between 1 and 23 AU, in *Solar Wind Six,* edited by V.J. Pizzo, T.E. Holzer, and D.G. Sime, p. 587, NCAR Technical Note 306 + Proc., Boulder, CO, 1988.

Winterhalter, D., E.J. Smith, J.H. Wolfe, and J.A. Slavin, Spatial gradients in the heliospheric magnetic field: Pioneer 11 observations between 1 AU and 24 AU, during solar cycle 21, *J. Geophys. Res.*, **95,** 1, 1990.

Woltjer, L., A theorem on force-free magnetic fields, *Proc. Nat. Acad. Sci., U.S.A.,* **44**:6, 389, 1958.

Yang, W.-H., Expanding force-free magnetized plasmoid, *Astrophys. J.*, **348,** L73, 1990.

Yeh, R., Magnetic structure of a flux rope, *Astrophys. J.*, **305,** 884, 1986.

Zhang, G., and L.F. Burlaga, Magnetic clouds, geomagnetic disturbances, and cosmic ray decreases, *J. Geophys. Res.*, **93,** 2511, 1988.

Zhao, X.-P., and A.J. Hundhausen, Organization of solar wind plasma properties in a tilted heliomagnetic coordinate system, *J. Geophys. Res.*, **86,** 5423, 1981.

Zhao, X.-P., and A.J. Hundhausen, Spatial structure of solar wind in 1976, *J. Geophys. Res.*, **88,** 451, 1983.

Zwickl, R.D., J.R. Asbridge, S.J. Bame, W.C. Feldman, J.T. Gosling, and E.J. Smith, Plasma properties of driver gas following interplanetary shocks observed by ISEE-3, in *Solar Wind Five,* edited by M. Neugebauer, p. 711, NASA Conf Publ. 2280, Washington, DC, 1983.

AUTHOR INDEX

Abraham-Schrauner, B., 74
Acuna, M.H, 74, 80, 82
Akasofu, S-I., 20–21, 41, 90, 162
Alexander, 85
Alfven, H., 20, 36–37, 89
Altschuler, M.D., 15
Anderson, 70
Anselmet, F.Y., 184
Argentiero, P.D., 74, 82
Arnold, V.I., 147, 151

Bame, S.J., 44, 113, 170–71
Barnes, A., 13, 30, 40–41, 43–44
Barouch, E., 29, 123–25, 167, 195
Bartels, J., 89, 115, 118
Bavassano, B., 65–67, 169
Bavassano-Cattaneo, M.B., 82
Behannon, K.W., 17, 24, 29, 51, 56, 58, 62, 65, 67, 96, 100, 119, 132–33, 158, 169
Belcher, J.W., 38, 54–55, 169, 202
Bieber, J.W., 29, 124
Billings, D E., 3, 118
Boltzmann, L., 47
Borgas, M.S., 186
Borrini, G., 42, 82
Boyd, T.J., 71, 82, 123
Bruno, R., 16, 19, 67
Buck, G.L., 96
Burlaga, L.F., 10, 17, 20–22, 28–33, 38, 43–47, 49, 51–52, 54–71, 74–77, 80, 82, 83, 85–86, 88, 91–102, 104–5, 107–10, 112–16, 119, 121–14, 137–43, 145–47, 149–53, 156–67, 169–72, 174–75, 177–78, 180–82, 185–89, 191, 194–97, 199, 204–7, 209, 212–13

Cahill, L.J., 51, 53
Cane, H.V., 85, 114
Chandrasekhar, S., 96

Chao, J.K., 83–88, 131
Chapman, S., 89, 90, 115, 118
Chen, J., 106, 127
Choudhuri, A.R., 105
Cliver, E.W., 167
Cocconi, G., 90
Colburn, D.S., 133
Coleman, P.J., Jr., 23, 204
Coles, W.A., 40, 43
Collard, H.R., 43, 134
Coniglio A., 193
Crooker, N.U., 94, 105, 112

Davis, L., 54–55, 169
Decker, R.B., 85
Dennison, P. A., 39–40
Denskat, K. U., 51, 169
Detman, T.R., 108
Diodato, L., 38
Dryer, M., 90, 134, 140

Fainberg, J., 110
Faraday, M., 12
Farrugia, C.J., 68, 95–96, 100–2, 105, 108, 110, 112, 114
Feldman, W.C., 36, 38, 44, 119
Ferraro, V.C.A., 45, 96, 115
Fitzenreiter, R.J., 63–64, 67
Forbush, S.E., 14
Formisano, V., 34, 37
Freeman, J.W., 39
Freeman, M.P., 101
Friere, G.F., 90
Frisch, U., 183
Fry, C.D., 20

Garren, D.A., 96, 106
Gauss, F., 12
Gazis, P.R., 13, 41, 43–44, 85, 133–34, 142–43, 151, 203, 205
Geranios, A., 91

Gibson, C.H., 184
Gilmore, R., 147
Gold, T., 71, 89–90, 96
Goldstein, B.E., 31, 43, 127, 203–4
Goldstein, H., 96
Goldstein, M.L., 169–71, 174–77, 210–11
Gosling, J.T., 37–38, 42, 58–59, 69, 71, 93, 95, 108, 113–14, 133–34, 170–71
Greene, J.M., 11
Gurnett, D.A., 109–10

Habbal, S.R., 88
Hada, T., 88
Hakamada, K., 21, 41, 162
Halsey, T.C., 194
Hansen, R.T., 16, 119
Harshiladze, A.F., 107
Harvey, J.W., 21, 121
Heineman, M.A., 85, 131
Hentschel, H.G.E., 193
Hewish, A., 39, 40
Hida, K., 113
Hildner, E., 26, 93
Hirshberg, J., 85, 131
Hoeksema, J.T., 16, 20, 26–27
Hollweg, J.V., 52
Holm, D., 11,
Holzer, T.E., 153
Howard, R.A., 16
Hoyle, F., 96
Hseih, K.C., 74, 80
Hu, Y.Q., 88
Hudson, P.D., 53
Hundausen, A.J., 16, 19–20, 36, 39–42, 58, 71, 84, 88–91, 93, 114–15, 118, 127–29, 133–34, 153, 170

Intriligator, D.S., 43
Ipavich, F., 114
Isenberg, P., 48–49, 66
Ivanov, K.G., 107

Jackson, B.V., 93
Jeffrey, A., 54, 70, 79, 82
Jokipii, R., 43, 127
Jones, J.A., 93–94

Kahler, S.W., 89, 93, 95–96
Kailasnath, P., 184, 186
Kakinuma, T., 41
Kayser, S.E., 43, 165
Kennel, C.F., 80, 88
King, J., 30, 172
Klein, L.W., 29–30, 33, 48–49, 66–67, 92–93, 96, 100–1, 104–5, 113, 132, 147, 157–58, 178
Koenigl, A., 105
Kojima, M., 41
Kolmogorov, A.N., 183–84, 191
Koomen, M.J., 16, 92
Kopp, R.A., 16, 120
Korzhov, N.P., 61
Krieger, A.S., 118, 121
Kupershmidt, B.A., 11

Lamb, Sir Horace, 205
Landau, L.D., 70, 82
Lazarus, A.J., 38, 43, 49, 101, 133–34, 142–43, 166, 201, 203, 205, 207, 209, 213
Lemaire, J., 65
Lepping, R. P., 51, 56, 58, 62, 65, 74, 80, 82, 85, 93–94, 99, 101, 107, 110, 112, 114, 131, 158
LeRoux, J.S., 26
Lie, S., 11
Lifshitz, E.M., 70, 82
Lindeman, F.A., 89
Liu, S., 182
Lockwood, J.A., 167
Lopez, R.E., 39
Lorentz, A.H., 12
Lovejoy, S., 182
Luhmann, J., 33
Lundquist, S., 97
Lust, R., 96

Maagoe, S., 40
Mandelbrot, B., 178, 182, 184
Mariani, F., 29, 55–56, 58, 62, 166
Marsch, E., 13, 88, 116, 169, 182
Marsden, R.G., 11, 113
Martin, 55
Marubashi, K., 96
Matthaeus, W.H., 171, 175
McComas, D.J., 69, 90, 108
McDonald, F.B.,152, 165–67, 180

McNutt, R.L., 33, 201, 203–4
Meakin, S., 182, 193–94
Meneveau, C., 182, 186, 188, 195
Mermin, N.D., 28
Meyer, B., 116
Michel, D.G., 28
Miggenrieder, H., 116
Mihalov, J.D., 134
Mish, W.H., 159, 177–78, 180, 190
Mitchel, D.G., 40
Monin, A.S., 183
Montgomery, D, 175
Montgomery, M.D., 35–36
Morrison, P., 11, 90, 167
Mulhauser, K.H., 116
Munakata, L., 41

Nakagawa, T., 68
Nerney, S.F., 204
Ness, N.F., 14–15, 17, 22, 28–29, 32–34, 37, 40, 43, 48, 51, 52, 54, 62, 80, 152, 165–66, 207, 212–13
Neubauer, F.M., 29, 66, 80, 82, 88, 169
Neugebauer, M., 10, 37, 43, 51, 58, 61–62, 115
Neupert, W.M., 118, 121
Newkirk, G., 15
Noci, G., 118
Nolte, J.T., 118

Ogilvie, K.W., 36, 38–39, 44, 47, 61, 74, 76, 85, 115–16, 125, 129–30, 133, 169, 171, 191
Olbert, S., 84, 86–87
Olmstead, C., 162
Osherovich, V.A., 91, 102–3, 105, 109–10

Paladin, G., 182, 191, 193–94
Parker, E.N., 3–6, 9, 27, 30–31, 33, 60, 83, 89, 107, 115, 120, 123, 129, 133, 146
Perko, J., 197, 199
Piddington, J.H., 90
Pietronero, L., 194
Pillip, W., 116
Pirraglia, J.A., 196–97, 199
Pizzo, V.J., 31, 115, 118, 121, 127–29, 134, 141–43, 203–4

Plumpton, C., 45, 96
Pneuman, G.W., 16–17, 40–42, 118–206
Potgieter, M., 26
Procaccia, J. 19

Raeder, J., 69
Reames, D., 6, 95
Rhodes, E.J., 40
Richardson, I., 101
Richter, A.K., 74, 80, 82, 88, 114
Rickett, B., 40, 42–43
Roberts, D.A., 50, 118, 169, 177, 210–11
Rosenau, P., 88
Rosenbauer, H., 80, 92, 116–17, 127
Rosenberg, R.L., 23
Rosenbluth, M.N., 107
Russell, C.T., 74, 85
Rust, D.M., 93, 97

Saito, T., 26
Sanderson, T.R., 113
Sarabhi, V., 123
Sari, J.W., 51
Saywer, C., 40
Schatten, K.H., 15
Schertzer, D., 182
Schluter, A., 96
Schulz, M., 16–17, 19
Schwenn, R., 13, 34–36, 38, 43, 58–59, 92–93, 115, 121–22, 126, 133, 140, 166
Scudder, J.D., 3, 35, 39, 44, 48–49, 66, 74, 78, 80, 82, 85, 131
Sen, A.K., 60
Serbu, G.P., 35
Sheeley, N., 21, 40, 42, 92, 118, 121
Siebesma, A.P., 194
Sime, D.G., 42
Siregar, E.D., 23, 209–11
Siscoe, G.L., 51, 55–56, 85, 131
Sittler, E.C., 39, 44
Slavin, J.A., 30
Smith, C., 29
Smith, E.J., 17, 20–21, 23, 26, 29–30, 40, 43, 54–58, 62, 84, 117, 133, 151, 157
Snyder, C.W., 37, 58, 115
Solodyna, C.V., 54–56
Sonett, C.P., 71, 73, 133

Sonnerup, B.U.O., 51, 53
Sreenivasan, K.R., 182–84, 186, 188, 193, 195
Stanley, H.E., 182, 193–94
Steinolfson, R.S., 88, 114, 140
Stone, E.C., 165
Stone, R., 82, 110
Suess, S.T., 4, 26, 31, 88, 100, 204
Suey, R.W., 51
Svalgaard, L., 20
Symonds, M.D., 40

Taniuti, T., 70, 79, 82
Taylor, H.E., 85
Taylor, J.B., 96
Tel, T., 182, 193–94
Thomas, B.T., 20, 26, 29–30
Tranquille, C.T., 113
Tsurutani, B.T., 55–58, 62, 82, 93
Turner, J.M., 62–63

van de Hulst, H.C., 3
Vandas, M., 90, 107–8, 112
Vellante, M., 49
Veselovsky, I.S., 207, 209

Villante, U., 19–20, 67
Vinas, A.F., 74, 78, 80
Voges, W., 116
Volkmer, P.M., 82
Vulpiani, A., 182, 191, 193

Watanabe, T., 42
Whang, Y.C., 40, 42–43, 73, 88, 118, 121, 127, 145–46, 153, 156, 160, 163–64, 195
Wilcox, J.M, 14–16, 20, 40
Wilson, R.M., 93
Winge, C.R., 204
Winterhalter, D., 30
Wolfe, J., 43, 84, 117, 133–34, 151

Yaglom, A.M., 183
Yang, W.-H., 100
Yun, S.H., 74

Zhang, G., 93, 113
Zhao, X.-P., 19, 41
Zink, S.H., 116
Zwickl, R.D., 112

SUBJECT INDEX

acceleration, 134
Alfven Mach number, 36–37
Alfven shock, 53
Alfven speed, 36
Alfvenic fluctuations, 169
alpha particles, 10
Archimedian spiral, 6, 20, 28
asymmetry, streams, 153, 155–56
autocorrelation coefficient, 170–71
autocorrelation length, 51
azimuthal magnetic field angle, 22–23

basic equations, 9–13
beta, 37, 121
bidirectional streaming, 95
  protons, 113
  suprathermal electrons, 113
binomial multiplicative process, 195–96
  two-scale, 199–200
boundary conditions, 7

catastrophe theory, 147–51
catastrophe
  A2, 148
  dual cusp A3, 148–49
  swallowtail A4, 149–51
central dilation, 5
chaos, 157
characteristic speed, 70
charge-exchange, 4
clusters, 200
CME, 91, 130
coalescence, 157
  interaction region, 133, 138, 156
  merged interaction regions, 145, 153
  shocks, 148, 161
coherence, 169
collision, streams, 123
comet, 4

competition, 126, 137, 160
compound streams, 129–33, 141, 162
compression, 123–25
conservation laws
  energy, 72
  magnetic flux, 12, 72
  mass, 11–12, 72
  normal momentum flux, 72
  tangential momentum flux, 72
constraints, 8
continuity equation, 11
control space, 148, 150
convective term, 10
convergence line, 121
convergence surface, 121
coordinates, 7–9
  azimuthal angle, 8
  elevation angle, 8
  heliographic coordinate system, 8–9
  inertial heliographic coordinate system, 7–8
core temperature, 35
corona, 3
coronal hole, 119–20
  density, 119
  equatorial, 119, 125, 130
  polar, 119, 121, 130
  streams, 119
coronal magnetic field, 40
corotate, 20
corotating interaction region, 130, 135, 175
corotating merged interaction region, 151–62, 180
  amplitude, 157
  definition, 138
  formation, 139–40, 146, 156–57, 160
  observations, 151–52
  spectra, 178
  width, 146–47, 157

corotating shocks, 43
  boundary condition, 118, 127
  coronal hole, 119
corotating streams, 115–37, 142, 170, 174
  boundaries, 121, 124–25
  density, 115
  destruction, 134
  equation of motion, 127
  existence, 115
  fluctuations, 191
  latitude variation, 122
  leading edge, 117
  magnetic cloud interaction, 131
  magnetic field strength, 118
  mesa, 117
  models, 115, 126–29, 134, 137
  observations, 115–16
  polarity, 115
  source density, 117
  source temperature, 115, 118
  source, 118–21
  temperature perturbation, 127
  trailing edge, 117
corotating flow systems
  correlation length, 175–76
  helicity length scale, 170
cosmic ray intensity, 5, 14, 132, 165–66, 176
  diffusion, 90
  drifts, 90
  long-term decrease, 176
  magnetic cloud, 90
  plateau, 176
critical point, 7, 148, 150

D-sheet, 62
declining phase, 152
deep-space probes, 13
defects, 28
degenerate critical point, 148, 150
density, 34–42
  coronal streamer, 42
  latitude variation, 42
  sector boundary, 42
  solar cycle variations, 38
  speed relation, 39
depletion layer, 114
destruction, 134
  pressure wave, 136

  shock, 148
  stream, 139
dipole field, 23, 40
directional discontinuity, 51, 54–58, 68
  definition, 54
  density change, 56
  direction change distribution, 56
  early observations, 54
  fast and slow wind, 55
  field strength change distribution, 55
  physical nature, 55
  radial gradient, 56–58
  separation 56
  thicknesses, 58
  time separations, 56–57
discontinuity, 7
displacement current, 10
displacements, 124
distribution function
  exponential, 171, 192, 198
  lognormal, 171–72, 184
  magnetic field strength, 198
  speed differences, 173, 191–92
  stretched exponential, 192
  tail, 170
draping, 68–69, 108
dynamical systems, 160

east-west deflections, 127
ejecta, 24, 89–90, 129–30, 160, 170
electrical conductivity, 12
energy minimum, 96
entropy, 12, 70
equation of motion, 10
equatorial plane, 5
equilibria, 11
erosion, 129, 156
  stream, 134, 137, 168
Euclidean Space, 5
evolution, 5
expansion, self-similar, 100, 103
Explorer 10, 13
Explorer 33, 39, 45, 51, 61, 76, 131

Faraday's law, 12
fast flows, classification, 129
fast shock. *See* shock
filaments, 67
filtering, 134, 153
flux deficit, 30–32

Subject Index

Forbush decrease, 90
force, 10
  Lorentz force, 106, 116
  magnetic curvature, 96
  magnetic pressure gradient, 106
  magnetic pressure, 96, 163
  magnetic tension, 106
force-free field, 7, 11
  constant alpha, 96, 98–99
  ejecta, 89
  expansion, 98–102
  field strength, 91
  flux rope, 96
  geometry, 94
  helical fields, 96–97, 99
  kinematic model, 100
  Lundquist's solution, 97
  spheroidal, 107–8
  stability, 105, 147
  toroidal, 107
formation
  local merged interaction region, 162
  magnetic field strength fluctuations, 195
  pressure wave, 118, 136–37
  shock pair, 137, 150–51
  shocks, 127, 148
  vortex street, 206, 209
frozen-field, 72, 123
four-sector pattern, 15–16, 19, 21, 27
fractals, 178
  magnetic field, 179
  speed, 179

galactic jets, 105
Gauss's law, 12
geomagnetic activity, 89, 115
geomagnetic disturbance, 89
geomagnetic field, 4
global merged interaction region, 138, 165–68
  definition, 165
  shell-structure, 166
group, 97
growth
  interaction region, 118, 129, 137, 176
  magnetic field strength fluctuation, 194–195
  pressure wave, 118
  turbulence, 175

Helios, 43, 91, 115–16, 121, 126, 182
Helios 1, 13, 17, 23, 34–35, 39, 42, 58, 61, 74, 85, 93–94, 117, 122, 127–28, 132, 134, 139, 140, 142, 167
Helios 2, 13, 17, 23, 34, 35, 42, 58, 61, 74, 85, 94, 117, 132, 142, 162
Hamiltonian dynamics, 11
handedness, 97
heating mechanism, 43
helicity, 96–97
  scale length, 176
heliopause, 4
heliosphere, 5
  amplitude, 20, 23
  asymmetry, 17–18, 20
  directrix, 20
  evolution, 26
  footpoints, 17
  generator, 20
  geometry, 19
heliospheric current sheet, 6, 10, 12, 16–20, 40, 121, 201, 205
  kinematic model, 25
  latitudinal extent, 17, 21, 23
  near equatorial, 17
  perturbations by ejecta, 20
  quadrupole distortions, 19
  radial variations, 24–27
  rotating plane model, 18
  shape, 20, 26
  solar cycle variations, 42
  speed, 26
  stream interactions, 22
  tilt angle, 20, 120
  velocity perturbations, 25
heliospheric plasma sheet, 41, 157
Heos-1, 34

ideal gas law, 10, 12
IMP, 44, 91, 132–33, 171, 187, 189, 190
IMP-1, 14, 39
IMP-3, 34
IMP-6, 38, 43, 44
IMP-7, 38, 43–44, 85, 121
IMP-8 38, 43–44, 94, 99, 121, 130, 134–35, 141–43, 153, 156, 174, 195, 205,
inertial effects, 10
inner-heliosphere probes, 13
integral curves, 28

interaction region, 116–18, 125, 128–29, 135
  corotating interaction region, 117
  creation, 118
  definition, 116
  electron temperature, 126
  equation of motion, 116
  expansion rate, 135
  growth, 118, 127, 129, 176
  kinematic steepening, 126
  pressure, 116, 126
  proton temperature, 126
  radial evolution, 153–54
  size, 129, 176
interactions
  corotating stream
    corotating pressure wave, 144–45
    corotating stream, 130
    magnetic cloud, 131–32
    shock, 130–31, 140–43
  fast stream
    slow stream, 139–40, 142, 153
  magnetic cloud
    corotating interaction region, 146–47
  shock
    tangential discontinuities, 85
    interaction region, 85
  twin stream, 140–41
interface, 132, 143
intermediate frequency fluctuations, 179
intermittency exponent, 184
intermittent turbulence, 182, 184–89
  binomial cascade model, 186, 188
  observations, 182–86
  theory, 186–89
interplanetary medium, 3
interplanetary scintillation, 39, 40–42
interstellar medium, 3–4
interstellar pickup protons. *See* pickup protons
invariance, scale, 175
invariant line, 5
invariant point, 5
invariants, 74
inverse cascade, 175, 180
ion acoustic instability, 110
ion acoustic waves, 109, 111
irregular variations, 129
irreversible, 137, 161

ISEE-3, 99, 132–33, 159, 174
intermediate frequency fluctuations, 169

jump-ramp approximation, 181

K-coronameter, 16
Kelvin-Helmholtz instability, 11, 60–62
  Alfven speed, 61
  directional discontinuity, 61–62
  nonlinear evolution, 209–13
  observations, 60–61
  stream interface, 67
  theory, 60
  velocity shear, 62
kinematic effects, 10, 76
  3-D, 124
  broadening, 122
  corotating interaction region, 123
  limitations of kinematic models, 126
  magnetic cloud, 100
  magnetic field strength fluctuations, 195
  shock distortion, 131
  spiral magnetic field, 28
  steepening, 58, 122, 124
  stream evolution, 164
  stream interface, 58
  streams, 123, 163
Kolmogorov turbulence, 175, 183, 184

Lagrangian acceleration, 10
Lagrangian equation, 123
large-scale fluctuations 157, 169–200
  definition, 169
  magnetic field strength fluctuations, 194
  radial evolution, 177
  solar cycle variations, 181
  spectra, 171–80
large-scale magnetic field, 14–33
latitude variations, 39–42
Lie groups, 11
local merged interaction region, 162–65
  definition, 138
  formation, 162
low frequency fluctuations, 169
low-energy protons, 95
Luna 2, 13
Luna 3, 13

M-regions, 119
magnetic barrier, 114
magnetic cloud, 7, 10, 132–33, 144
  acceleration, 106
  asymmetry, 102, 103, 105
  axis, 97
  bidirectional streaming of protons, 113
  boundary 93, 112
  CME, 93
  definition, 91
  disappearing filaments, 93, 112
  discontinuities, 112
  draping, 108
  electron polytropic index, 109
  electrons, 96
  electron temperature, 109
  expansion, 147
  field strength, 91
  flares, 93
  flow direction, 103, 105
  flux rope, 96
  force-free field model, 96–98
  force-free magnetic field lines, 96, 99
  geometry, 94
  ion acoustic waves, 109–11
  kink instability, 105
  linear dynamical model, 100
  low energy particles, 112
  magnetic barrier, 114
  magnetic field strength asymmetry, 102–3
  magnetic holes, 112
  magnetic pressure, 96
  observations, 92, 95
  origin, 93
  oscillations, 106
  polytropic law, 96
  proton beta, 112
  proton temperature, 109
  rotation, 103, 105
  shock pair, 114
  shocks, 113
  solar filaments, 97
  solar flare, 95
  solar particle event, 95
  speed, 101
  spheroid, 95
  stability, 105, 147
  temperature, 91
  topology, 95–96
  torus, 95
  turbulence, 110, 112
  work, 101
magnetic energy transfer, 175
magnetic equator, 16
magnetic field lines
  coronal hole, 121
  magnetic tongue, 90
magnetic field strength,
  1 AU, 36
  kinematic 3-D model, 125
  latitudinal variations, 33
  Parker's model, 29
  radial variations, 29–33
  speed dependence, 30
  temporal variation, 30
magnetic flux, 195, 192
magnetic force, 10
magnetic hole, 12, 62–67
  complex, 65–66
  current sheets, 64–65
  definition, 62
  directional discontinuity, 62
  electric field force, 65
  linear, 63, 65
  magnetic cloud, 112
  observations, 63–64, 66, 68
  occurrence rate, 62
  radial variation, 66
  reconnection, 62
  sector boundary, 65–66
  size, 66
  theory, 63, 65
magnetic latitude, 41
magnetic permeability, 10
magnetic pinch, 106
magnetic pressure force, 11, 31, 37, 127
magnetic solar cycle, 6
magnetic storms, 71
magnetic tension force, 11
magnetic tongue, 90
magnetized plasma cloud, 89
magnetoacoustic speed, 134
magnetosphere, 89
magnetic helicity, 175
manifold, 7–8
mapping solar fields to heliosphere
  neutral line, 23
  sectors, 15
  source surface field, 20

Mariner 2, 13, 37, 39, 42
Mariner 5, 40, 86
Mariner 10, 44
Mars, 11
maximum brightness contour, 16–17, 20
Maxwell set, 148, 151
Maxwell's field equations, 12
Maxwellian distribution, 35
memory loss, 137, 141, 161
meridional flow, 31, 204
merged interaction regions, 133, 138–68
   catastrophe theory, 147–51
   classification, 138, 147
   definition, 138
   formation, 158, 159, 176
   radial variations, 162
   size, 162
merging shocks, 161
meridional flux transport, 30–31
mesa, 17
mesoscale, 116
MHD models
   1-D, 47, 146, 156
   2-D 141, 142
   3-D, 203
MHD shock. *See* shock
MHD versus gas dynamical models, 129
MHD waves, 7
microscale, 47
minimum variance analysis, 51
minor ions, 10
models
   deterministic, 170
   statistical, 170
moments, 193
moment temperature, 35
momentum flux, 126
multifractal fluctuations, 181–200
multifractal magnetic field strength fluctuations, 192–200
   binomial model, 195
   existence, 196
   multifractal spectrum, 181, 191, 194, 196
   scaling laws, 192–93
   solar cycle var, 197–200
multifractal speed fluctuations, 182–92
   corotating streams, 189,
   structure function, 183
multiple neutral lines, 26

multispacecraft observations, 85

near-Earth spacecraft, 13
neutral line, 16–18, 20, 120
   amplitude, 24
   evolution, 26
   latitudinal extent, 21
   singular, 16
   solar cycle evolution, 26–27
   temporal variations, 20–21
non-linear system, 137, 160
nondissipative, 11
north-south flows, 201–3
   observations, 201–2
   pressure gradients, 203–4
number density, 10

oblique shock, 82–85
   definition, 82
   normal, 82–83
   observations, 82, 86
orange-segment hypothesis, 16–17
order, 157
orientation, 5
outer heliosphere dynamics, 137
outer-heliosphere probes, 13
overtaking, 123, 139, 147, 153

Pioneer 6, 56, 67, 86, 88
Pioneer 8, 55–56
Pioneer 10, 23, 30–31, 43–44, 57, 151, 153, 164, 166, 190–91
Pioneer 11, 23, 30–31, 39, 43, 57, 133, 153, 156, 203, 205
parallel shock, 79–82
   abundance, 82
   classification, 79
   definition, 79
   existence, 79
   velocity, 79
   waves, 82
Parker's model, 27–28, 33
period doubling, 153–54, 156, 159
perpendicular shock, 75–78
   field direction 75
   field magnitude and density, 75–76
   normal energy flux, 75
   normal momentum flux, 75–76
   observations, 76–77
   structure, 75
phase, 169

phenemenology, 7
photoemission, 35
photosphere, 3
photospheric magnetic field, 15
physical objects, 7
pickup protons, 4, 10, 45
Pioneer 7, 131
Pioneer 9, 42
Pioneer 10, 13
Pioneer 11, 13, 17
Pioneer-Venus Orbiter, 13, 24, 33
planar structure, 66–67, 69
plasma characteristics, 34–44
plasma clouds, 89
plasmoid, 95
point defect, sun, 28
polar coronal hole, 141
polar magnetic field, 16–17
   reversal, 26
polarity, 15–17, 20–24, 121, 130
   corotating stream, 115
   evolution, 23
   latitudinal variation, 23
   magnetic, 6
   radial variation, 24
polynomials, 147
polytropic exponents, 12
polytropic law, 1 AU electrons, 39
polytropic relations, 12, 44, 125
   electrons, 39
   exponents, 12
   magnetic cloud, 109–10
potential field model, 15–16, 18, 119–20
pressure balance, 4
pressure balanced structure, 45–69
   density, 47
   forces, 45
   observations, 45–50
   temperature, 47
   theory, 45
pressure gradient force, 4, 10
pressure wave, 118
   birth, 118, 136
   definition, 135
   destruction, 136
   formation, 118
   growth, 118
   observations, 135
pressure versus scale, 47–48, 80, 169
proton beta and fluctuations, 112

quadrupole distortion, 16
quasi-perpendicular forward fast shock, 77–78
quiet wind, 129

radial expansion, 5
radial variations, 42–44
   density, 43
   proton temperature, 43
Rankine vortex, 105
Rankine-Hugoniot conditions, 71–72, 160
rarefaction, 123–124
reconection
   magnetic cloud, 90
   magnetic hole, 62
   tearing mode, 66
recurrent geomagnetic storms, 119
reverse parallel shock, 76–77
   definition, 80
   observations, 80–81
   signature, 80
reversible, 11
rotation axis, 5
rotational discontinuities, 52–54
   Alfven wave, 52
   angular momentum, 54
   density and temperature, 53
   normal magnetic field component, 52
   observation, 54
   polarization, 54
   pressure, 52
   speed, 53
   vorticity, 54
ruled surface, 20

Sakagake, 68
shock normals, 131
scalar field, 10
scale invariance, 175
scintillation, 42
sector boundary, 120
   displacement, 24
   ejecta, 24
   radial variation, 24–27
   temporal variation, 24
sectors, 6, 15–16, 18, 24, 102, 120
   ascending activity, 25
   definition, 15
   existence 14, 15
   solar cycle evolution, 25–27

self-affine, 178, 186
sensitive dependence, 141
separatrix, 148
shear, 11, 47, 67, 121–22, 127
shear layer, 204–5
  linear instability, 206
  nonlinear instability, 209
sheath, 132
shell, 166–167
shock, 7, 11, 10–89, 132
  classification, 70–71, 73
  corotating, 84, 129
  corotating stream interaction, 130
  destruction, 148
  energy, 70
  entropy, 70
  evolution, 148
  fast shock, 71
  formation, 127, 148
  forward, 71, 133, 143, 145–46
  kinematic distortion, 131
  merging, 161
  normal, 73–74
  oblique, 71
  parallel, 71
  perpendicular 71
  polynomials, 147
  quasi-parallel, 71
  quasi-perpendicular, 71
  reverse, 71, 133, 143, 145–46
  slow shock, 71
  speed, 73
  transient, 84
shock pair, 139, 140, 146, 148, 161
  formation, 137, 148, 150–51
  corotating, 133
  model, 133–34
shock surface
  mesoscale distortions, 85
  ripples, 85
  shape, 83–86
  transient, 85
SI^+–SI^- pair, 133
simple streams, 129
singular point, 12
singular surface, 12
singularities, 5–6, 11–12, 148
  magnetic hole, 66–67
  multiplicity, 148
  sun, 28

vortices, 204
slow shocks, 85–88
  CME, 88
  formation, 88
  observations, 86–88
  signature, 86
  standing, 88
  theory, 85, 86
shock normal
  Acuna-Lepping method, 74
  coplanarity theorem, 73
  Lepping-Argentiero method, 74, 78
  magnetic coplanarity method, 73, 77
  multiple-spacecraft methods, 74, 76
  parallel shock, 79
  velocity coplanarity method 74
  Vinas-Scudder method, 74, 78
solar activity cycle, 5, 14–15
solar flare, 4, 89
solar maximum, 21
solar minimum, 17, 21, 41, 152
Solar Probe, 88
solar rotation, 124
solar rotation axis, 5
solar wind, 4–5
  density, 13
  temperature, 13
  magnetic field, 6
source surface, 15, 40, 120
spacecraft potential, 35
spectra, 175
  $1/f$, 175
  corotating merged interaction region, 178
  exponent, 184
  f^{-2}, 177–79, 181, 191
  large-scale fluctuations, 177, 180
  radial evolution, 177
  shocks, 180
  solar cycle variation, 181
  speed of corotating stream, 191
  speed, 177
speed
  1 AU, 34
  distribution function, electron, 35
  distribution function, proton, 35
  fluctuation, smoothing-out, 134
  global view, 42
  heliospheric current sheet, 38, 40–41
  latitude variation, 124

radial variation, 206, 208
sector boundaries, 40–41
solar cycle variations, 37–38
spheromak, 107
spiral angle, 28
spiral magnetic field, 10, 27–33
azimuthal component, 28
north-south asymmetry, 28
radial component, 28
stationary, 171
steepening, 171
steplike decreases, 167
stirring, 181
stirring scale, 175–76
strahl, 35
stream erosion, 134
stream interface, 58–60, 128–30, 135, 141–42
definition, 58
formation, 58
heliospheric plasma sheet, 58
motion, 129
observation, 58–59
origin, 58–59
stream
definition, 130
erosion, 137
evolution, 136
filtering, 134
interactions, 130, 139–40, 142
latitude boundary, 40
transient, 174
width, 153
stream remnant, 135
strict stationarity, 171
sub-Alfvenic solar wind, 37
subharmonic, 159–60
sudden commencement, 71, 89
sudden impulse, 133
Sun, magnetic field, 5
sunspot number, 14, 153
supermagnetoacoustic, 37
suprathermal electrons, 95
symmetry, 7, 10
system of transient flows, 167–68
axial, 5
scaling, 193

tangential discontinuities, 50–52, 133
Alfven waves, 52
curvature, 51
multispacecraft observations, 52
normal, 51
pressures, 51
tangent space, 8
temperature, 34–36, 35, 47
radial variation, 38, 44
solar cycle variations, 38
temperature-speed relation, 38–39
termination shock, 4–5, 170
thermal energy transport, 12, 36, 39
thermal pressure force, 10, 37
tilt angle, 17, 27
tilted solar magnetic dipole, 16, 203
tokamak, 107
topology, 5
trailing edge, 123
transformation
central dilation, 6
Lorentz, 12
projective, 123
rotation, 6
spiral similarity transformation, 6
transient flow systems
correlation length, 176
helicity length scale, 176
turbulence, 10–11, 169–70, 172, 178, 180
ejecta, 90
exponent, 175
growth, 175
heating, 11
intermittent. *See* intermittent turbulence
magnetic cloud, 110, 112
origin, 181
radial evolution, 172
twin streams, 140–41, 157
two-sector pattern, 16, 21, 27

Ulysses, 44

Voyager 1, 13, 17, 21, 23–24, 30–31, 33, 41–44, 76, 85, 91, 94, 100, 135, 140, 142, 145, 151–52, 157, 159–60, 162, 165, 167, 171–72, 174, 175, 177–80, 182
Voyager 2, 13, 22–24, 30–31, 33, 41, 43–44, 47, 80, 85, 94, 110, 135, 142, 145–46, 151, 152, 165, 167, 178, 180, 196–97, 199, 201, 203, 205

vector field, 9
Vela 3, 34, 39, 126
velocity field, 5
velocity structure function, 183
Venus, 3, 13
viscous force, 11
vortex, 7
  rotation direction, 204–5
  size, 204–205
  velocity, 205–206, 211
vortex street, 201–13
  analytical model, 204–7
  density, 211
  flux deficit, 212–13
  formation, 206, 209
  nonlinear model, 209–13
  speed profile, 206
  stability, 205, 213
  vorticity, 210

Walen's equation, 123
  solution, 124

x-rays, 121